Write Like a Chemist

Write Like a Chemist

A Guide and Resource

Second Edition

Marin S. Robinson
Northern Arizona University
Fredricka L. Stoller
Northern Arizona University

With contributions by Molly S. Costanza-Robinson, Middlebury College, and James K. Jones

OXFORD
UNIVERSITY PRESS

OXFORD
UNIVERSITY PRESS

Oxford University Press is a department of the University of Oxford. It furthers the University's objective of excellence in research, scholarship, and education by publishing worldwide. Oxford is a registered trade mark of Oxford University Press in the UK and certain other countries.

Published in the United States of America by Oxford University Press
198 Madison Avenue, New York, NY 10016, United States of America.

Library of Congress Cataloging-in-Publication Data
Names: Robinson, Marin S., author. | Stoller, Fredricka L., author.
Title: Write like a chemist : a guide and resource / Marin S. Robinson, Fredricka L. Stoller.
Description: Second edition. | New York : Oxford University Press, 2022. |
Includes bibliographical references and index.
Identifiers: LCCN 2021051253 (print) | LCCN 2021051254 (ebook) |
ISBN 9780190098957 (paperback) | ISBN 9780190098940 (hardback) |
ISBN 9780190098971 (epub)
Subjects: LCSH: Chemistry—Authorship—Textbooks. | Communication in chemistry—Textbooks. | Technical writing—Textbooks.
Classification: LCC QD9.15 .R63 2022 (print) | LCC QD9.15 (ebook) |
DDC 808.06/654—dc23/eng/20211201
LC record available at https://lccn.loc.gov/2021051253
LC ebook record available at https://lccn.loc.gov/2021051254

DOI: 10.1093/oso/9780190098940.001.0001

Dedicated to
Jeff and Bill,
Chuck and Ronnie,
and Kara

Contents

List of Figures and Table with Typical Move Structures (i.e., Organizational Patterns)

Journal Article

Conference Abstract

Poster

Research Proposal

List of Tables with Common Verb Tense–Voice Combinations and Their Functions

Journal Article

Conference Abstract

Poster

Research Proposal

Preface

Write Like a Chemist, now in its second edition, is designed to be used as a *textbook* in upper-division and graduate-level university chemistry classes and as a *resource* for chemistry students, post-docs, faculty, and other professionals who want to perfect their chemistry-specific writing. To this end, *Write Like a Chemist* focuses on four types of writing:

- the journal article;
- the conference abstract;
- the scientific poster;
- the research proposal.

Each type of writing, or genre, is directed toward a distinct audience and written for a different purpose. For example, writing a journal article requires a style and organizational format that are quite different from a research proposal. Thus, to write like a chemist requires learning to write for multiple audiences and purposes.

One of the best ways to learn to *write* like a chemist is to *read* like a chemist. Indeed, many of today's chemists taught themselves to write by reading others' works and using those works as templates for their own writing. Corrections (often in red pen) from peers, mentors, reviewers, and editors along the way also played an integral role in the learning-to-write process. Although ultimately successful, this approach was often painful and inefficient for all involved. The goal of *Write Like a Chemist* is to teach writing in a more efficient way. Because reading is integral to writing, we use the chemical literature (and other examples of discipline-specific writing) to model conventional writing practices; indeed, excerpts from 9 ACS conference abstracts, 22 NSF CAREER proposals, 6 NSF Graduate Research Fellowship Program proposals, and more than 175 excerpts from ACS journal articles are included in the book. But we do not stop there. *Write Like a Chemist* endorses a *read–analyze–write* approach that combines the *reading* of authentic passages with the *analysis* of those passages to gain insights into the writing conventions of the targeted genres. Reading and analysis activities are supplemented by exercises that reinforce disciplinary writing conventions. Step-by-step Writing On Your Own tasks, posted on the *Write Like a Chemist* (2nd ed.) website, guide users of the book in using disciplinary conventions to write on their own, using excerpts as models.

Because it will take more than a single encounter with *Write Like a Chemist* to become a skilled writer, the book can serve as a rich resource over the years when the goal is to communicate effectively in writing with chemists, other scientists, and funding agencies.

Second Edition of *Write Like a Chemist*

The first edition of *Write Like a Chemist* was developed and piloted with the support of two awards from the National Science Foundation. The second edition continues and updates these efforts. The second edition (like the first) targets multiple instructional settings (including writing-dedicated courses, labs, and seminars) and a broad cross section of users (including chemistry students, post-docs, faculty, and other professionals from multicultural and multiple-language backgrounds). Although intended primarily for chemists, the book also benefits chemistry majors who ultimately decide to pursue other careers. Indeed, learning to write in a well-organized and concise manner requires writing skills that are highly coveted across many fields. Moreover, the read–analyze–write approach featured in *Write Like a Chemist* is readily transferable to other disciplines.

The second edition, updated to reflect current chemistry-specific writing practices, highlights disciplinary conventions set forth in the *ACS Guide to Scholarly Communication* (Banik et al., 2020). The second edition also includes (1) new excerpts from the primary literature and other chemistry genres; (2) updated tables and figures that can be consulted for chemistry-specific writing patterns and practices; (3) infographics developed by *inChemistry* to convey essentials for writing conference abstracts and preparing scientific posters; (4) a scientific poster template and corresponding model poster that can be used for poster creation; and (5) updated exercises and task types (that can be assigned in chemistry-writing classes and seminars, in addition to being used by others for self-study). The volume has been streamlined, without sacrificing thoroughness. Numerous additional resources have been placed on the *Write Like a Chemist* (2nd ed.) website (see pp xvii–xviii), as itemized later in the preface.

Unique Features of *Write Like a Chemist*

Write Like a Chemist is unique in many ways, not the least of which is the fact that it was conceived by a chemist (M.S.R.) and an applied linguist (F.L.S.) and developed early on with input from faculty and students in both disciplines. As part of our interdisciplinary effort, we analyzed chemistry-specific writing practices using tools from applied linguistics and corpus linguistics,[1] the latter a discipline that investigates language empirically through computer-based analyses of large collections of authentic texts known as corpora (or corpus, singular). Early in the project, a 1.5-million-word corpus of chemistry texts was created, comprising largely journal articles. This database was used to identify common, generalizable patterns in the language of chemistry. The ACS Journals Search was also used extensively to accomplish many of the same aims. Findings from both sets of analyses—centered on common chemistry-specific grammatical patterns, verb tense–voice combinations, word usage, phrasal patterns, and more—are included throughout the book.[2]

Writing conventions from the *ACS Guide to Scholarly Communication* (Banik et al., 2020) have been modeled throughout the volume, not only in excerpts from the primary literature. Yet, because writing conventions vary, and are likely to change with time, we urge readers to consult journals of their choice in addition to conference abstract, poster, and research proposal specifications as they prepare their written work.

Organization of the Book

Following the introductory chapter (Chapter 1), Section 1 of *Write Like a Chemist* is divided into three modules:

Module 1	Journal Article (Chapters 2–7)
Module 2	Conference Abstract and Scientific Poster (Chapters 8–9)
Module 3	Research Proposal (Chapters 10–13)

Each module combines authentic readings with exercises to introduce and reinforce chemistry-specific writing skills for the four targeted genres. *Writing on Your Own* tasks for each chapter (posted on the *Write Like a Chemist* (2nd ed.) website) guide writers in completing the type of writing emphasized in each module.

Section 2 of *Write Like a Chemist* focuses on writing conventions that run across different types of chemistry writing, including those genres highlighted in Section 1:

Chapter 14	Figures and Tables
Chapter 15	Citations and References
Chapter 16	Language Tips

The Language Tips chapter provides helpful hints about language areas that often prove troublesome for writers (e.g., conciseness; word choice; chemistry-specific writing conventions as related to abbreviations, capitalization, compound labels, numbers and units; grammar; punctuation; commonly confused words; and overall writing effectiveness). To facilitate self-study, language-tip review exercises and answer keys, in addition to proofreading recommendations, are posted on the *Write Like a Chemist* companion website.

Write Like a Chemist (2nd ed.) Website

Accompanying the book is the *Write Like a Chemist* (2nd ed.) companion website. The website includes student resources and password-protected instructor resources (see pp xvii–xviii).

Icons Used in *Write Like a Chemist*

Icons, each one with a special meaning, are used throughout the book:

Exercises

Indication of items posted on the *Write Like a Chemist* (2nd ed.) website

Definition of a key term or concept, elaboration of important points, and reminders about skilled writing

Reference to a later part of the book for additional information and/or practice

A useful principle (i.e., rule of thumb) with broad applications for skilled writing

In closing, we wish the users of *Write Like a Chemist* good luck. With this book and hard work, you will learn to write like a chemist!

Marin S. Robinson and Fredricka L. Stoller

Notes

1. Our early analyses of the language of chemistry were influenced by Biber, Conrad, and Reppen (1994, 1998), Hyland (1998), and Swales (1990, 2004), among others.
2. Conventional linguistic terminology is not always used when presenting language-related matters in the book. Rather, we use terminology that best reaches our intended audience (i.e., chemists).

Companion Website

A companion website is available for readers (i.e., students and instructors). The URL and login credentials for students are noted on the back cover. Instructor resources can be accessed by contacting Customer Service at Oxford University Press.

Student Resources (open to all)

- Writing On Your Own: Step-by-step supplemental guidance is provided for
  - writing journal article sections (Chapters 2–7), a conference abstract (Chapter 8), and research proposal sections (Chapters 10–13). Embedded into Writing On Your Own tasks are peer-review activities.
  - preparing a scientific poster (Chapter 9).
  - finalizing all written work.
- Peer-review checklists, to accompany peer-review activities.
- Self-study exercises, answer keys, and proofreading recommendations for each language tip in Chapter 16.
- A full-color, downloadable version of a scientific poster template in PowerPoint (Figure 9.1 in Chapter 9) that can be used for formatting scientific posters.
- A full-color, downloadable version of a scientific poster (Figure 9.6 in Chapter 9) that can serve as a model for creating a poster.
- "Canned" research projects. We recognize that not all users of *Write Like a Chemist* will have a sufficiently robust research project to write about. Four "canned" research projects, all based on authentic research, are available to fill this need. Each provides sufficient data and background information for a mock journal article or poster.

Instructor Resources (password required):

- Instructor Guidelines (with an orientation to the book and general suggestions for teaching writing).
- Answer keys for exercises in the book.
- Practice peer-review exercises (with suggested feedback) that can be assigned.
- Analytical and holistic scoring rubrics for major writing assignments (journal-quality paper, conference abstract, poster, and research proposal).
- Answer keys for canned research projects.

Acknowledgments

We gratefully acknowledge the U.S. National Science Foundation for supporting the initial Write Like a Chemist project (DUE 0087570 and DUE 0230913), which resulted in the first edition of *Write Like a Chemist* (2008, Oxford University Press). It was that experience, and the feedback received from those who used the first edition, that motivated us to write this second edition. We are indebted to Jeremy Lewis, Acquisitions Editor at Oxford University Press, for his decades-long support of our work.

We are also grateful for the support of the Departments of Chemistry & Biochemistry and English at Northern Arizona University (NAU), as well as our NAU colleagues and students. In particular, we thank Maria M. Bolar-Gastony, PhD student in Applied Physics and Materials Science at NAU, who assisted us with the answer key (posted on the *Write Like a Chemist* (2nd ed.) website for faculty consultation).

We are particularly indebted to the chemistry faculty who piloted early drafts of *Write Like a Chemist* at their home institutions. Their efforts not only influenced the first edition of the book, but also this second edition. (If faculty affiliations have changed, we note pilot institution in parentheses.) We also thank those pilot faculty (indicated with an asterisk aside their names) who contributed a quote to the book.

Frances Blanco-Yu*	Penn State New Kensington (Seton Hill University)
David Collins	Brigham Young University, Idaho (Colorado State University, Pueblo)
Ellen R. Fisher*	Colorado State University
Brian Gilbert	Linfield College
Alex Grushow*	Rider University
Angela Hoffman	University of Portland
Timm Knoerzer*	United States Air Force Academy (Nazareth College)
Daphne Norton	University of Georgia (Emory University)
Donald Paulson*	California State University, Los Angeles
Dan Philen	Emory University

Jennifer N. Shepherd	Gonzaga University
Betty H. Stewart*	University of Texas at Dallas (Austin College)
Joe Vitt	University of South Dakota
Carl Wamser	Portland State University
Barry L. Westcott	Central Connecticut State University

We also acknowledge colleagues and professionals who contributed one or more quotes to the book.

Joseph H. Aldstadt	University of Wisconsin, Milwaukee
Joan Curry	University of Arizona
Robert Damrauer	University of Colorado Denver
Charles H. DePuy	University of Colorado Boulder
David B. Knaff	Texas Tech University
Richard Malkin	University of California, Berkeley
Sharon R. Neufeldt	Montana State University
Bradley F. Schwartz	Southern Illinois University School of Medicine
James B. Weissman	Dicerna Pharmaceuticals
Thomas J. Wenzel	Bates College

We gratefully acknowledge American Chemical Society (ACS) Publications Support personnel, the ACS, and its partner Copyright Clearance Center's RightsLink for granting us permissions to reprint excerpts (including phrases, sentences, one or more paragraphs, titles, figures, and tables) from journal articles published by the ACS. An in-text citation accompanies each selection that is reprinted with ACS copyright permissions (e.g., "From Díaz-Mula et al., 2019" or "Adapted from Díaz-Mula et al., 2019"), and the corresponding bibliographic information (including a direct link to each article) is included in the reference list of cited works at the end of the book. Readers interested in further permissions to these materials should contact ACS. We also thank *inChemistry*, the ACS student member magazine, for permissions to reprint infographics on preparing conference abstracts and posters.

We gratefully acknowledge the following individuals who granted us permission to use excerpts from their successful NSF research proposals or ACS conference abstracts:

Recipients of NSF Graduate Research Fellowship Program (GRFP) Awards

Caitlin Nicole Cain	University of Washington, Seattle
Tucker Ryan Huffman	The Scripps Research Institute
Olivia Pearson (née George)	Georgia Institute of Technology
Fionna Morningstar Drozda Samuels	Colorado State University
Elizabeth Eileen Wait	University of Texas at Austin
Hannah Weaver	University of California, Berkeley

Primary Investigators of NSF CAREER Awards

Diana Aga	University of Buffalo
Daniel J. Dyer	Southern Illinois University, Carbondale
Howard Fairbrother	Johns Hopkins University
Nathaniel Finney	Tianjin University
Anna D. Gudmundsdottir	University of Cincinnati
Karen S. Harpp	Colgate University
Paul Hergenrother	University of Illinois, Urbana-Champaign
Robert P. Houser	University of Northern Colorado
Jeffrey S. Johnson	University of North Carolina, Chapel Hill
Gary R. Kinsel	Southern Illinois University Carbondale
Amnon Kohen	University of Iowa
Jeehiun Katherine Lee	Rutgers University
Gary A. Lorigan	Miami University
L. Andrew Lyon	Chapman University
David L. Patrick	Western Washington University
Christoph G. Rose-Petruck	Brown University
Andrei Sanov	University of Arizona
Eileen M. Spain	Occidental College
Mark E. Tuckerman	New York University
James R. Vyvyan	Western Washington University
Robert A. Walker	Montana State University
Timothy H. Warren	Georgetown University

Authors of American Chemical Society Conference Abstracts
(grouped by abstract)

Xu-Liang Cao	HEALTH CANADA, Ottawa, Canada
Betty Galarreta	Pontificia Universidad Catolica del Peru, Lima
Yulán Hernández	Pontificia Universidad Catolica del Peru, Lima
Lesly Lagos	Pontificia Universidad Catolica del Peru, Lima
Lorena Veliz	Pontificia Universidad Catolica del Peru, Lima
Amanda Garner	University of Michigan
Hankyeol Jung	University of California San Diego
Benjamin Zhou	University of California San Diego
Alina Schimpf	University of California San Diego
Paul Mahaffy	NASA Goddard Space Flight Center, Greenbelt, MD
Luke Whitaker	University of Strathclyde, Hitchin, UK
Katherine Wheelhouse	GSK, Stevenage, UK
William Kerr	University of Strathclyde, Hitchin, UK
Marc Reid	University of Strathclyde, Hitchin, UK
Yichao Yan	University of Michigan
Sophia Robinson	University of Utah
Koen Hendriks	University of Michigan
Matthew Sigman	University of Utah
Melanie Sanford	University of Michigan
David Yingzhao Zhao	Northeastern University, Malden
Roman Manetsch	Northeastern University, Malden
Xiaolin Zhu	San Diego State University
Yixiong Lin	San Diego State University
Yue Sun	San Diego State University
Dian Zhu	San Diego State University
Yong Yan	San Diego State University

We thank Dr. Roald Hoffmann for permission to reproduce his poem "Next Slide Please" from *The Metamict State* (1987; Orlando: University of Central Florida Press, pp. 51–52).

Closer to home, we express our heartfelt appreciation to Jeff Robinson and Bill Grabe for their patience and ongoing support of our decades-long Write Like a Chemist project and this second edition.

The opinions, findings, conclusions, and recommendations expressed in this book are those of the authors and do not necessarily reflect the views of the National Science Foundation, the American Chemical Society, or authors whose works are included in *Write Like a Chemist*.

Marin S. Robinson and Fredricka L. Stoller

Section 1

Writing Modules

1 *Learning to Write Like a Chemist*

Writing, more than any other skill developed as a chemistry student, has enabled me to advance my career.
—James B. Weissman, Dicerna Pharmaceuticals

Welcome to *Write Like a Chemist*. If you are reading this book, we suspect that you are a chemistry major and rather than finding our title humorous (write like a *chemist*, really?), you want to do just that—learn to write in chemistry. Good for you! In this chapter, designed to get you started, we introduce our basic approach to writing in chemistry and offer an overview of the book's contents. By the end of this chapter, you should be able to do the following:

- Identify common writing genres in chemistry and those covered in this book
- Describe the five essential components of genre analysis and explain why genre analysis is so useful for developing writers
- Explain what is meant by audience and identify the audiences addressed in this book
- Differentiate between broad and fine organizational structures
- Explain the meaning and significance of a move and a move structure
- Make use of the book's approach to improve your chemistry writing skills

Many effective writers develop their discipline-specific writing skills by reading and analyzing the works of others in their fields. Learning to write in chemistry is no exception; chemistry-specific writing skills are developed by reading and analyzing the writing of chemists. We coined the phrase "read–analyze–write" to describe this approach, which is modeled throughout the book. In this chapter, we lay the foundation for the read–analyze–write approach by analyzing a few common, nonscientific examples of writing. We use

Write Like a Chemist. Marin S. Robinson and Fredricka L. Stoller, Oxford University Press. © Oxford University Press 2022.
DOI: 10.1093/oso/9780190098940.003.0001

these everyday examples (e.g., recipes, letters, and jokes) to introduce you to the process of analyzing writing and to share with you the tools that you will need to analyze chemistry writing in subsequent chapters.

Genre

I had no idea how much time I would spend writing in my career as a chemistry professor at an undergraduate institution. With course materials, grant proposals, and research papers, I am always writing something.

—Thomas J. Wenzel, Bates College

Unless you are reading this chapter very early in the morning, you have likely already encountered several different types of writing today. Posts on social-media platforms, emails, text messages, articles, and news briefs are just a few examples of writing that you might read on a daily basis. You may have also glanced at some chemistry-specific writing in lab manuals, course notes, reference books, or chemical catalogs. Each of these types of writing is unique and distinguishable from the others. This is true even if they share overlapping content. For example, information about the chemical properties of ethanol is presented differently in an organic chemistry textbook, chemical catalog, and online chemical dictionary.

The word used to describe these different types of writing is **genre**. For example, there are different genres in literature (e.g., poems, short stories, romance novels) and in film (e.g., comedy, horror, mystery). Different genres exist in chemistry, too. Although the word may sound a little funny at first, you will soon learn that recognizing a chemistry genre is the first step toward writing successfully in that genre.

Genre

A type of writing that is distinguished from other types of writing because of differences in content, form, style, audience, purpose, and context.

This book focuses on four genres commonly read and written by chemists; these genres are addressed in three modules:

- Journal Article (Module 1)
- Conference Abstract and Scientific Poster (Module 2)
- Research Proposal (Module 3)

Exercise 1.1.

Make a list of four to five genres that a college student majoring in chemistry might read or write. Make a second list with three to five genres that a professional chemist in academia, industry, or a government lab might read and write. How do these genres differ from one another?

At the core of the read–analyze–write approach is **genre analysis**, a systematic way to read and analyze writing. Through genre analysis, you will identify and examine essential components of a genre, thereby facilitating your ability to write that genre effectively. This book focuses on five such components: audience and purpose, organization, writing conventions, grammar and punctuation, and science content. As shown in Table 1.1, each component can be further divided into two or more subcomponents. Our goal is to teach you to analyze chemistry-specific writing according to these components and subcomponents. To get you started, we begin by identifying these components in some familiar types of writing (not in chemistry).

Genre Analysis

A systematic way of analyzing a genre to identify its distinguishing features.

Table 1.1. Components of Genre Analysis Addressed in *Write Like a Chemist*

audience and purpose	organization	writing conventions	grammar and punctuation	science content
conciseness level of detail level of formality word choice	broad structure fine structure ("moves")	abbreviations and acronyms capitalization formatting verb tense voice	commas, colons, semicolons parallelism subject–verb agreement word usage	graphics text

Audience and Purpose

Writing is a process, not a one-time event.

—Ellen R. Fisher, Colorado State University

Before you begin to write, you must decide who will read your work (your **audience**) and why you are writing (your **purpose**). In turn, your audience and purpose will influence the levels of detail, formality, and conciseness that you use in your writing and the words that you choose to use. To illustrate, consider two everyday genres: a recipe and a shopping list. The recipe is written to instruct a hopeful chef (audience) how to prepare a particular type of food (purpose); the shopping list is written to remind a shopper (audience) what items to buy (purpose). Because of their different audiences and purposes, detailed instructions are needed in the recipe but not in the shopping list. (Imagine how useless a recipe would be if it included only a list of ingredients or how unwieldy a shopping list would be if it included instructions for locating each item in the store!) Moreover, because recipes are often published, the writing is more formal, with titles, headings, lists of ingredients presented in a parallel fashion, and unambiguous, fully punctuated sentences (e.g., "Melt 2 tsp. butter in a small saucepan."). Shopping lists, on the other hand, are scrawled out in personal shorthand (e.g., "choc, OJ, mlk") with no titles, headings, or punctuation. Thus, we can see how audience and purpose influence levels of detail, formality, and conciseness in particular genres.

Audience and Purpose

Audience refers to the people who will most likely read a specific piece of writing. **Purpose** refers to the aims, goals, or intentions of the writer.

As a second example, consider two genres of letters: a cover letter for a job or internship and a thank you note to a friend. These two types of correspondence are sent to different individuals (audience) with whom the writer has different relationships and for entirely different reasons (purpose). These differences are reflected not only in tone (i.e., the cover letter is formal and professional, whereas the thank you note is personal and friendly) but also in characteristic phrases. A potential employer would be quite surprised to read a cover letter signed "Hugs, Mario" as would a friend reading a thank you note beginning with "To Whom It May Concern." **Word choices** such as these are anticipated by readers. Choosing the right word(s) requires careful consideration (Figure 1.1). If expected words are missing, or a wrong phrase is used, readers will have a difficult time recognizing the genre and even understanding the author's purpose. Hence, by learning words and phrases that are characteristic of a genre, you can make your own writing sound more like a typical example of that genre.

Figure 1.1. Even Snoopy struggles for just the right words.

Word Choice

Readers expect characteristic words and phrases to be used in a genre (e.g., the word *Discussion* is used to demark the start of a journal article Discussion section). Effective writers must learn to incorporate these words into their writing.

Keep in mind that both audience and purpose define a genre. Two pieces of writing with the same intended audience may be written very differently if they have different purposes. For example, a university catalog and a university student newspaper are both written for a student audience, but the two publications are distinct from one another in many ways (e.g., organization and content).

With these everyday examples in mind, let's consider audience and purpose for chemistry-specific genres. We begin with audience. Chemists write for many different audiences, including students, teachers, professional chemists, and funding-agency proposal reviewers, to name only a few. Thus, it is instructive to divide audience into different categories. For our purposes, we consider four categories: expert audience, scientific audience, student audience, and general audience. The **expert audience** includes professional chemists with advanced knowledge in a subdiscipline of chemistry, such as biochemistry, analytical chemistry, or organic chemistry. The subdiscipline is often reflected in names of journals written for experts in that field (e.g., *Biochemistry*, *Analytical Chemistry*, or *The Journal of Organic Chemistry*). The **scientific audience** comprises readers with scientific backgrounds but not necessarily in the authors' field or subdiscipline. For example, a biologist or geologist asked to review a chemist's research proposal would be considered a scientific audience. The **student audience** consists of individuals who are reading to learn chemistry at any level, such as in high school (with an introductory chemistry book) or in graduate school (with a book on quantum mechanics). The **general audience** includes readers who are interested in a chemistry topic but with little to no formal training in chemistry, such as an English or history teacher reading *Science News* or *Popular Science*.

Audiences for Chemistry Writing

Expert	Readers with expert-level knowledge in a specific area of chemistry.
Scientific	Readers with significant scientific knowledge but not in the specific area targeted in the written work.
Student	Readers learning chemistry.
General	Readers with little or no chemistry knowledge.

Together, these four audiences form a continuum that spans a wide range of expertise in chemistry. In general, journal articles are written for an expert audience, research proposals and scientific posters for a scientific audience, textbooks for a student audience, and popular science articles for a general audience. A genre can change position on the continuum if an audience is expected to have more (or less) chemistry-specific knowledge. For example, a poster presented at a highly technical conference should address an expert audience, but a poster presented at an undergraduate research symposium should target a student or general audience. Moreover, a single genre often addresses more than one audience. Although a journal article is written primarily for an expert audience, parts of its Introduction are often written for a scientific audience. You can see that determining your audience is an integral part of the writing process.

In this book, we focus on two audiences: the expert audience in Module 1 (Journal Article), the scientific audience in Module 2 (Conference Abstract and Scientific Poster), and both expert and scientific audiences in Module 3 (Research Proposal). Other genres, such as technical memos and reports, target these same audiences, but they are not covered in this book. An important goal of this book is to help you move beyond writing student genres (e.g., undergraduate lab reports) and begin writing for expert and scientific audiences.

Closely linked to audience is purpose; a genre is also influenced by the purpose for the writing. Several different purposes for chemistry-specific writing are listed here (with representative genres shown in parentheses). This book addresses only the first two of these purposes:

- To present research results or convey new scientific insights (journal articles, conference abstracts, and posters)
- To persuade funding agencies to provide financial support (research proposals)
- To teach or instruct (textbooks)
- To convey instructions (lab or operating manuals)
- To provide chemical information (safety data sheets)
- To communicate with colleagues (memos or emails)

Exercise 1.2.

Look back at the lists of chemistry genres that you created in exercise 1.1. Who is the primary audience for each genre: general, student, scientific, and/or expert? Some genres will target a single audience; others will target multiple audiences.

Exercise 1.3.

Many scientific journals publish articles for more general audiences in addition to the articles written for experts. For example, the *Journal of the American Chemical Society* (*JACS*) includes a *Spotlights* section in which science writers "translate" for non-experts the scientific discoveries reported in recent *JACS* articles. In this exercise, we ask you to compare the writing in a *Spotlights* article (McGuire, 2020) with the writing in the corresponding journal article written for experts (Li et al., 2020).

a. Compare the titles of the two articles. What difference(s) do you notice?

Spotlights article	Cryogenic Adhesive: Just Add Water
Journal article	Supramolecular Adhesion at Extremely Low Temperatures: A Combined Experimental and Theoretical Investigation

b. Now read and compare excerpts from the two articles. Identify at least two differences in the writing styles for the two audiences.

Spotlights section	At low temperatures, many types of adhesives become brittle or loosen their grip. Now, Gai Zhao, Shengyi Dong, and co-workers have developed a strong, long-lasting, low-temperature copolymer adhesive made from an organic monomer and water.
Journal article	Herein, we report the application of supramolecular polymers as adhesive materials at low temperatures. Tough and reversible adhesion effects were observed when the supramolecular polymers were used at extremely low temperatures, with strong and durable adhesion performance being retained even in liquid nitrogen (−196 °C).

Organization

The second essential component of genre analysis is organization. When you decide to write in a particular genre, you implicitly agree to follow the genre's organizational structure. Such is the case for romance novelists, Disney scriptwriters, and "whodunit" mystery writers; all must adhere to a time-honored formula (or risk having their work remain unpublished). Indeed, one of the best ways to learn to write in a new genre is to analyze its organizational structure.

Organizational structure can be divided into broad and fine features. Broad structural features are indicated, for instance, by readily identifiable sections or headings (e.g., Introduction, Results, and Discussion); fine structural features are identified by patterns of organization within paragraphs and within sections (e.g., from more general to more specific) and by transitions between paragraphs. Both sets of features contribute to the readability and flow of the written piece. A joke book, for example, can be broadly organized by type (e.g., knock-knock jokes, light bulb jokes, genie jokes), but a single joke can also be organized into finer segments (e.g., opening line, punch line). We can think of these finer structural features as the many steps (or **moves**) that writers take to progress from the beginning to the end of each section of their writing, always with the purpose of communicating clearly with their intended audience. Writers who make use of conventional moves in their written work meet the organizational expectations of their intended audience. (Although the move concept, like the term *genre*, may sound odd, its utility will become clearer as you progress through the book.)

A Move

A step taken by writers to achieve part of their overall purpose. Writers who use conventional moves in their written work meet the organizational expectations of their readers.

The move concept is a bit easier to illustrate with an example; hence, we do this with a joke. The overall purpose of a joke is to make the reader (or listener) laugh. The smaller parts of the joke—the moves—serve as building blocks to help the writer (teller) achieve the overall purpose of the joke. Our sample joke is a variation of a "guy-walks-into-a-bar" joke. (We could not resist this joke because it pokes fun at incorrect punctuation.) In Table 1.2, the joke itself is in the left-hand column, the moves are in the center column, and the sentences that accomplish the moves are in the right-hand column. The joke is told in seven moves: the joke setup, a four-step action/reaction sequence between the guy (panda) and the bartender, the punch-line setup, and the punch-line delivery. The sequencing of moves plays an important role in achieving the purpose of the joke; if the moves were sequenced differently (e.g., if the punch line were given

Table 1.2. **Variation on a Guy-Walks-into-a-Bar Joke and Its Moves**

joke[a]	moves	illustration of moves
A panda walks into a bar and orders a drink. When he's done, he draws a gun and fires two shots into the air. "What was that for?" asks the confused bartender. The panda produces a badly punctuated wildlife manual and tosses it over his shoulder. "I'm a panda," he says, at the door. "Look it up." The bartender turns to the relevant entry and, sure enough, finds an explanation. "**Panda**. Large, black-and-white bear-like mammal, native to China. Eats, shoots, and leaves."	Set up the joke	A panda walks into a bar and orders a drink.
	Describe guy action #1	When he's done, he draws a gun and fires two shots into the air.
	Describe bartender reaction #1	"What was that for?" asks the confused bartender.
	Describe guy action #2	The panda produces a badly punctuated wildlife manual and tosses it over his shoulder. "I'm a panda," he says, at the door. "Look it up."
	Describe bartender reaction #2	The bartender turns to the relevant entry and, sure enough, finds an explanation.
	Set up the punch line	"**Panda**. Large, black-and-white bear-like mammal, native to China.
	Deliver the punch line	Eats, shoots, and leaves."

[a] Joke adapted from Truss (2003).

first), the joke would no longer be successful. Thus, the appropriate moves must be present and sequenced in the correct order.

In addition to listing the moves, as we did in Table 1.2, moves can also be represented graphically using a diagram similar to a flow chart, which we refer to as a **move structure**. A move structure illustrates required moves, optional moves (when appropriate), and the sequence of moves, including any repeated move patterns. A move structure for the panda joke is illustrated in Figure 1.2.

A Move Structure

A flowchart-like representation of the moves within a genre. The diagram visually depicts required and optional moves, repeated moves, and the sequencing of moves.

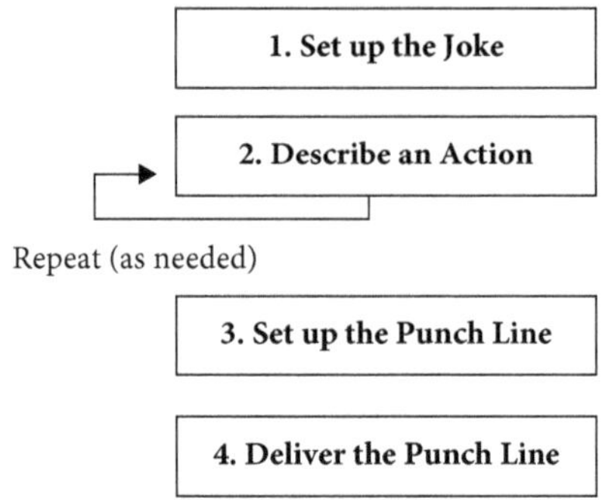

Figure 1.2. A visual representation of the move structure for the panda joke in Table 1.2.

Analogous move structures are used throughout this book to illustrate major sections of the journal article, conference abstract, poster, and research proposal. The move structures are meant to guide you in reading and writing these genres. Important features of the move structures in this book include the following:

- Moves are numbered to convey their conventional order in a genre (e.g., move 2).
- Some moves are divided into submoves. If numbered (e.g., submoves 2.1 and 2.2), the numbering conveys their conventional order in the genre. If unnumbered, they can be addressed in any order.
- Each move and submove begins with an action verb (e.g., state, identify, provide).
- Unless stated otherwise, a move or submove is required; without the move or submove, the genre would be incomplete and ineffective (e.g., imagine a joke that omits the punch line).

Some moves or submoves are not required in all instances, indicated in this book by the words *if applicable* or *optional* (in parentheses). *If applicable* means that the move is required only when appropriate. For example, in the Methods section of a journal article, the move "Describe Numerical Methods" is labeled *if applicable* (see Figure 3.1 in Chapter 3) because not all authors use numerical methods in their work. Similarly, the submove "preview key findings" in the Introduction section is labeled *optional* (see Figure 6.1 in Chapter 6) because it is the author's prerogative to include that move or not. Of course, we cannot possibly know which moves or submoves are required for your written work, so, as a writer, you will need to determine which moves and submoves are applicable or optional for your own purposes.

Required Moves

Most moves are required; that is, the genre would be incomplete or unrecognizable without them. Occasionally, moves or submoves are required only in some instances or are left to the discretion of the writer. We label such moves *if applicable* and *optional*, respectively.

Exercise 1.4.

You are likely already familiar with the Acknowledgments section of a journal article. This short section is commonly included at the end of an article, just before the references. Select an American Chemical Society journal of your choice and browse through several different acknowledgments. Using the move structure in Figure 1.2 as a guide, propose a move structure for this section. Begin each move with an action verb. Number your moves to convey their order in the genre. Read enough different acknowledgments to identify at least one required move and one optional move (i.e., a move that does not appear in all examples). Be sure to indicate which move(s) are optional in your move structure.

The Right Answer?

Exercise 1.4, like many exercises in this book, is designed to get you thinking about writing; hence, it will have several "right" answers (although some answers may be better than others).

Writing Conventions

Every discipline has its own writing conventions (the third essential component of genre analysis), and chemistry is no exception. **Writing conventions**, as the name implies, are generally accepted (and expected) practices in a given discipline; they are not "right" in the absolute sense (unlike most rules of grammar and punctuation). The American Chemical Society (ACS) documents generally accepted discipline-specific writing practices in its *ACS Guide to Scholarly Communication*. Your goal should be to recognize and learn chemistry-specific writing conventions and use them in your writing properly.

Writing Conventions

Discipline-specific rules of writing, established by major professional associations and journals. Examples include how to format graphics, how to cite references, what to capitalize, and whether to use past or present tense.

American Chemical Society (ACS)

The premier American professional organization for chemists, chemical engineers, and other professionals interested in chemistry.

In chemistry-specific genres, there are **formatting** conventions, as one example, for tables, figures, in-text citations, references, and number–unit combinations, to name only a few. Formatting rules reflect common practices and readers' expectations in areas such as font size (e.g., in poster titles), bolding (e.g., in labeling, where **1** can be used to represent a chemical compound), italics (e.g., in references, where volume numbers are italicized), and placement (e.g., in citations, where numbers are superscripted). The rules also dictate whether to include a space between a number and its unit. For example, which is correct: 10mm or 10 mm, 100° C or 100 °C? Formatting conventions will help you answer this question.

Formatting

Writing conventions specific to a discipline (like chemistry) that dictate the appearance and physical placement of written elements in, for example, tables, figures, references, headings, and number–unit combinations.

The use of **abbreviations** and **acronyms** is another writing convention that spans the genres of chemistry. In every genre, we see abbreviations and acronyms for words and phrases that are used repeatedly in that genre (and often, these abbreviated terms appear to be a "foreign language" to newcomers to that genre). To write effectively in chemistry, you need to learn standard abbreviated terms. Although you may already be familiar with "m" for meters and "g" for grams, you may be less familiar with "µg" for micrograms and "h" for hours. With abbreviations such as these, chemists omit the "s" for plural units ("g" not "gs" for grams) and seldom use periods ("min" not "min." for minutes).

Genres also vary by their conventional uses of verb tense (past, present, or future) and voice (active or passive voice). For example, most jokes are told in present tense (e.g., "A panda *walks* into a bar and *orders* a drink" as opposed to "A panda *walked* into a bar and *ordered* a drink"). Present tense is used to make the joke more vivid for the listener or reader. Jokes also tend to be told in active rather than passive voice:

Active	A panda walks into a bar and orders a drink.
Passive	A bar is entered by a panda and a drink is ordered.

In this book, we examine how tense and voice are used in chemistry journal articles, conference abstracts, posters, and proposals. As we will see, different tense–voice combinations are used, depending on which genre, or section of a genre, is being written.

Abbreviations and Acronyms

Abbreviations and **acronyms** are agreed-upon short forms for commonly used words and units. When spoken aloud, abbreviations are often pronounced letter by letter (e.g., A-C-S), whereas acronyms form a pronounceable word (e.g., NASA).

Writing conventions may seem a bit picky to you at this point; however, by adhering to chemistry writing conventions, you take an important step toward sounding like an expert chemist. If you submit a journal article, for example, with improperly formatted units and figures, incorrect abbreviations, and inappropriate verb tenses (e.g., present tense in sentences that are conventionally written in past tense), readers may judge you as a careless scientist and dismiss your work.

Grammar and Punctuation

The fourth component of genre analysis addressed in this book relates to **grammar** and **punctuation**. Unlike writing conventions, the rules of grammar and punctuation apply across the disciplines, to many formal written genres. (The rules may be altered intentionally in creative writing genres such as poetry or fiction writing.) Although grammar and punctuation are not the focus of this book, we do point out common pitfalls experienced by novice writers, such as errors with colons and semicolons, subject–verb agreement, and **word usage** (e.g., affect vs. effect).

Grammar and Punctuation

Grammar	Rules for combining words into sentences (e.g., subject–verb agreement).
Punctuation	Written marks (e.g., commas, colons, semicolons, hyphens) governed by rules and used for communicating clearly with one's audience.

In general, rules of grammar and punctuation are followed across formal written genres (and academic disciplines).

Word Usage

The term *word usage* refers to correct and incorrect uses of words and phrases. For example, there is a right and wrong way to use such words as *affect/effect* and *spectra/spectrum*.

In contrast, the term *word choice* refers to decision(s) that writers make when choosing among several conventionally accepted words and phrases for a particular audience.

The panda joke (Table 1.2), adapted from *Eats, Shoots & Leaves: The Zero Tolerance Approach to Punctuation* (Truss, 2003), illustrates the importance of correct punctuation (specifically, the troublesome comma). How should the punch line be punctuated in the panda joke to provide a proper definition of a panda? Here is what we recommend:

Punch line	**Panda** Large, black-and-white bearlike mammal, native to China. Eats, shoots, and leaves.
Corrected	**Panda** Large, black-and-white bearlike mammal, native to China. Eats shoots and leaves.

Like punctuation, the misuse of commonly confused words (e.g., affect/effect, compose/comprise, fewer/less, its/it's) can result in miscommunication and undermine the message conveyed in your writing.

Exercise 1.5.

Consider these sentences. Choose the correct word in each. What rule guided your choice? See Chapter 16, "Language Tips," Part 6: Word Usage, for help, if needed.

a. The human wrist is *comprised/composed* of eight bones, but the ankle has only seven.

b. *Fewer/Less* samples were used in the original series.

c. The new procedure *affected/effected* the yield.

d. The instrument was chosen for *its/it's* detection limits.

Science Content

It goes without saying that having a crystal-clear understanding of a subject is a prerequisite to effectively writing about it.

—Joseph H. Aldstadt, University of Wisconsin–Milwaukee

The fifth and last essential component of genre analysis addressed in this book is science **content**. It is impossible to write clearly and effectively if you lack a clear understanding of the chemistry involved; understanding the chemistry and writing about that chemistry go hand in hand. Writing is also a useful tool for learning chemistry. Chemists often "think through their hands" (i.e., through writing). You will find that as you write about your science, you will gain deeper insights into and knowledge of that science.

Content

The topic(s) covered in a given genre; content is expressed through both text and graphics.

Every genre has rules (often unspoken) restricting content for that genre. A joke restricts content by appropriateness; depending on the audience, some content may be viewed as offensive rather than funny. Journal articles also restrict content; a chemistry-related journal article must be written about novel research in a subfield of chemistry. Although a chemist could write a paper that describes how a cake is baked ("After mixing, the ingredients were heated in an oven for 60 min at 176 °C."), we all know that such a paper would never be published in a chemistry journal, even if it adhered to all other defining characteristics of that genre. Written work will not be recognizable as an instance of a genre if the content differs from the expectations of the genre. Thus, a genre requires not only appropriate organization and language, but also appropriate content.

Chemistry content is typically expressed in two ways: text (prose, written language) and graphics (tables, figures, and schemes). One key to effective chemistry writing is the appropriate use of both forms of expression. In this book, we illustrate how chemists use text and graphics to communicate content with their intended audience(s) and how authors weave back and forth between the two to tell a story of scientific discovery.

As you analyze the ways in which chemists communicate content, don't be surprised if you learn some new chemistry, too. Although the primary focus of this book is writing, we believe that your chemistry knowledge will also expand as you read, analyze, and write.

What to Expect

My graduate advisor told me something many times that I now tell my students: Writing is thinking. To truly communicate a scientific idea in a precise written form, one really needs to think long and hard about the best way to accomplish that goal.

—Alexander Grushow, Rider University

Each module in Section 1 of this book begins with an introduction to the targeted genre, including an overview of the sections that commonly characterize the genre. The remainder of the module examines each section in greater depth. For example, Module 1 (Journal Article) includes an introductory chapter followed by individual chapters dedicated to the different sections of a journal article: Abstract, Introduction, Methods, Results, and Discussion (although not in that order). Most chapters begin with an authentic example of the targeted section from the chemistry literature (i.e., from ACS journals). You are asked to read the example (multiple times) and analyze it for its essential components (audience and purpose, organization, writing conventions, grammar and punctuation, and science content). The remainder of the module includes excerpts from the chemistry literature, explanations, and exercises designed to strengthen your ability to read, analyze, and write in that genre. Module 2 (Conference Abstract and Scientific Poster) and Module 3 (Research Proposal) are organized similarly to Module 1; they include introductions to the genre, models of the genre, a discussion of each section of the genre, and exercises. Through the combination of reading, analyzing, and writing in Section 1, you will learn to recognize the defining characteristics of four important genres in the field of chemistry and to incorporate those characteristics into your own writing of those genres.

Section 2 includes chapters that are relevant to the four genres covered in Section 1. In these chapters, you will learn to format figures and tables (Chapter 14) as well as citations and references (Chapter 15). The last chapter (Chapter 16, "Language Tips") provides additional details about many of the language features and chemistry conventions highlighted throughout the book. Explanatory notes and illustrative sentences and passages from the chemistry literature are included. Practice exercises, answers, and proofreading suggestions are posted on the *Write Like a Chemist* (2nd ed.) website to facilitate self-study and review.

Write Like a Chemist (2nd ed.) Website

The *Write Like a Chemist* (2nd ed.) website includes Writing on Your Own tasks that guide you, step by step, in (a) writing and revising sections of each target genre addressed in Section 1, (b) finalizing your written work, (c) formatting figures and tables, and (d) formatting citations and references.

Exercise 1.6.

Skim the poem "Next Slide, Please," written by Roald Hoffmann (Hoffmann, 1987, pp 51–52), who in 1981 shared the Nobel Prize in Chemistry with Kenichi Fukui. In this poem, Hoffmann pokes fun at seminar presentations. The poem serves as an example of a genre that differs in so many ways from the scientific papers written by this world-renowned chemist.

a. Who is Hoffmann's audience?
b. What was his purpose for writing this poem?
c. Although you have probably not read any of Hoffmann's many scientific papers, speculate on ways in which this poem is different in organization, writing conventions, grammar and punctuation, and science content from the many journal articles that he has written for expert audiences.

Next Slide, Please

there was no question that the reaction worked
but transient colors were seen
in the slurry of sodium methoxide in dichloromethane
and we got a whole lot of products
for which we can't sort out the kinetics
the next slide will show
the most important part
very rapidly
within two minutes
and I forgot to say on further warming
we get in fact the ketone
you can't read it on the slides
but I refer to the structure you saw before
the low temperature infrared spectrum
as I say
gives very direct evidence
so does the NMR
we calculated it

throwing away the geminal coupling
which is of course wrong
there's a difference of 0.9 parts per million
and it is a singlet
and sharp
which means two things
either
you're doing the NMR in excess methoxide
and it's exchanging
or
I would hazard a guess
that certainly in these nucleophilic conditions
there could well be
an alternative path
to the enone you see there
it's difficult to see
you could monitor this quite well in the infrared
I'm sorry in the NMR
my time is up I see
well this is a brief summary of our work
not all of which
I've had time to go into
in as much detail as I wanted
today.

Chapter Review

As a review of what you've learned in this chapter, define each of these terms, in the context of this chapter, for a friend or colleague who is new to the field of chemistry:

ACS	genre	optional move	word choice
audience	genre analysis	purpose	word usage
content	move	required move	writing conventions
formatting	move structure		

Similarly, explain the following to a friend or colleague who has not yet given much thought to chemistry genres geared toward expert and scientific audiences:

- Genres commonly used by chemists at various levels in their training
- Five components of genre analysis and how they facilitate the read–analyze–write approach to writing in chemistry

- Audiences that scientific writing typically addresses
- Common purposes for scientific writing
- Differences between broad and fine organization
- Relationship between a genre's move structure and its organization
- Examples of formatting, word choice, and word usage that are both appropriate and inappropriate in chemistry-specific writing genres
- Two different means of communicating science content

Module 1 *Journal Article*

2 *Overview of the Journal Article*

An author should recognize that journal space is a precious resource created at considerable cost. An author therefore has an obligation to use it wisely and economically.

—American Chemical Society, *Ethical Guidelines to Publication in Chemical Research*

This chapter introduces the journal article module. The chapter describes some defining characteristics of journal articles while emphasizing concise writing and organization. By the end of this chapter, you should be able to do the following:

- Recognize the importance of concise writing
- Identify the broad organizational structure of journal articles
- Explain what is meant by targeted reading

Writing on Your Own: Getting Started

Visit the *Write Like a Chemist* (2nd ed.) website for step-by-step guidance in getting started on writing your own journal-quality paper.

Module 1 focuses on writing a journal-quality paper, a paper suitable for submission to a refereed chemistry journal. **Refereed journals** only include articles that have made it through a rigorous peer-review process. In this process, a submitted manuscript is critically reviewed by two or more anonymous reviewers. The reviewers are asked to judge the manuscript's scientific merit and writing quality. Authors are often required to revise their work before it is accepted for publication.

Write Like a Chemist. Marin S. Robinson and Fredricka L. Stoller, Oxford University Press. © Oxford University Press 2022.
DOI: 10.1093/oso/9780190098940.003.0002

The entire review process can take six months or longer. An account of the review process typically appears in the published article, for example:

Received for review Jan. 12, 2021. Revised manuscript received March 5, 2021. Accepted for publication April 2, 2021.

Once published, the journal article becomes part of the **primary literature** of chemistry. The primary literature is a permanent and public record of all scientific works, many of which are refereed journal articles.

Refereed Journals

Refereed journals only publish papers that have gone through a rigorous peer-review process. Submitted manuscripts are evaluated by experts (peers) for quality and originality. Based on reviewers' remarks, journal editors decide to accept, accept with revisions, or reject each submission.

Primary Literature

The **primary literature** comprises peer-reviewed publications that report and interpret results of original research. In general, these publications are the first and most authoritative record of the work.

In this chapter, we take a bird's eye view of the journal article. We consider the journal article's audience and purpose, stress the importance of concise writing throughout the journal article, and examine the broad organizational structure of the journal article. In subsequent chapters, we will examine sections of the journal article in more depth.

Audience and Purpose

The major purpose for writing a chemistry journal article is to share the results of original research with other chemists. The primary audience for a journal article is an expert one; readers are typically well educated and highly experienced in the subfield of chemistry addressed in the article. Because journal articles are written largely for experts, newcomers to the field (e.g., students or chemists exploring a new research area) are often frustrated by the advanced level of these articles. Details are often omitted that the non-expert reader would find useful. (If you find yourself in this situation, we recommend that you consult related works written for less experienced audiences—textbooks, review articles, general science articles—to help you work your way through the journal article.)

Although the bulk of the journal article is written for experts, a few sections are often accessible to less sophisticated readers. For example, general or summative remarks in the abstract, Introduction, and conclusions are often accessible to a scientific audience, allowing those readers to grasp key concepts of the work. Similarly, many chemistry journals include features (e.g., book reviews, editorials, and news articles) that are written specifically for scientific and general audiences.

Exercise 2.1.

Browse through a research article in three different ACS journals (*Analytical Chemistry*, *Biochemistry*, and *Environmental Science & Technology*) and complete these tasks:

a. Find two or three sentences that are easy to understand in each section of the article (abstract, Introduction, Experimental, Results, and Discussion). Next, find two or three sentences that are difficult to understand in each section. What differences, if any, do you notice in the readability of these sections? Which sections are the easiest to understand? Why do you think some sections are easier to read than others?
b. Read one of your three articles more carefully. What makes the authors sound like experts? Jot down a few examples of expert-like writing.

Exercise 2.2.

Browse through the Table of Contents of several issues of *Environmental Science & Technology* or *ACS Nano*. In most issues, you will see that, in addition to research articles, the journal also contains features such as editorials, comments, viewpoints, and reviews. Glance through the pages of two such items. For each item, identify the intended audience and purpose of the entry.

Conciseness

I find that there is nothing more tedious than papers that go on and on, with no obvious point.

—Richard Malkin, University of California–Berkeley

Recall from Table 1.1 in Chapter 1 that audience and purpose are communicated through four subcomponents (conciseness, level of detail, level of formality, and word choice). The first of these, conciseness, is a hallmark of writing in chemistry. Chemistry readers (experts and non-experts alike) want crisp, clean sentences that say what needs to be said and no more. They do not want to be bogged down in words that fail to advance or, worse, that confuse meaning. Because concise writing is important in every section of the journal article, we address it here, in Chapter 2, and throughout the module.

Novice writers often equate wordy writing with expert writing. They adapt a wordy and pretentious writing style because (1) they want to make their papers longer (not at all a goal in scientific writing!) and (2) they want to sound more professional. However, most professionals (particularly scientists) prefer a more concise, direct style. As an example, going all the way back to 1978, scientists attending a British Ecological Society conference were asked to read two texts that presented the same information, although one text was considerably wordier than the other (Turk, 1978). Overall, the scientists rated the concise version as being easier to read and more appropriate than the wordy version. The scientists also asserted that the more concise author was more objective, had a more organized mind, and inspired more confidence in the work. Views about the importance of conciseness remain the same today. Thus, wordy language, rather than making you sound more professional, can obscure your message and discredit your objectivity.

Wordiness

The *ACS Guide to Scholarly Communication* (Banik et al., 2020, section 5.1.3) recommends that these phrases be omitted from papers because they are vacuous and contribute to wordiness:

- As already stated
- It has been found that
- It has long been known that
- It is interesting to note that
- It is worth mentioning at this point
- It may be said that
- It was demonstrated that

Removing Unnecessary Words to Achieve Conciseness

The ability to write concisely is a coveted skill among chemists and an important step toward sounding like an expert. The key is to say only what needs to be said, deleting unnecessary words (i.e., words that add little substance, state the obvious, or can be inferred by other words in the sentence). For example, compare these two sentences. The wordy sentence contains so many unnecessary words that the authors' message is nearly lost.

Wordy	In a paper published by Bonderic et al.,[2] experiments were described that led to similar results.
Concise	Bonderic et al.[2] reported similar results.

(It is clear to the reader that Bonderic et al. published a paper because of the in-line citation. It is also obvious that "experiments were described" in that paper.)

Sound Like an Expert

Learning to write concisely will help you sound like an expert. This skill requires that you delete words, as you revise and edit your work, that add little substance or state the obvious.

Exercise 2.3.

Read this wordy passage (adapted from Guo et al., 2019) and suggest ways to make the italicized phrases more concise or delete them entirely. (As a guideline, our answer has 113 words.)

As we already mentioned above, we speculated that oxygen plasma treatment can not only remove the residual CTAB molecules on the Au NPLs but can also cause irreversible damage to the morphology of the NPLs.[34] *In order for us to be able to confirm our speculation*, we *very carefully* imaged the surface morphology of the same Au NPLs after the plasma treatment for different periods of time. By overlapping the SEM images of the same NPL before and after plasma treatment for 5 min, *we were able to find* that the lateral size of the Au NPLs was reduced by a few nanometers. As a result, *this led us to discover the important result that* the plasma-treated Au NPLs exhibited a smaller average red shift than those without plasma treatment upon surface modification with the same PEG–SH molecules. (142 words)

There are several ways to make the adapted passage in Exercise 2.3 more concise. One technique is to replace a group of words with a single word that has the same meaning. For example,

Wordy	This finding *makes it apparent* that . . . (6 words)
More concise	This finding *demonstrates* that . . . (4 words)
Wordy	We *come to the conclusion* that . . . (6 words)
More concise	We *conclude* that. . . (3 words)

The *ACS Guide to Scholarly Communication* provides many useful suggestions for converting wordy multiple-word phrases into more concise alternatives (see Table 2.1 for a few examples).

Table 2.1. Suggestions for Concise Writing[a]

wordy	more concise
a number of	many, several
are known to be	are
at the present time	now
based on the fact that	because
despite the fact that	although
for the reason that	because
if it is assumed that	if
in order to	to
is known to be	is
it is clear that	clearly
prior to	before
reported in the literature	reported

[a] From the *ACS Guide to Scholarly Communication* (Banik et al., 2020, section 5.1.4).

Contractions?

Even though contractions seem to make writing more concise, do not use them in your papers; contractions are not appropriate in an academic style.

Concise Writing

See Chapter 16, "Language Tips," Part 1: Conciseness, for additional tips on concise writing.

Exercise 2.4.

Using Table 2.1 for guidance, suggest a more concise alternative for each italicized phrase.

a. *In spite of the fact that* the mechanism of the stability of UFBs remains under intensive investigation, industrial applications of UFBs have recently arisen in industries such as agriculture, fishing, and medical therapy. (Adapted from Hirai et al., 2019)

b. These proteins *are known to be* well-studied typical globular proteins with subunit or subdomain structure. (Adapted from Hirai et al., 2019)

c. The saturated vapor inside the UFBs is ~7.7×10^{-3} eÅ^{-3} *due to the fact that* the saturated vapor content is 23 g^{-3} at 25 °C. (Adapted from Hirai et al., 2019)

d. *It is clear that* these data indicate that PL and EL sample the distribution of CT states in these blends differently. (Adapted from Dimitrov et al., 2019)

Another way to be concise is to eliminate words that are redundant or that provide information that the reader can assume. For instance, the phrase "In our judgment" (in the wordy example) can be removed because the reader can assume that the conclusions being drawn are based on the researchers' judgments:

Wordy	In our judgment, this finding clearly demonstrates . . . (7 words)
More concise	This finding clearly demonstrates . . . (4 words)

(Some would argue that "clearly" could also be eliminated, but the word was used in the original article and is a favorite among chemists.)

Conciseness can also be achieved with parentheses, eliminating information that is superfluous:

Wordy	The results, as illustrated in Table 3, suggest that . . . (9 words)
More concise	The results (Table 3) suggest that . . . (6 words)
Wordy	Ethanol (ACS grade and purchased from MilliporeSigma) was added . . . (9 words)
More concise	Ethanol (ACS grade, MilliporeSigma) was added . . . (6 words)

Exercise 2.5.

Revise these sentences. Identify the parts that you consider to be too wordy. Make those parts more concise or delete them entirely.

Example

Original	The lipid-binding potential was *observed to be* independent of pH; *as a consequence*, the results *presented in this paper* do not support hydrophobic interactions.
Revised	The lipid-binding potential was independent of pH; hence, the results do not support hydrophobic interactions.

a. The headspace of the vial was flushed with N_2 and CO_2 (80%/20%) for 30 min to be sure all of the air had been removed.
b. In consequence of this fact, it can be assumed that $ArCl^+$ does not interfere with the quantification of arsenic in the soil extracts.
c. In a large number of cases, stem and root lengths, plant dry weight, and the number of floral branches were significantly enhanced as we show in Table 1. (Adapted from Marra et al., 2019)
d. Despite the fact that we have observed performance improvements in organic photovoltaics, the mechanism of photoinduced electron–hole separation at organic donor–acceptor interfaces remains poorly understood. (Adapted from Dimitrov et al., 2019)
e. All of the chemicals were research grade, and they were all purchased from Fisher, which is located in Pittsburgh, PA.

Using Nominalizations to Achieve Conciseness

Another technique employed by chemists to achieve conciseness is to use **nominalizations**. Nominalizations are nouns that are formed from other parts of speech, usually by adding endings such as -tion, -sion, -ment, -ity, -sis, and -ence. For example,

solubility (noun, from the adjective *soluble*)
distillation (noun, from the verb *distill*)

Nominalizations often allow several words to be combined into a single word.

Without a nominalization	After we distilled the liquid, it was colorless. (8 words)
With a nominalization	After distillation, the liquid was colorless. (6 words)

By using a nominalization in this last example, the word "we" could be removed from the sentence, making it sound more objective. Table 2.2 lists common nominalizations used in chemistry journal articles.

Table 2.2. Common Nominalizations Used in Chemistry Writing

absorption	concentration	extraction	presence
activation	conductivity	formation	purification
addition	conversion	intensity	reaction
aggregation	dependence	interaction	reactivity
agreement	diffusion	luminescence	rearrangement
analysis	efficiency	measurement	reduction
calculation	emission	oxidation	synthesis
comparison	excitation	preparation	treatment

Nominalizations

A noun formed from a verb or an adjective, usually by adding a word ending (e.g., -tion, -ment). For example,

extraction (noun, from the verb *extract*)
efficiency (noun, from the adjective *efficient*)

Nominalizations often result in more concise writing. See Part I of Chapter 16 for additional tips on using nominalizations.

Keep words that convey important content; delete words with little substance.

Exercise 2.6.

Consider these sentences from chemistry journal articles. Identify instances of nominalizations using Table 2.2 as a guide.

Example

Potential *conversion* of the N1 adduct to its N^6 derivative was made possible through a Dimroth *rearrangement*, although the *efficiency* of this process is highly dependent on *reaction* conditions and adduct structure (Scheme 2).[26] (From Veldhuyzen et al., 2001)

a. In aqueous colloidal dispersions, van der Waals attraction and electrostatic repulsion serve as major long-range interactions.[4] (Adapted from Wang et al., 2018)
b. The production of isomer II over isomer I could be achieved by increasing the stagnation pressure of the supersonic expansion used to generate the clusters. (From Wang et al., 2018)
c. Colloidal lead halide perovskite nanocrystals (NCs) have high tunability in the visible light region and high photoluminescence quantum yields (PL QYs) for green and red emissions, but bright blue emission is still a challenge. (From Zhang et al., 2019)
d. The relative intensity of this signal permits a comparison of the strength of the interaction between oxides in dependence on the preparation conditions (milling time, calcination temperature and time, presence or absence of water). (From Spengler et al., 2001)
e. In addition, the presence of OVs in combination with the spatial separation structure makes the Au/end-CeO_2 an excellent candidate for plasmon-induced NIR N_2 photofixation. (From Jia et al., 2019)

Exercise 2.7.

Try rewriting sentence (d) in Exercise 2.6 without using any nominalizations. How does your revision compare to the original in terms of conciseness and clarity?

Nominalizations usually make writing more concise. One important exception, however, is using nominalizations with forms of the verb *do* (i.e., *did* or *was done*). In such cases, nominalizations can make the sentence wordier and should be avoided. Consider these examples:

Wordy	Komiyama et al.[2] did an analysis . . .
Concise	Komiyama et al.[2] analyzed . . .
Wordy	A synthesis was done by Martinez et al.[7] . . .
Concise	Martinez et al.[7] synthesized . . .

et al.

This is an abbreviation (for the Latin *et alia*) that means "and others." It is used when referring to authors of a publication that has three or more authors. The first author's last name is listed, followed by et al. (with the period).

Exercise 2.8.

Rewrite these sentences to make them more concise. (Do not delete citations, indicated by superscript numbers.)

a. At this point in time, there exist only a small number of reported examples of the synthesis of carboranes from eneynes,[20] the first of these being the synthesis of 1-isopropenylcarborane.[21] (Adapted from Valliant et al., 2002)

b. Table 1 accurately summarizes the reaction products from the two reactions that were performed: the hydrogenation reaction and the oxidation reaction.

c. Polyadducts of C_{60} with well-defined three-dimensional structures are of great importance, based on the fact that they possess interesting biological[1] and material properties.[2] (Adapted from Mas-Torrent et al., 2002)

Organization

In Chapter 1, we learned that genres have both broad and fine organizational structures. In this chapter, we focus on the *broad* organizational structure of the journal article, indicated by identifiable sections and headings. In general, journal articles are divided into four major sections. These sections have familiar names: Introduction, Methods, Results, and Discussion. Collectively, this organizational structure is referred to as the IMRD format. Some journals expect a separate Conclusions section; in other journals, conclusions are placed at the end of the Discussion section. In addition to these four major sections, journal articles contain a title, abstract, acknowledgments, references, and supporting information. (Acknowledgments are required for works supported by a funding agency; otherwise, this section is optional.) Chapters 3–7 explore most of these sections in detail; in this chapter, we briefly highlight the IMRD sections.

IMRD

The typical broad organization of a journal article:

Introduction
Methods (or Experimental)
Results
Discussion

The *Introduction* of a journal article identifies the research area, explains the importance of the research, provides background information, cites and summarizes key literature in the field, points out what still needs to be studied, and introduces the reader to the work presented in the article. The *Methods* section—formally known as Materials and Methods or Experimental (section)—describes how the study was conducted. The *Results* section summarizes quantitative (and possibly qualitative) data collected during the study. In the *Discussion* section, authors interpret their data and suggest the larger implications and/or applications of their results. Each major section can be further divided into moves, as we will see in subsequent chapters.

In recent years, variations have appeared in the traditional IMRD format. For example, some journals include explicit headings for all four divisions, some use fewer than four explicit headings, and some use no headings at all. The sequencing of the sections also varies. For example, *The Journal of Organic Chemistry* typically places the Experimental section at the end of the article, whereas the *Journal of the American Chemical Society* places the Experimental section in Supporting Information. Interestingly, in the past, chemists were discouraged from combining the Results section (presentation of data) with the Discussion section (interpretation of data), yet today a combined Results and Discussion section is commonplace. Because every journal has different requirements, it is important to read the **Author Guidelines** for your targeted journal when preparing a manuscript for electronic submission.

Author Guidelines

A set of journal-specific instructions for hopeful authors that includes the following:

length specifications
section requirements
appropriate research areas
manuscript preparation and submission guidelines

Exercise 2.9.

Browse through *The Journal of Organic Chemistry*, *Journal of the American Chemical Society*, and *Organic Letters*. What variations in sectional divisions do you see? Are all headings shown explicitly? Repeat this exercise with two or three non-ACS chemistry journals (e.g., *Applied Surface Science*, *Chemical Physics*, and *Tetrahedron*, all published by Elsevier).

Hourglass Structure

The IMRD format creates what is sometimes called an hourglass structure, a feature common to journal articles across many fields of academic research. The hourglass depicts the way in which the scope or specificity of the paper changes throughout its sections, as shown in Figure 2.1. The Introduction begins with a broad overview of the research area but narrows as the authors mention specifics about their presented work. This specificity is maintained throughout the Methods and Results sections and then broadens again at the end of the Discussion, where research findings are described in a broader context. An hourglass icon is used throughout Module 1 to remind you where you are in the hourglass structure.

Exercise 2.10.

Browse through three articles in *Chemical Research in Toxicology* or the *Journal of Agricultural and Food Chemistry*. For each article, find one sentence from each IMRD section that is consistent with the hourglass structure. Examine each group of four sentences. To what extent is the hourglass structure apparent in these four-sentence sets?

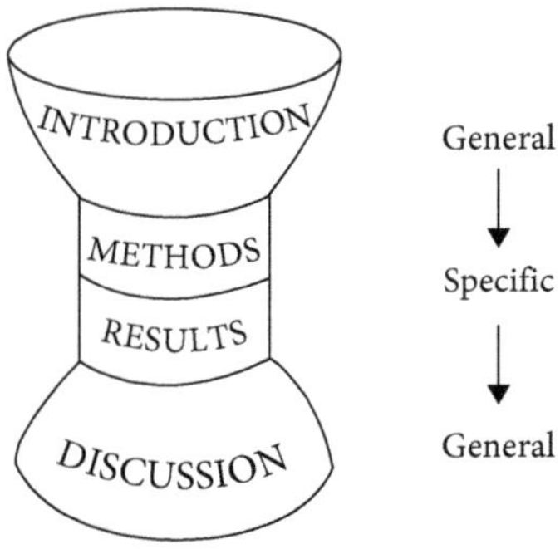

Figure 2.1. The hourglass structure of the IMRD format.
Adapted from Hill et al. (1982).

Targeted Reading

Before ending this chapter, an additional purpose of the IMRD format is worth noting: the IMRD structure promotes **targeted reading**, allowing readers to locate science content in an expedient manner. You might be surprised to learn that chemists do not typically read every word in every article from start to finish. Rather, most chemists read selectively, targeting sections most pertinent to their interests. For example, most chemists glance first at an article's title, abstract, and keywords. This cursory glance is often enough to decide if the article is sufficiently relevant to merit a closer look. If the content appears promising, they next target a specific section of the article. For example, a chemist interested in planning a synthesis will read the Methods section; a reader wanting to learn more about potential uses of a novel compound will read the Introduction and Discussion sections. Sections considered less relevant are skimmed or skipped entirely. Only a few articles, those most pertinent to the reader's interests, will be read in their entirety (and then usually many times).

Targeted Reading

Students are generally accustomed to reading texts from start to finish, with the intent of understanding everything. You may have read a textbook chapter in this manner.

Scientists rarely read research articles like this. Generally, scientists *target* a given section to look for specific information. For example, a chemist might target the Methods section for the sole purpose of finding out what brand of instrument was used.

The IMRD format, together with the finer organizational structure (moves) and language of the journal article, helps readers quickly locate the content that they seek. As a developing writer, it is important that you learn to present your content in these expected ways. As readers scan your paper, they should be able to quickly locate your topic, the nature of your work, the methods you used, and your conclusions.

Exercise 2.11.

Browse through three research articles in an ACS journal of your choice. With only a cursory glance at the title, abstract, and the IMRD sections, determine the research topic, methods used, and major conclusions. (Note: You should be able to identify the topic and methods, even if you do not understand them.)

Writing is a lot like composing a piece of music. Random notes do not make a symphony, nor do random assemblies of words tell a cogent story. It is critical that writers learn how to put ideas together, through the very final stages of composing, using tried and true methods that are common to the field of chemistry. Only in this way can their ideas unfold clearly for the reader.

—Timm Knoerzer, United States Air Force Academy

Chapter Review

As a self-test of what you've learned in this chapter, define each of these terms for a friend or colleague who is new to the field:

Author Guidelines	et al.	primary literature
conciseness	hourglass icon	refereed journal article
	IMRD format	

Also, explain the following to a friend or colleague who has not yet given much thought to writing a journal article:

- Main purpose of a journal article
- Audience of a journal article
- Importance of concise writing
- Techniques for writing concisely
- Broad organizational structure of a typical journal article and its variations
- Purpose(s) of each section (IMRD) of a journal article

3 *Writing the Methods Section*

Tell me how you did it, but be concise. A long-winded step-by-step Methods section sidetracks and irritates the expert reader.
—Betty H. Stewart, University of Texas at Dallas

In this chapter, we focus on writing a Methods section for a journal-quality paper. We begin with the Methods section because it is the section that many chemists write first, in part because this section describes what they know best: the procedures they have repeated (many times) to conduct their work. Moreover, most research groups use similar methodologies for several years; hence, previously written Methods sections can serve as models for writing new Methods sections. Together, these factors make the Methods section one of the easier sections to write and an excellent place to begin our writing instruction.

That said, Methods sections can vary significantly across different subdisciplines of chemistry. For example, the section has several different names. It is typically called Experimental or Methods in physical and analytical chemistry journals, Materials and Methods in environmental and biochemistry journals, and Experimental Section in organic and inorganic chemistry journals. For brevity in this book, we call it simply the Methods section.

Methods Section

What we call the Methods section has different labels in different areas of chemistry

- Experimental or Methods (e.g., physical and analytical chemistry)
- Materials and Methods (e.g., environmental chemistry and biochemistry)
- Experimental Section (e.g., organic and inorganic chemistry)

Write Like a Chemist. Marin S. Robinson and Fredricka L. Stoller, Oxford University Press. © Oxford University Press 2022.
DOI: 10.1093/oso/9780190098940.003.0003

Chemistry journals also locate the Methods section differently in articles. Although many journals follow the IMRD format and begin the Methods section immediately after the Introduction, organic journals place it after the Conclusions (i.e., Introduction, Results and Discussion, Conclusions, Experimental Section). In journals with *Letters* in their titles (e.g., *Organic Letters*, *Nano Letters*), the Methods section is placed in Supporting Information, not in the actual article. We examine several additional differences across Methods sections throughout this chapter. By the end of this chapter, you will be able to do the following:

- Know how to address the intended audience of a Methods section
- Identify which details to include in and exclude from a Methods section
- Recognize differences in Methods sections across different chemistry subdisciplines
- Organize a Methods section following standard moves
- Use capitalization, abbreviations, and parentheses appropriately
- Format numbers and units correctly
- Use verb tense and voice in conventional manners

Writing on Your Own: Methods Section

Visit the *Write Like a Chemist* (2nd ed.) website for step-by-step guidance in writing your own Methods section.

The purpose of the Methods section is to explain *how* your research was conducted. Relevant information about instrumentation and experimental and/or numerical procedures is described. The goal is to provide enough detailed information so that an expert (not a novice) could repeat the work. Because many of you have written a Methods section previously for a chemistry lab course, we begin with an exercise to test your current knowledge about what details to include in a Methods section.

Exercise 3.1.

What have you learned about writing Methods sections from other writing courses and labs? Place a "Yes" next to those items that you think should be included in a Methods section of a journal article. Place a "No" next to items that you think should be omitted. Use a question mark (?) if you are unsure.

_____ Table of chemicals with their physical properties (e.g., mp, MW, ρ)
_____ Amounts of reagents used in a synthesis (e.g., mg, mmol)

____ Directions for preparing a stock solution
____ Quality (grade) of chemicals used
____ Names and locations of chemical vendors
____ Brand names of commercial instrumentation used
____ Diagram of a distillation apparatus
____ Illustration of a novel or custom-built apparatus
____ Operating conditions for a gas chromatograph
____ Equations used to calculate percent yield or dilution ratios
____ List of disposable supplies (e.g., test tubes, gloves, tissues)
____ List of routine equipment and glassware (e.g., heating mantel, Erlenmeyer flask)
____ Step-by-step instructions of the procedure
____ Warnings to other scientists about unusual hazards
____ Quantitative statements of reaction times and temperatures
____ Descriptions of the physical appearances of synthesis products
____ IR or NMR data confirming product purity
____ Statistical packages used (including the name of the software)
____ Reports of other software used to keep track of data (e.g., Excel)

Reading and Analyzing Writing

We formally begin this chapter by asking you to read and analyze an excerpt from a Methods section published in the *Journal of Agricultural and Food Chemistry* pertaining to the chemical analysis of pomegranate fruit and juice. (You will eventually read additional excerpts from this article, section by section, throughout the module.) The study analyzes different parts of pomegranate fruit for proanthocyanidins (PAs), a group of phenolic compounds recognized for their antioxidant capacity and ability to prevent illnesses such as cardiovascular disease and cancer. The authors use phloroglucinolysis combined with liquid chromatography–electrospray ionization–mass spectrometry (LC–ESI–MS) to investigate the PA profile. Phloroglucinolysis uses acidic conditions to depolymerize the PAs and form adducts that help identify PA composition. LC–ESI–MS is a form of mass spectrometry that uses liquid chromatography to separate compounds, electrospray to create an ionized aerosol, and mass spectrometry to determine product mass. The original Methods section comprised five sections. In excerpt 3A, we include only the first three; however, these three illustrate how experienced chemists describe materials (chemicals and pomegranate samples) and methods (phloroglucinolysis and LC–ESI– MS) in a Methods section.

Exercise 3.2.

Read excerpt 3A (a Methods section formally labeled Materials and Methods). Consider the audience, organization, writing conventions, grammar, and punctuation used in the excerpt as you answer these questions:

a. Who is the authors' intended audience? How is this audience reflected in level of detail, level of formality, and word choice?
b. How do the authors make their writing concise?
c. How do the authors organize their information?
d. What writing conventions do you notice (i.e., units, capitalization, formatting, abbreviations, verb tense, voice)?
e. Which 2–4 segments of the excerpt illustrate correct uses of parallelism (e.g., among section headings), commas, and subject–verb agreement?
f. What instrumentation did the authors use, and how is it described?
g. Based on your response to exercise 3.1, what information did you expect to see in this Methods section that was not included?

Excerpt 3A (adapted from Díaz-Mula et al., 2019)

MATERIALS AND METHODS

Chemicals. Phloroglucinol and standards of proanthocyanidin B2, (+)-catechin (C), (–)-epicatechin (EC), (–)-gallocatechin (GC), (–)-epigallocatechin (EG), (–)-gallocatechin gallate (GG), (–)-epigallocatechin gallate (EGCG), (–)-catechin gallate (CG), (–)-epicatechin gallate (ECG), ellagic acid, and cyanidin 3-O-glucoside were purchased from Sigma-Aldrich (St. Louis, MO, U.S.). Stock solutions for each compound at a concentration of 2000 mg/L were first prepared in methanol, except for ellagic acid that was prepared in dimethyl sulfoxide (DMSO), and then serially diluted to working concentrations. Acetonitrile and methanol of LC–MS grade were from J.T. Baker (Deventer, Netherlands), DMSO was from LabScan (Dublin, Ireland), and formic acid, hydrochloric acid, and sodium acetate were from Panreac (Barcelona, Spain). Ascorbic acid was from Acros Organics (Geel, Belgium). All other chemicals and reagents were of analytical grade. Milli-Q system (Millipore Corp., Bedford, MA, U.S.) ultrapure water was used throughout the study.

Pomegranate Samples. Pomegranates (four samples) of cultivar Mollar Elche (Elche, Spain) harvested in October 2016 were obtained from a local market. The fruit was manually separated into husk, membranes, and arils (Figure 1). Husk and membranes were weighed, freeze-dried, and ground with a mixer grinder (Oster Professional BPST02-B), and the powders were kept in a desiccator before analysis. Because of their high sugar content, arils were not able to be freeze-dried and were analyzed directly. The arils were pressed with a hand-press juicer to obtain the press cake and the juice, which was subsequently filtered through cheesecloth to remove solid particles. Seeds were separated from the press cake remaining after pressing the arils. The seeds, press cake, and pellet obtained after filtration of the juice were freeze-dried and ground with the mixer grinder. Clarified juice was analyzed directly for the determination of anthocyanins and ellagitannins but was passed through Sep-Pak cartridges before the phloroglucinolysis reaction. Commercial pomegranate juices (also from Mollar cultivar), obtained by pressure extraction of the whole fruit, were provided by a juice processor (AMC, Murcia, Spain). Juices were centrifuged at 3500g for 10 min, and the pellet obtained was freeze-dried and ground. The moisture content was determined from the weight of the samples before and after freeze drying.

Analysis of Proanthocyanidins by Phloroglucinolysis. *Sample Preparation.* Solid samples (husk, membranes, seeds, press cake, and pellets) were processed directly after freeze drying. . . . [*The remainder of this subsection is omitted.*]

Phloroglucinolysis Reaction and Analysis by LC–ESI–MS. PAs were quantified as previously reported by Kennedy and Jones,[13] with some modifications.[21] A solution of 0.1 M HCl in methanol containing 50 g/L phloroglucinol and 10 g/L ascorbic acid was prepared. A total of 50 mg of dried samples was mixed with the phloroglucinolysis reagent (0.8 mL) and incubated in a water bath for 20 min at 50 °C. The reaction was stopped by placing the vials in ice and adding 1 mL of 40 mM sodium acetate. The samples were centrifuged at 3500g for 10 min and filtered with a 0.22 µm polyvinylidene fluoride (PVDF) filter prior to the injection. The same reaction was applied to 1 mg of proanthocyanidin B2 using proportional amounts of reagents. Samples were analyzed using an Agilent HPLC 1200 series instrument equipped with a diode array detector (Agilent Technologies, Waldbronn, Germany) and an ion-trap mass spectrometer detector in series (Bruker Daltonics, Bremen, Germany). Separation of phenolic compounds was achieved on a reverse-phase C18 Pursuit XRs column (250 × 4 mm, 5 µm particle size, Agilent Technologies, Waldbronn, Germany), operating at room temperature and a flow rate of 0.8 mL/min. A total of 10 µL of sample was injected. The mobile phases were water with acetic acid (2.5%) (phase A) and acetonitrile (phase B). The gradient changed according to the following conditions: 0 min, 3% B; from 0 to 5 min, 3–9% B; from 5 to 15 min, 9–16% B; from 15 to 45 min, 16–50% B; from 45 to 47 min, 50–90% B; maintained at 90% B for 1 min; decreased to the initial conditions in 3 min; and maintained for 5 min (total time 57 min). The UV–vis chromatograms were recorded at 280 nm. The optimal conditions of the electrospray interface were as follows: gas temperature, 325 °C; drying gas, 11 L/min; and nebulizer pressure, 65 psi. Spectra were acquired in the range of *m/z* 50–800 (target mass of 400) in the negative ionization mode. Compound stability was set at 75%. Flavan-3-ol monomers and phloroglucinol adducts obtained after the phloroglucinolysis reaction were identified using their elution order, absorbance spectra, and mass fragmentation patterns (MS/MS) compared to available standards when possible. Quantification was developed in MS, using the area of the extracted ion chromatograms (EICs) for each compound. Flavan-3-ol monomers were quantified with their own commercial standards, and phloroglucinol adducts were quantified with their corresponding free flavan-3-ol, assuming higher response in MS for adducts (twice more, as indicated below from the phoroglucinolysis of proanthocyanidin B2). The concentrations of flavan-3-ols in the products analyzed were expressed in micrograms per gram of fresh weight (FW) in solid parts and milligrams per liter in juices, taking into account the moisture of the samples. To calculate the apparent mean degree of polymerization (mDP), the sum of all subunits (flavan-3-ol monomer and phloroglucinol adducts, in moles) was divided by the sum of all flavan-3-ol monomers (in moles).

Analyzing Audience and Purpose

The major purpose of a Methods section is to describe, for other chemists (an expert audience), the procedures that were used to obtain the results presented in the article. A well-written Methods section serves as a resource for expert chemists who wish to (1) develop similar procedures, (2) compare their own procedures with those presented, or (3) familiarize themselves with procedures in a branch of chemistry other than their own.

Because journal-article Methods sections are written largely for experts, they are not at all like Methods sections in chemistry lab manuals. Lab manuals are written largely for instructional purposes, to reinforce new techniques and help students carry out experiments successfully. As such, they tend to include details (e.g., lists of equipment, safety precautions, and step-by-step directions) not needed (or wanted) by expert readers. Leaving out such details makes the writing more concise. Concise writing is important because, unlike a single lab experiment, methods sections in journal articles describe multifaceted works that took months or years to complete. Similarly, lab manuals often include language inappropriate for journal articles. For example, in a lab manual, a student might be directed to "Stir the mixture. Heat to reflux." In a journal article, however, this would be restated in past tense and passive voice as "The mixture was stirred and heated to reflux," making the writing more formal.

Experiment vs. Work

Because of the multifaceted nature of a research project, authors seldom use the word *experiment*, which connotes a simplistic activity that could be accomplished in a single day. Words such as *project* or *work* are preferred.

To see for yourself how chemists write Methods sections for journal articles, we have included excerpts from various chemistry journals throughout this chapter. These excerpts illustrate appropriate levels of detail, formality, and conciseness for expert audiences. We encourage you to use these excerpts (rather than lab manuals or lab reports) as models for your writing.

Exercise 3.3.

Access the ACS website and browse through Articles (not Communications, Viewpoints, Additions and Corrections, etc.) of the Current Issue (not Articles ASAP, Just Published, etc.) of the following journals: *Analytical Chemistry*, *Biochemistry*, *Inorganic Chemistry*, *Journal of the American Chemistry Society*, *The Journal of Organic Chemistry*, and *The Journal of Physical Chemistry A*. For each journal, determine where the Methods section is located and how it is labeled.

Analyzing Organization

Different subdisciplines of chemistry (e.g., analytical and organic chemistry) use different organizational approaches in their Methods sections. To address these differences, we propose two organizational structures for the Methods section. The first pattern applies largely to analysis articles, where the primary focus is on analyzing a sample or developing a new method. This approach is commonly observed in journals such as *The Journal of Analytical Chemistry* and *Environmental Science & Technology*. The second pattern applies largely to synthesis articles, where the primary focus is on synthesizing a new compound. This pattern is commonly observed in journals such as *The Journal of Organic Chemistry* and *The Journal of Inorganic Chemistry*. Of course, overlap between these patterns exists because analysis articles can include syntheses (and vice versa). Nevertheless, these two structures provide a useful starting place for newcomers to chemistry writing. The Methods section, of both types, is written primarily for an expert audience and is in the narrowest part of the IMRD hourglass structure.

Organization in Analysis Articles

We begin with the organizational structure of Methods sections in analysis articles (Figure 3.1). In move 1, authors tell readers about materials used in their work. The term *materials* refers to chemical reagents, samples (e.g., soil, water, or food), biological media (e.g., cell cultures), actual materials (e.g., tests kits), and/or other tangible items used to conduct the work. In field studies, the sampling site is also described in move 1. When different types of materials are reported, subheadings (in bolded font) are commonly used to set them apart (e.g., the subheadings **Chemicals** and **Pomegranate Samples** in excerpt 3A). Examples of

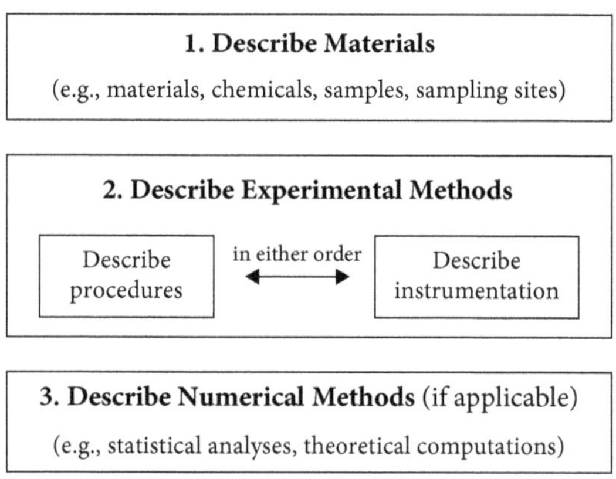

Figure 3.1. A visual representation of the move structure for a typical Methods section (aka Experimental Section) in an analysis paper.

common subheadings in move 1 include the following: **Chemicals, Reagents and Materials**, **Samples**, **Cell Cultures**, and **Site Description**.

In move 2, Describe Experimental Methods (Figure 3.1), authors describe the procedures and specialized instrumentation used to collect data. The move involves two submoves: describe procedures and describe instrumentation. The ordering of the two submoves, although interchangeable, should reflect the sequence of events in the study. Thus, if derivatization preceded GC analysis, derivatization should be mentioned first.

When describing procedures, authors identify the analytical or instrumental methods used in their work. Subheadings (in bold font) are used to designate each procedure, for example, **Analysis of Proanthocyanidins by Phloroglucinolysis** (a procedure in excerpt 3A). Procedures associated with method validation or quality control are also described here, with subheadings such as **Method Validation** or **Quality Assurance/Quality Control (QA/QC)**. If the work involved a synthesis, synthetic procedures are also described in this submove, for example, **General Procedure for the Phosphitylation** (from Ghodke et al., 2019, not shown).

QA/QC

This abbreviation refers to quality assurance/quality control. QA/QC procedures are standardized methods used to verify the quality (accuracy and precision) of data.

Sub-subheadings are also common when describing procedures. To distinguish them from subheadings, italic font is often used. For example, in excerpt 3A, two sub-subheadings were used (both italicized). Similarly, following the general procedure for phosphitylation in Ghodke et al. (2019, not shown), four sub-subheadings were used (all in italics). Each sub-subheading gave the name of the new compound, followed by a brief description of the synthesis. Subheadings (and sub-subheadings) are useful in Methods sections because they help organize multiple procedures. They also assist readers in quickly locating a method, procedure, or compound of interest. (Scientists often read papers selectively, looking only for information about a particular procedure, instrument, or compound.)

When describing instrumentation, authors identify the scientific apparatus used in the study. Both custom-built and commercially available instruments are described. Subheadings (and sub-subheadings) are also common in this submove (e.g., *Phloroglucinolysis Reaction and Analysis by LC–ESI–MS* in excerpt 3A). Ordinary instrumentation, such as a rotary evaporator or a vacuum pump, is not described.

The last move of the Methods section, Describe Numerical Methods, is included when numerical or mathematical procedures (e.g., statistical analyses) were used to analyze, derive, or model data presented in the paper. In such cases, the experimental methods are described first (move 2), and the numerical methods are described last (move 3). Subheadings used to demark move 3 include **Statistical Methods**, **Theoretical Methods**, or **Data Analysis**.

Exercise 3.4.

We have intentionally scrambled the order of the subheadings in Sets 1 and 2. Rearrange the subheadings in each set to adhere to the move structure in Figure 3.1. (In some cases, more than one ordering is correct.)

Set 1	**Statistical Analysis.**
	FTIR Measurements.
	Starch Samples.
	FT–Raman Measurements.
Set 2	**Time-of-Flight Mass Spectrometry (TOF).**
	Chemicals.
	SPE and Cleanup.
	Sample Pretreatment.

Organization in Synthesis Articles

In synthesis articles, the Methods section is typically called the Experimental Section. The move structure for this section is shown in Figure 3.2. Note that move 1 does not begin (as it would in an analysis article) with a description of the chemicals used and their sources. Instead, move 1 focuses on general experimental methods with a section entitled **General Information** or **General Conditions**. Chemicals are not named individually but are referred to generically (e.g., "All reagents and solvents were commercially available and used without further purification."). Sources of chemicals are

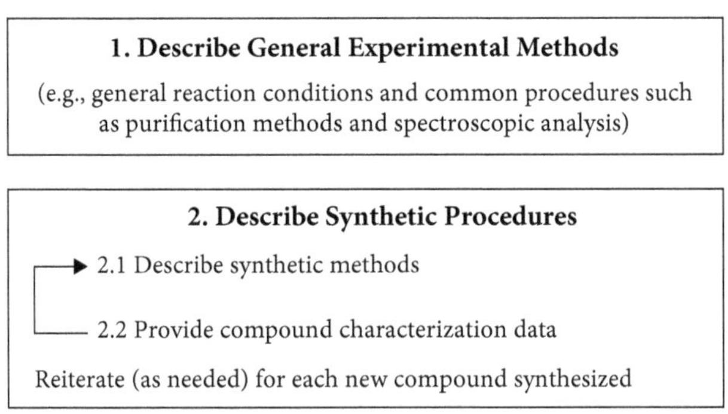

Figure 3.2. A visual representation of the move structure for a typical Methods section (aka Experimental Section) in a synthesis paper.

omitted entirely, unless they were difficult to obtain, critical to the work, or catalysts. Methods common to all syntheses in the work are also briefly described, such as how solvents were removed, how compounds were purified, or how compounds were identified (e.g., by NMR spectroscopy or gas chromatography).

Specific synthetic procedures are described in move 2, which has two submoves. In submove 2.1, the methods used to prepare each compound are described. Enough detail is included so that experienced chemists can repeat the work. In submove 2.2, characteristics of the compound (e.g., yield, color, melting point) and data that verify the compound's identity (e.g., NMR results) are presented. These two submoves are repeated for each compound synthesized. It is not uncommon for 10 or more new compounds to be described in a single paper.

Noticeably absent in Figure 3.2 is a description of numerical methods (corresponding to move 3 in analysis papers, Figure 3.1). Numerical methods (e.g., theoretical computations and statistical analyses) are often included in synthesis papers; however, they are presented in Supporting Information rather than in the Experimental Section. Hence, they are not addressed here. Consult Author Guidelines in your targeted journal for information on how to describe computational methods in Supporting Information.

Exercise 3.5.

Browse through a synthesis article in *The Journal of Organic Chemistry*, an environmental measurements article in *Environmental Science & Technology*, and a theory article in *Journal of Physical Chemistry A*. What different procedures are described in each Methods section? What headings and subheadings are used? How well do these Methods sections follow the moves in Figures 3.1 and 3.2? What variations do you notice?

Analyzing Excerpts

There is no question that, in every part of my career, clear and simple writing has helped me communicate what I wanted said. In research, I often have had comments about the clarity of my writing. I cannot but think that this makes a favorable impression on reviewers. Although clarity cannot supplant content, it certainly "encourages" reviewers, and then readers, to spend some time with our work. Surely, that is what we want when we bring something to press.

—Robert Damrauer, University of Colorado Denver

With general audience and organizational considerations in mind, we now examine excerpts, move by move, from Methods sections in analysis and synthesis articles (in Parts 1 and 2, respectively). Later in the chapter, we examine writing conventions that run across Methods sections.

Part 1: Analyzing Excerpts Move by Move in Analysis Articles

We begin with excerpts that illustrate the three moves of Methods sections in analysis papers (Figure 3.1).

Move 1: Describe Materials

The purpose of move 1 is to describe the materials used in the work for an expert audience (chemists). Excerpts 3B and 3C illustrate how chemicals are described in analysis journals. (See also excerpt 3A.) As you read these excerpts, consider what details about chemicals are included and excluded. Note, too, the different headings used in these Methods sections.

Excerpt 3B (adapted from Zhang et al., 2019)

EXPERIMENTAL SECTION

Chemicals and Materials. Sodium citrate, (3-aminopropyl)trimethoxysilane (APTMS), sodium silicate solution, 4-mercaptobenzoic acid (MBA), 2-naphthalenethiol (NT), and 5,5′-dithiobis (2-nitrobenzoic acid (DTNB) were purchased from Alfa Aesar. $HAuCl_4$, HCl, and HNO_3 were purchased from Shanghai Reagent Corporation, China. All chemicals were used without further purification. Milli-Q water (18.2 MΩ cm) was used throughout the study, and all glassware was cleaned with aqua regia before use.

Excerpt 3C (adapted from Bekker et al., 2018)

MATERIALS AND METHODS

Chemicals. All chemicals were of analytical reagent grade. Water was obtained from a Milli-Q purification system (Millipore, North Ryde, NSW, Australia). All solvents were of high-pressure liquid chromatography (HPLC) grade and purchased from Rowe Scientific (Lonsdale, SA, Australia). L-Ascorbic acid (99%), (+)-catechin hydrate (98%), L-cysteine hydrochloride (99%), ethylmethyl sulfide (EMS, 96%), formic acid (98%), potassium metabisulfite (PMS, 98%), and sodium sulfide nonahydrate ($Na_2S{\cdot}9H_2O$, 98%) were obtained from Sigma-Aldrich (Castle Hill, NSW, Australia). Iron(III) sulfate hydrate (97%) was purchased from Acros Organics (Thermo Fisher, Scoresby, VIC, Australia). Ethanol (HPLC grade), copper(II) sulfate (99%), tartaric acid (99.5%), and sodium chloride (99.5%) were obtained from Merck (Frenchs Forest, NSW, Australia).

Several features are worth noting in excerpts 3B and 3C. First, the excerpts are written in paragraphs, with complete sentences (not lists with sentence fragments). Although commonplace in lab reports, lists should be avoided in journal articles.

Second, chemical names are not capitalized; instead they are written in lower case (e.g., sodium sulfide). The first letter of a chemical name is capitalized only at the start of a sentence or in a chemical formula (e.g., Na_2S). Also note that information about chemical purity (e.g., 98%) is reported in parentheses and without the word *purity* (which is understood by chemists). Reagent grade is also reported in parentheses unless it is the same for all chemicals; then it can be stated in a single sentence (e.g., "All chemicals were of analytical reagent grade" in excerpt 3C). Consider these wordy and concise examples:

Wordy	**Chemicals.** The following solvents were purchased from Aldrich: 99.8% purity methanol, 97% purity ethanol, and 99% purity 1-pentanol.
Concise	**Chemicals.** Methanol (99.8%), ethanol (97%), and 1-pentanol (99%) were purchased from Aldrich.

Fragments

Fragments are incomplete sentences (e.g., Methanol from Aldrich.). Journal articles have only complete sentences, not fragments (e.g., Methanol was purchased from Aldrich.). Exceptions include titles, section headings, figure captions, and table titles.

The sources of chemicals (i.e., vendors) are also included in analysis articles (excerpts 3B and 3C). When multiple vendors are involved, chemicals are grouped by vendor, thereby eliminating the need to repeat the same vendor's name. In some journals, the vendor's location (city, country) is also included the first time the vendor is mentioned, especially if the vendor is not well known. Typically, the location is reported in parentheses at the end of the sentence, as shown in excerpt 3C.

Information *Not* to Include about Chemicals

As illustrated in excerpts 3B and 3C, expert chemists omit information about chemical reagents that is readily available from other sources, such as physical data (boiling point, density, molecular weight), general hazard information (e.g., flammability or reactivity), and disposal instructions. This information is appropriate for lab reports but not journal articles.

Chemicals that are used as reference or calibration standards are also described in move 1. Note that the final concentration of any stock standard solution should be mentioned (in addition to any other dilute solutions prepared from the stock solution). Do not explain how a stock solution (or any solution) was prepared; simply state the final concentration and solvent (e.g., 100 μg L^{-1} in ethanol), as illustrated here and in excerpt 3D:

Incorrect	The internal standard phenanthrene-d10 was prepared by adding 1 mL of a 2000 μg mL^{-1} solution to a 50 mL volumetric flask and diluting to the mark with hexane (final concentration 40 ng μL^{-1}).
Correct	Phenanthrene-d10 was used as the internal standard (40 ng μL^{-1} in hexane).

Excerpt 3D (adapted from Chirollo et al., 2018)

MATERIALS AND METHODS

Calibration. Primary stock standard solutions for DHS and penG were prepared in water (1 mg mL^{-1}). Solutions were stored frozen (−20 °C) and were stable for 1 month. Eight point working solutions of 1–1000 ng mL^{-1} DHS and 0.1–500 ng mL^{-1} penG were assembled by dilution of the respective stock solutions in deionized water. The working standard solutions were stored at −20 °C in small plastic tubes to allow the use of only one cap during each extraction day and to avoid degradation. Negative buffalo milk samples were fortified, to obtain calibration curves, with penG and DHS standard solutions to give matrix final concentrations of 1, 10, 20, 50, 100, 200, 500, and 1000 μg kg^{-1} for DHS and 0.1, 1, 2, 4, 10, 50, 100, and 500 μg kg^{-1} for penG.

---

mg L^{-1} or mg/L?

Both ways to express "per" appear in the literature, but journals generally prefer one form or the other. Check your journal's Author Guidelines to see which one to use.

Deionized ≠ DI

Novice writers often ask if they can use DI for deionized water or DD for double-distilled water. Both terms are generally written out fully rather than abbreviated.

Exercise 3.6.

Imagine that you prepared a stock standard solution of arsenic in your research project. You purchased an arsenic concentrated standard (1000 μg mL^{-1}) from Spex Industries in Hoboken, New Jersey. You prepared a 100 μg mL^{-1} stock standard solution by adding 10 mL of the concentrated standard to a 100 mL volumetric flask and diluting to the mark with deionized water. How would you report this information in the Methods section of an article for *Analytical Chemistry*?

Chemists also frequently collect and analyze samples in their work. In such cases, move 1 is used to share information about the samples, sample source, and/or collection sites. For example, pomegranate samples are described in except 3A. Additional examples are shown in excerpts 3E and 3F. Once again, complete sentences are used, and vendor information is placed in parentheses. If sample collection is part of a long-term sampling campaign (e.g., the Integrated Atmospheric Deposition Network in excerpt 3E) or employs a Standard Method (e.g., Standard Method 3125 B in excerpt 3F), that should be noted in the text.

Excerpt 3E (adapted from Wang et al., 2018)

MATERIALS AND METHODS

Air Sampling. Six Integrated Atmospheric Deposition Network (IADN) sites, including two urban sites (Chicago, IL, and Cleveland, OH), two remote sites (Eagle Harbor, MI, and Sleeping Bear Dunes, MI), and two rural sites (Sturgeon Point, NY, and Point Petre, Ontario) were selected as the sampling sites. Particulate air samples were collected with a high-volume air sampler (General Metal Works, model GS2310) for 24 h every 12 days from 1 January to 31 December 2015 at five IADN sites. At Point Petre, samples were collected every 36 days. The high-volume air sampler used a 0.3 μm quartz fiber filter (Whatman QM-A, 20.3 × 25.4 cm) for collection of the particulate phase followed by a cartridge filled with XAD-2 resin to collect the vapor phase. All the samples were kept at −20 °C until analyzed. Both particulate and vapor-phase samples were collected and analyzed for legacy pollutants, including polychlorinated biphenyls (PCBs), polycyclic aromatic hydrocarbon (PAHs), and organochlorine pesticides (OCPs) as part of IADN, which is a long-term project funded by the U.S. EPA's Great Lakes National Program Office.

Abbreviating U.S. and U.S. State Names

U.S. (without a space) is commonly used for United States; US (without periods) is also sometimes observed. Two-letter abbreviations are used for states (e.g., OH for Ohio).

Excerpt 3F (adapted from Pieper et al., 2018)

MATERIALS AND METHODS

Water Lead Sampling. Each water lead kit prepared at Virginia Tech included sampling instructions and three sampling bottles. In addition, a demonstration video of the sampling protocol was made available. After a minimum 6 h stagnation, residents were instructed to choose one drinking water tap (e.g., kitchen or bathroom faucet) to (1) collect 1 L of cold water at a normal flow (first draw sample); (2) flush the sample tap for 45 s and collect a 500 mL sample (1 min flush sample); and (3) flush the sample tap for an additional 2 min and collect a 125 mL sample (3 min flush sample). Along with water samples, residents provided sampling address, which was aggregated by the five primary zip codes for further analyses.

After shipment to Virginia Tech, all samples were acidified with nitric acid (2%) and digested for a minimum of 16 h before metals analysis via ICP-MS per Standard Method 3125 B.[26] For data quality assurance and quality control, blanks and spikes of known concentrations were measured every 10–15 samples. Residents were offered $20 per sampling event after the first round of sampling in August 2015.

Abbreviating Units of Time

Seconds (s), minutes (min), and hours (h) are abbreviated, but day, week, month, and year are not abbreviated (e.g., 3 days, 5 weeks, 4 months, 10 years). See Chapter 16, "Language Tips," Part 3: Writing Conventions, for more details on numbers and units.

Exercise 3.7.

Excerpts 3E and 3F include many abbreviations, numbers, units of measure and time, and parentheses. Scan the two excerpts to complete a–d.

a. Give the correct abbreviations for seconds, minutes, hours, and days.
b. Which is correct in each pair?
   10min or 10 min   25 °C or 25° C   three sites or 3 sites   10% or 10 %
c. Based on excerpt 3E, list three different types of information that are placed inside parentheses.
d. Look over excerpt 3E again. Can you assume that the authors have *not* referred to PCBs, PAHs, and OCPs earlier in the article? How do you know?

Move 2: Describe Experimental Methods

Move 2 is typically the longest move in the Methods section. Its purpose is to provide information about procedures and instrumentation in enough detail that an experienced chemist could repeat the work (often to the frustration of a graduate student who would like more detail). Move 2 involves two submoves in which authors describe procedures and instrumentation (Figure 3.1). Each submove is described separately in the following discussion.

Submove: Describe Procedures

We first consider how procedures are described in analysis articles. There are literally hundreds of procedures in such articles; the excerpts included here are intended to provide a few illustrative examples. Because multiple procedures are typically involved, descriptive subheadings are used to denote each procedure. Following the subheading, the procedure is described in concise, complete sentences, including only those details needed by an expert audience. Such details are often quantitative in nature. Step-by-step directions (e.g., "first, zero the balance") are omitted, as is information that is available elsewhere (e.g., instructions from an operator's manual).

Let's begin with excerpt 3G from *Environmental Science & Technology*. The authors use a procedure called solid phase extraction (SPE). After the subheading, the authors describe the procedure. Note what details they include, such as the volumes of methanol and deionized water, the temperature of the water bath, and the flow rate of the SPE process. Also note how the authors convey the steps and their order in the SPE process.

Excerpt 3G (adapted from Coffin et al., 2019)

MATERIALS AND METHODS

Solid Phase Extraction. Following the incubation of plastic items in simulated digestion fluid, the samples were filtered and immediately extracted using SPE based on standardized protocol.[42] Prior to loading filtered simulated digestion fluid, each SPE column was preconditioned with 10 mL of methanol and 10 mL of deionized water. Each column was washed with 10 mL of deionized water, and samples were eluted with 10 mL of methanol into a clean glass scintillation vial. Throughout the SPE process, the flow rate was ≤1 mL/min. Following elution, the samples were concentrated to 1 mL under a gentle stream of nitrogen gas in a 40 °C water bath. The samples were not evaporated to dryness to avoid loss of volatile chemicals. The samples were syringe filtered (0.20 µm nylon) and stored at −20 °C until analyzed.

Exercise 3.8.

Re-read excerpt 3G and answer these questions:

a. What details do the authors include regarding the solid phase extraction (SPE) process? What quantitative values are included?

b. How do the authors indicate the order of events? What did they do first, second, third, etc.? To what extent is ordinal language (words such as first, second, then, and next) used to convey this sequence?

Descriptions of procedures often include numbers and units, formatted according to chemistry writing conventions. For example, notice how the numbers are formatted in excerpts 3H and 3I. Find examples of numbers in both numerical and word forms. Also notice that, in nearly all cases, there is a space between the number and its unit (e.g., 100 µM). One exception is 10,000g where the g (in italics) refers to acceleration due to gravity. Another exception is the percent sign (e.g., 5% not 5 %).

Excerpt 3H (adapted from Richardson et al., 2018)

Zn(II)-Uptake Assay. HEK293T cells (~500,000) were plated in a six-well dish. After 2 days, the medium was aspirated and ^{70}Zn(II)-enriched medium (1.0 mL) was added to each appropriate well. Cells were harvested by washing each well with phosphate-buffered saline (PBS) containing EDTA (100 µM, 2 × 2.0 mL), adding Milli-Q water (18.2 MΩ cm, 1.00 mL) to each well, and incubating for 2–4 min. A portion of the resulting cell lysate (0.90 mL) was collected and transferred to an Eppendorf safe-lock tube, snap-frozen in liquid nitrogen, and thawed at room temperature for a total of three freeze–thaw cycles. The insoluble lystate was pelleted (10 min at 10,000g), and the soluble protein concentration of each sample was determined by using the Pierce Protein Assay (660 nm) in a 96-well plate (200 µL of reagent, 14 µL of sample, bovine serum albumin standard curve). Samples were frozen and lyophilized overnight, and the resulting white, flaky solid was taken up in concentrated nitric acid (0.20 mL, ultrapure for trace metal analysis, BDH Aristar) and heated at 70 °C for 1 h. A portion of each sample (142 µL) was diluted with water (2.00 mL) and a Tb internal standard (Agilent 5190-8590, stock at 10,000 ppb in 10% nitric acid diluted to 100 ppb in Milli-Q water for addition to samples, 2 ppb final concentration), after which the metal content of the samples was assessed by ICP–MS.

Excerpt 3I (adapted from Neufeld et al., 2018)

MATERIALS AND METHODS

Experimental Methods. *Resazurin Assay.* The deviations from the negative control (resazurin with PBS) were determined by monitoring the absorbance values associated with the formation of the resorufin complex at 570 nm using a 96-well-plate assay. Stock solutions (100 mM) of all 19 species were made in PBS and subsequently diluted five times to obtain six different concentrations. Aliquots (100 µL) of each concentration were added to a 96-well plate followed by the addition of 20 µL of the resazurin dye (already in solution at a fixed concentration as prepared by the manufacturer). The 96-well plate was placed in a plate reader and allowed to shake for 10 s before measuring the absorbance at 570 nm every hour for a total of 4 h (Synergy 2 Multi-Mode Reader, BioTek, Winooski, VT, U.S.). The plate reader was maintained at 37 °C to mimic experimental conditions. The data are represented as the average percent deviation from the negative control. This was calculated by the difference between the sample and control value, divided by the control value, and multiplied to obtain a percent value ($n \geq 3$).

Formatting Numbers and Units

Segments of excerpts 3H and 3I are double underlined to call your attention to correctly formatted numbers and units. A few rules are summarized here.

- include a space between the number and unit (e.g., 100 µM) except for % (10%)
- use *numbers* with units of time and measure (e.g., 3 min, 15 ppb, 100 µM)

- use *words* with units other than time and measure (e.g., six-well dish) for numbers less than 10; for 10 and greater, use *numbers* (e.g., 19 species, 96-well dish)
- include a space before °C (e.g., 37 °C)

See Chapter 16, "Language Tips," Part 3: Writing Conventions, for additional tips on formatting numbers and units.

Exercise 3.9.

Using excerpts 3H and 3I as well as Chapter 16, "Language Tips," Part 3: Writing Conventions: Numbers and Units, for guidance, select the correct answer in a–j.

a. (five or 5) days
b. (ninety-six- or 96-) well plate
c. (six- or 6-) well plate
d. (20,000 or 20000) ppb
e. (twelve or 12) species
f. 100° C or 100 °C
g. 200 mL or 200mL
h. 0.20 µL or .20 µL
i. 10% or 10 %
j. ($n \geq 0.01$) or ($n{\geq}0.01$)

We end our analysis of this submove (describing procedures) by observing how authors describe quality assurance/quality control (QA/QC) procedures in their Methods sections. A common approach is to provide only brief information about QA/QC in the text and to refer readers to Supporting Information (SI) for more details. An example is shown in excerpt 3J.

Standards and Blanks

Blanks, spikes, surrogates, and internal standards are all terms associated with QA/QC procedures. Collectively, they are used to measure sample contamination, analyte recovery, and analyte relative abundance.

Excerpt 3J (adapted from Cao et al., 2019)

2.3. Quality Assurance and Quality Control (QA/QC). Sample recovery was determined by spiking 10, 50, and 100 ng of 10 standard OPFRs into a quartz fiber filter (n = 5) and a forehead wipe (n = 5), respectively, and analyzed using the same method as the samples. One procedural blank and one field blank were included for every batch of eight particulate samples or eight forehead wipe samples. The recovery of triaryl phosphate (TAP) and triphenyl phosphate-d15 (TPHP-d15) (surrogate standards) for all samples averaged 89 ± 19% and 93 ± 25% (mean ± SD), respectively. A value of three times of the deviation of all blank samples was used as the method detection limit (MDL) for each compound detected in blank samples. For compounds undetected in the blanks, a signal of 10 times the noise level was used as the MDL (SI Table S3). More detailed information and data on QA/QC are provided in SI (Section S2.3).

Submove: Describe Instrumentation

Authors must also describe the instrumentation or scientific apparatus used in their work. (Routine equipment, e.g., a distillation apparatus or a rotary evaporator, should *not* be described.) Instrumentation generally falls into two categories: custom-built or commercial. Custom-built instrumentation includes novel or hand-built chambers, devices, or instruments. The first publication that describes a custom-built instrument offers the most detail and often includes a diagram. Excerpt 3K, as an example, includes a diagram of a new reagent-ion source and focusing ion-molecule reactor for use in proton-transfer-reaction mass spectrometry. In subsequent publications, however, only the essential features of the apparatus are described and the reader is referred to the original article for more information (excerpt 3L).

Plural of Apparatus

The word *apparatus* has two plural forms: apparatus and apparatuses; most chemistry journals prefer *apparatus*.

Excerpt 3K (adapted from Krechmer et al., 2018)

A schematic drawing of the Vocus is shown in Figure 1. Reagent ions are produced in a discharge ion source. The focusing ion–molecule reactor (FIMR) is composed of a glass tube with four quadrupole rods mounted radially on the outside. More detailed descriptions of the individual components follow below. [*section omitted*]

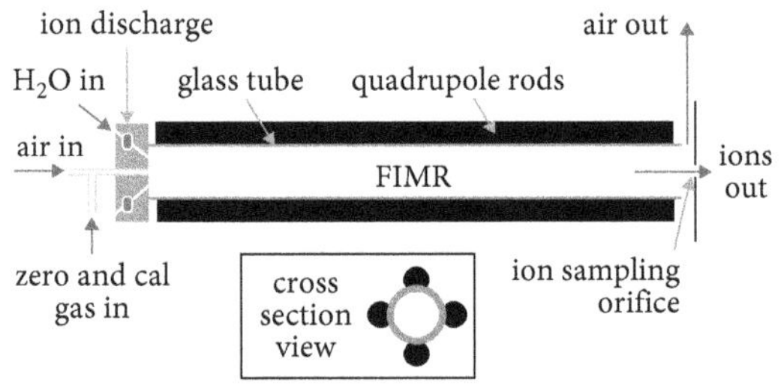

Figure 1. Design of the Vocus consisting of a discharge reagent-ion source and a focusing ion–molecule reactor (FIMR).

Excerpt 3L (from Chang et al., 2019)

A modified Smalley-type[31] laser ablation metal beam source was used to produce a precursor V atom sample in the form of a pulsed supersonically cooled gas jet. The detailed experimental arrangement and procedures for using this laser ablation metal atom beam source have been described previously.[32,33]

Descriptions of Instrumentation

Typically, instrumentation should be described in detail only if it is novel or custom-built. With most standard and commercially available instrumentation, stating a company name and model number (in parentheses) is sufficient.

What about commercial instrumentation? Nearly all chemists use some form of commercial instrumentation in their research such as gas chromatography, mass spectrometry, infrared spectroscopy, and/or nuclear magnetic resonance spectroscopy. A list of common instrumental techniques and their abbreviations are shown in Table 3.1. Because these instruments are used so frequently, sometimes only the instrument type is mentioned (e.g., a 500 MHz spectrometer). More often, however, authors also include the company name and model number of the instrument (in parentheses). Authors should also describe the parameters (i.e., settings) under which the instrument was operated, particularly if the parameters were optimized for a specific outcome. Because parameters can affect the results or reproducibility of the data, they should be reported. Characteristic ways to report parameters have been developed for many types of instruments. In passages P1–P3, we provide examples for FTIR, GC–MS/MS, and HPLC–MS–ESI instrumentation, respectively. Note that in P3, the authors report details about their instrument in Supporting Information rather than in the actual paper, a practice that is becoming increasingly common. Check the guidelines of your targeted journal to see what is preferred.

Table 3.1. Abbreviations for Common Instrumental Techniques
(adapted from Petronella, 2020, section 5.3.8)

instrumental technique	abbreviation
atomic absorption spectroscopy	AAS
electron spin resonance	ESR
Fourier transform infrared	FTIR
high-performance liquid chromatography	HPLC
high-pressure liquid chromatography	HPLC
inductively coupled plasma	ICP
mass spectrometry; mass spectrum	MS
scanning electron microscopy	SEM
secondary-ion mass spectrometry	SIMS
solid-phase microextraction	SPME
transmission electron microscopy	TEM
thin-layer chromatography	TLC
X-ray diffraction	XRD
X-ray fluorescence	XRF

P1 **FTIR Analysis.** Freeze-dried samples were ground with high-purity KBr and mechanically pressed to form translucent pellets. FTIR analysis was performed on a Shimadzu IR Affinity-1 Fourier transform infrared spectrophotometer equipped with a multiple transmission reflection (MTR) source and a RT-DLaTGS detector. The spectrophotometer was set to scan from 4000 to 400 cm^{-1}, averaging 64 scans with a 2.0 cm^{-1} interval. The KnowitAll software package was used to analyze the spectra. (Adapted from Yang et al., 2019)

P2 **Apparatus and Analytical Conditions.** The gas chromatography system used to analyze the pesticide residues consisted of a 7693 autosampler, 7890 B GC, and 7000 C GC–MS/MS triple quadrupole system (Agilent Technologies, Santa Clara, CA). Data acquisition and processing were implemented using Agilent Mass Hunter Quantitative Analysis B07.00 software.[13] Two ultra-inert columns were used, 15.0 m × 0.25 mm i.d., 0.25 µm, HP-5 ms (Agilent J&W, Santa Clara, CA). They were connected by an electronic pressure controller to enable a 5 min post-run backflush. The samples were injected using a multimode injector inlet in splitless mode through an ultra-inert inlet liner with a glass wool plug (Agilent Technologies, Santa Clara, CA). The injection volume was 3 µL.

The injector port was programmed so that the initial injection temperature was 70 °C for 0.02 min, ramped up at the rate of 850 °C/min to 320 °C, kept at this temperature for 5 min, then cooled down at a rate of 20 °C/min to 280 °C. Helium (99.999%) was used as the carrier gas and nitrogen (99.999%) as the collision gas. The oven temperature was held at 70 °C for 1 min, ramped to 150 °C at a rate of 50 °C/min, increased to 200 °C at 6 °C/min, then followed by a final ramp to 280 °C at 16 °C/min. The total run time was 24 min. The transfer line was maintained at 280 °C. The retention time lock setting used chlorpyrifosmethyl as the locking compound at retention time of 10.83 min. The instrument worked at a constant flow (1.25 mL/min in column 1 and 1.45 mL/min in column 2). (Adapted from Maestroni et al., 2018)

P3 **MS Analysis.** The enriched samples were analyzed using an Agilent 1200 high performance liquid chromatography (HPLC) coupled with an Agilent 6410 triple quadrupole mass spectrometer (MS) with an electrospray ionization (ESI) source (Agilent Technologies, U.S.). The instrument was operated in negative ionization and full scan acquisition (m/z 100–1000) mode (details in Supporting Information). (Adapted from Yang et al., 2019)

FTIR *What?*

Novice writers often say "FTIR was used," but this is incorrect. FTIR is the abbreviation for "Fourier transform infrared" (all serving as adjectives). Thus, you need to include a noun following FTIR, such as the following: FTIR analysis, FTIR measurements, FTIR spectroscopy, an FTIR spectrum, or several FTIR spectra. Likewise, a noun should be included after UV–vis and other similar techniques.

Exercise 3.10.

Reconsider passages P2 and P3. What operational parameters should you report in your Methods section (or Supporting Information) if you use an FTIR spectrometer? A GC–MS/MS system?

Exercise 3.11.

Consider the sentence "Samples were analyzed using UV–vis." Rewrite the sentence so that it is appropriate for a journal article.

Exercise 3.12.

Find three journal articles that describe an instrument that you have used. Compare the parameters reported in each article.

Move 3: Describe Numerical Methods

The final move of the Methods section of an analysis paper involves the description of statistical or other mathematical methods used to derive or analyze data. This move is required only if numerical methods were part of the work. Excerpts 3M and 3N demonstrate how some common types of statistical methods, including the Student's *t* test and analysis of variance (ANOVA), are described. Once again, subheadings are used to guide the reader's attention to this information. If specialized statistical software is used, the name and version number of the software package are generally reported. (Software packages used to analyze infrared and mass spectra are also reported, as in P1 and P2, respectively.) Important statistical parameters that affect the outcome of the statistical test (e.g., significance level) may be reported here or in the Results section. Note that routine software such as Microsoft Word or Excel should not be reported in this section (or anywhere else in the journal article).

Excerpt 3M (adapted from Lichtenberg et al., 2019)

Statistical Analysis. Statistical analysis was conducted using PROC GLM in SAS 9.4 for the fluorescence and reproduction assays. In each case, Dunnett's test was used to test for differences between treatment and control.[42] For the length assay, comparisons were made between treatments at the same concentration using a Student's *t* test.

Excerpt 3N (adapted from Besser et al., 2004)

Statistical Analysis. Analysis of variance (ANOVA) of toxicity data was conducted using SAS/STAT software (version 8.2; SAS Institute, Cary, NC). All toxicity data were transformed (square root, log, or rank) before ANOVA. Comparisons among multiple treatment means were made by Fisher's LSD procedure, and differences between individual treatments and controls were determined by one-tailed Dunnett's or Wilcoxon tests. Statements of statistical significance refer to a probability of type I error of 5% or less ($p \leq 0.05$). Median lethal concentrations (LC_{50}) were determined by the Trimmed Spearman-Karber method using TOXSTAT software (version 3.5; Lincoln Software Associations, Bisbee, AZ).

In excerpt 3O, the authors describe theoretical calculations performed with computational software based on molecular **ab initio** or **semiempirical** electronic structure theory. Computational parameters (e.g., the basis set and level of theory) are included in the descriptions. In theoretical papers, this section can be extensive. Here we include only a brief example.

Ab initio vs. Semiempirical

The Latin term *ab initio* means "from the beginning" or "from first principles." Ab initio calculations involve no experimental (empirical) data; they are derived solely from theory. (Note that, in ACS journals, ab initio is not italicized or enclosed within quotation marks.) Semiempirical calculations combine empirical (experimental) and nonempirical data.

Excerpt 30 (adapted from Yang et al., 2019)

Theoretical Computation. The computation of the optimized geometry and frequency was performed on the basis of DFT at the B3LYP/6-311+G (d, p) level of theory.[53] Vibrational analysis indicates all structures were at the minimum points on the potential energy surfaces. Transition states were characterized with one imaginary vibrational frequency. Intrinsic reaction coordinate (IRC) computations were performed to verify that transition states were connected to reactants and intermediates on the potential energy surfaces of reaction. All computations were performed using the Gaussian 09 software package.

Part 2: Analyzing Excerpts Move by Move in Synthesis Articles

We now examine excerpts that illustrate moves in Methods sections of synthesis papers (Figure 3.2). Many of you have already described synthetic procedures in your organic chemistry lab reports. Most likely, you included items such as the equation that you used to calculate percent yield or the steps that you followed to complete a synthesis (e.g., "Heat to reflux." or "Stir constantly for 10 min."). Are such items also appropriate in a journal article? To answer this question, we analyze several excerpts, each describing common chemical procedures.

Move 1: Describe General Experimental Methods

The Methods section of synthesis articles begins with a heading (e.g., **EXPERIMENTAL SECTION**). The first section after this heading typically describes general information about the procedures used in the work. This approach is the norm in journals such as *The Journal of Organic Chemistry* (excerpts 3P and 3Q) but can also be used in analysis journals such as *Analytical Chemistry* (excerpt 3R). A common sentence in this section is "All reagents were used as received," indicating that chemicals were not further purified. The purity (or grade) of the chemicals is also often specified (e.g., analytical grade). Note that the names of vendors are omitted (unlike in analysis articles). General conditions (e.g., that reactions were conducted under an argon atmosphere or that deionized water was used) are described. Instrumentation used to verify product identity and purity (e.g., chromatographic or spectroscopic methods) is also briefly mentioned in move 1.

Excerpt 3P (adapted from Murai and Mutoh, 2016)

General Remarks. The ^{1}H and ^{13}C NMR spectra were recorded in $CDCl_3$. Chemical shifts of protons are reported in δ values referenced to tetramethylsilane as an internal standard in $CDCl_3$, and the following abbreviations are used: s: singlet, d: doublet, t: triplet, m: multiplet. The HRMS spectra were recorded on a double-focusing mass spectrometer (EI). Column chromatography was performed on silica gel 60 N (spherical neutral) 100–210 µm. Flash column chromatography was performed on silica gel 60 N (spherical neutral) 40–50 µm. All manipulations were carried out under an argon atmosphere.

Excerpt 3Q (adapted from Tanaka et al., 2018)

General Information. All reagents and solvents were used as received from commercial suppliers. IR spectra were recorded on an FTIR spectrometer. ^{1}H NMR spectra were recorded in $CDCl_3$ containing tetramethylsilane as an internal standard and were acquired on a 500 MHz spectrometer. $^{13}C\{1H\}$ NMR spectra were acquired on a 125 MHz spectrometer. High-resolution mass spectra were obtained using a double-focusing magnetic sector mass spectrometer coupled with FAB. The light source was a high-pressure (100 W) mercury arcs and blue LED (18 W, 405 nm). Column chromatography was performed on Wakogel C-300, particle size 45–75 µm.

Excerpt 3R (adapted from Cheng et al., 2019)

Materials and Instrumentation. All reagents were of analytical grade and were used as received. The water used was deionized water. UV–visible absorption spectra were recorded on a Shimadzu UV-2600 spectrophotometer. Fluorescence titration experiments were performed on a Horiba Fluoromax-4 fluorescence spectrophotometer with a quartz cuvette (path length 1 cm).

Exercise 3.13.

Compare the information given about ^{1}NMR spectra in excerpts 3P and 3Q. What information is included in both excerpts?

Move 2: Describe Synthetic Procedures

We now examine how synthetic procedures are described in synthesis articles (move 2, Figure 3.2). The move has two submoves. In submove 2.1, synthetic methods are described; in submove 2.2, compound characterization data are presented. These two submoves are reiterated for each new compound synthesized. Both submoves are typically accomplished in the same (single) paragraph.

Let's begin by examining move 2 in two excerpts from *The Journal of Organic Chemistry* (excerpts 3S and 3T). Multiple compounds were synthesized in each project, but for brevity we include only one compound from each paper. We have also demarked the start of each submove (e.g., [*submove 2.1*]) in each excerpt. As illustrated in excerpt 3S, submove 2.1 begins with a subheading that states the compound's name and label (**6a**). The names and labels for compounds **1**, **3**, and **5a** also appear in this excerpt. These **compound labels** were first introduced, sequentially, in the Results & Discussion. (Recall the Results & Discussion section precedes the Methods section in *The Journal of Organic Chemistry*.) The authors reintroduce them again here so that readers do not need to refer back to earlier sections of the paper.

Compound Labels

Numerical or alphanumerical abbreviations for compounds (e.g., **2** and **6a**, respectively). They are bolded to differentiate them from other numbers. Compound labels are typically placed inside parentheses when first introduced, like other abbreviations; thereafter, they are used without parentheses. See Chapter 16, "Language Tips," Part 3: Writing Conventions, for more details on compound labels.

Excerpt 3S (adapted from Murai and Mutoh, 2016)

N,N-Dimethyl-1-phenyl-1-(phenylsulfanyl)propan-2-amine (6a). [*Submove 2.1*] To a solution of benzylphenylsulfane (**1**) (0.5 mmol, 0.1001 g) in THF (2.0 mL) was added *n*-BuLi (1.05 equiv) at −78 °C, and the mixture was stirred for 1 h. To the resulting solution was added a solution of *N,N*-dimethylthioformamide (**3**) (1.2 equiv, 0.051 mL) in THF (1.5 mL), and the mixture was stirred for 15 min at room temperature. To this was added methylmagnesium bromide (**5a**) (1.5 equiv), and this mixture was stirred at reflux for 2 h. The resulting mixture was poured into a saturated aqueous solution of NH_4Cl and extracted with Et_2O. The organic layer was dried over $MgSO_4$, filtered, and concentrated in vacuo. The residue was purified by flash column chromatography (SiO_2, hexane/EtOAc/Et_3N = 10:1:0.01) to give the amine **6a**: [*submove 2.2*] a pale yellow oil, 66% yield (0.089 g), dr = 91:9; ^{1}H NMR ($CDCl_3$) δ 0.75 (d, *J* = 6.3 Hz, 3H), 2.37 (s, 6H), 3.07–3.14 (dq, *J* = 6.3 and 10.7 Hz, 1H), 4.17 (d, *J* = 10.7 Hz, 1H), 7.02–7.49 (m, 10H); ^{13}C NMR ($CDCl_3$) δ 8.9, 40.4, 59.5, 63.7, 126.8, 127.1, 128.2, 128.4, 129.0, 133.3, 135.7, 141.1.

Excerpt 3T (adapted from Swenson et al., 2002)

2-(*p*-Toluenesulfonyl)-4′-methoxyacetophenone (2a). [*Submove 2.1*] A mixture of 2-bromo-4′-methoxyaceto-phenone (45.8 g, 200 mmol) and *p*-toluenesulfinic acid sodium hydrate (35.6 g, 200 mmol) in ethanol (1 L) was heated at reflux for 1.5 h. The mixture was stirred and cooled to room temperature, and the resulting solid was collected, washed with ethanol (2 × 50 mL), and dried to give pure **2a**: [*submove 2.2*] 54.6 g (90%), mp 126.0–127.0 °C; IR (cm^{-1}): 2951, 2906, 1676, 1599, 1572; ^{1}H NMR (400 MHz, $CDCl_3$): δ 2.45 (s, 3H), 3.90 (s, 3H), 4.67 (s, 2H), 6.95 (d, *J* = 8.8 Hz, 2H), 7.34 (d, *J* = 8.2 Hz, 2H), 7.76 (d, *J* = 8.2 Hz, 2H), 7.95 (d, *J* = 8.8 Hz, 2H); ^{13}C NMR ($CDCl_3$): δ 20.9, 55.1, 62.5, 113.4 (2C), 127.7 (2C), 128.3, 129.1 (2C), 131.1 (2C), 135.8, 144.3, 163.7, 186.0. Anal. Calcd for $C_{16}H_{16}O_4S$: C, 63.14; H 5.30; S, 10.54. Found: C, 63.49; H, 5.35; S, 10.33.

Submove 2.1: Describe Synthetic Methods

Let's return to submove 2.1 in excerpts 3S and 3T. You might have expected the syntheses to be written more like a lab manual with cookbook-like instructions for readers to follow, for example: (1) combine 2 mg of X and 4 mg of Y, (2) add 10 mL ethanol, and (3) stir for 5 min. But no lists are used, and the steps are written without imperatives (verbs such as *combine*, *add*, and *stir*). Instead, sentences are written in past tense and passive voice (e.g., *was heated*, *was stirred*) and amounts are placed in parentheses. For example, the three preceding tasks would likely be written as follows: A mixture of X (2 mg) and Y (4 mg) in ethanol (10 mL) was stirred for 5 min.

The language used by chemists to describe synthetic procedures may sound awkward at first, but over time, you will realize that this approach is both clear and concise. To help you learn this style of writing, we offer three "formulas" that you can use to describe a few common synthetic procedures. These formulas also follow the rules for formatting numbers and units. By using these formulas, you will sound more like an expert. (Note: For instructional purposes, we have italicized passages and amounts in these examples that will vary with each synthesis.)

Formula 1: To describe how to prepare a mixture and heat at reflux

- A mixture of *X* (*5 g, 0.01 mmol*), *Y* (*5 g, 0.02 mmol*), and *Z* (*10 g, 0.02 mmol*) in *A* (*10 mL*) was heated at reflux for *2 h* and cooled to *room temperature*.
- To a mixture of *X* (*5 g, 0.01 mmol*) and *Y* (*5 g, 0.02 mmol*) in *A* (*10 mL*) was added *Z* (*10 g, 0.02 mmol*). The mixture was heated at reflux for *2 h* and cooled to *room temperature*.

Formula 2: To describe how to continue and stop a reaction

- After *15 min*, a solution of *X* (*5 g, 0.01 mmol*) and *Y* (*5 g, 0.02 mmol*) in *A* (*10 mL*) was added to the mixture. The mixture was stirred at *10 °C* for *1 h*, warmed to *room temperature*, and quenched with *saturated aqueous ammonium chloride* (*5 mL*).

Formula 3: To describe how to extract, wash, and purify a product

- The water phase was extracted with *X* (2 × *10 mL*). The combined organic layers were washed with *Y* (2 × *10 mL*) and *Z* (2 × *10 mL*), dried with $MgSO_4$, and evaporated in vacuo. The crude product was recrystallized from *ethanol* at *10 °C* and further purified using *column chromatography*.

Units and Verb Agreement

Use a singular verb when one unit of measure is mentioned. Use a plural verb when two separate units of measure are mentioned.

Incorrect	To the mixture *were* added 8.5 g of X.
Correct	To the mixture *was* added 8.5 g of X.
Correct	To the mixture *was* added X (8.5 g).
Correct	To the mixture were added 8.5 g of X and 4.5 g of Y.

Describing a Synthesis

Phrases commonly used when describing a synthesis include these:

- heated to reflux
- heated at reflux
- cooled to room temperature
- the organic layer
- the aqueous layer
- the/an aqueous solution
- the reaction mixture
- the resulting mixture
- the crude product

Exercise 3.14.

Rewrite the following synthetic procedure (adapted from D'hooghe et al., 2004) using formulas 1–3 for guidance. The five steps listed (a–e) are from a single procedure involving multiple steps. Note the use of compound labels in steps d and e.

a. Make a solution containing 15 mmol (3.60 g) of 2-(bromomethyl)-1-((4-methylphenyl)methyl)aziridine in 50 milliliters of acetonitrile.
b. Add 2.56 g (15 mmol) of benzyl bromide.
c. Heat the solution for 5 hours, refluxing.
d. Remove the solvent using a vacuum to produce *N*-benzyl-*N*-(2,3-dibromopropyl)-*N*-((4-methylphenyl)methyl)amine (**4**).
e. Purify the crude product using column chromatography on silica gel, eluting with a mixture of hexane and ethyl acetate. A typical yield of purified **4** is 85%.

Exercise 3.15.

Errors in the use of numbers and units were intentionally added to the following excerpt. Identify at least 10 mistakes and correct them.

3-Benzyloxy-6-bromo-4-methoxyphenethyliminophosphorane (5a). To a solution of LAH (250mg) in THF (15ml) was added a solution of 3-benzyloxy-6-bromo-4-methoxy-β-nitrostyrene (500 mg, 1.37 mmol) in THF (15 mL). The mixture was boiled under reflux for eight hours. Addition of several mL of water followed by extraction with ether and drying the ether layer (K_2CO_3) gave the corresponding phenethylamine, which was purified by column chromatography on silica gel (hexane-CH_2Cl_2, 6:4 v/v) to give **5a**: yellow oil (330 mg, 72 %). The amine (160 mg, .48 mmol) and triphenylphosphine (130 mg, .48 mmol) in 100 mgs of CCl_4 and CH_2Cl_2 (7 mL) were stirred at 40° C for 72 hours under dry nitrogen. (Adapted from Rodrigues et al., 2004)

We call your attention to details included in (and excluded from) excerpts 3S and 3T. The quantities of reagents were included (e.g., mmol for solid reagents and mL for solvents). Stirring times, reaction temperatures, and procedures used to wash, dry, and purify the product were also reported. Because these factors directly influence product yield and purity, they must be mentioned. Routine procedures that do not directly affect product yield and purity can be omitted, such as the types of glassware used, weighing procedures, or dilution steps. Authors sometimes do mention specialized glassware such as a three-neck round-bottom flask or an oven-dried Schlenk tube, but routine glassware (Erlenmeyer flasks, graduated cylinders, etc.) and any disposable items (pipettes, gloves, etc.) should not be mentioned. Although there are no hard and fast rules, you should omit common details unless you want to draw attention to them (see Figure 3.3).

Figure 3.3. "Every little detail" is too much detail for experts and Snoopy!

Exercise 3.16.

The three excerpts here describe the process of recrystallization for product purification. The first excerpt is from a lab manual; the last two are from articles in *The Journal of Organic Chemistry (JOC)*. Read and compare the excerpts as you answer these questions:

a. What details are included for experts?
b. What details are included for students but omitted for experts?

Lab manual	The residue obtained in evaporation of the methylene chloride is next recrystallized by the mixed-solvent method. Using a steam bath or hotplate, dissolve the residue in a small quantity (about 2 mL) of hot acetone and add dropwise just enough low-boiling (bp 30–60 °C) petroleum ether to turn the solution faintly cloudy. Cool the solution and collect the crystalline product by vacuum filtration, using a small Buchner funnel. (From Pavia et al., 1998)
JOC	The mixture was evaporated to dryness and recrystallized from the methanol/chloroform solution. (Adapted from Frydrych et al., 2019)
JOC	The major diastereomer (–)-**5a** was purified by recrystallization from hexanes/CH_2Cl_2 = 3:1 to give 3.30 g of this compound as a white solid (60%). (From Gunther et al., 2019)

Exercise 3.17.

Re-read excerpts 3S and 3T and answer these questions:

a. Different units are used to describe amounts of reactants, solvents, and products. What are those units? Include correct punctuation and capitalization.
b. For what procedures are time and temperature reported?
c. What details are included for refluxing, washing, and drying the products?

We have listed details that are commonly included in and excluded from descriptions of syntheses in Table 3.2. From these lists, it is clear that journal articles are not intended to teach novices how to conduct a synthesis; instead, they offer a blueprint for the experienced chemist. In other words, the recipe is included, but basic cooking instructions are not.

Table 3.2. Details Commonly Included and Omitted When Describing a Synthesis

details commonly included	details commonly omitted
• a specialized type of glassware • the mass and moles of solid reagents or volume of liquid reagents (often in parentheses) • information on the reaction time and temperature • information on stirring • information on how the product was filtered, washed, and dried (including solvents used and their amounts) • the final product mass and percent yield • a description of the final product (a solid, a white powder) • analytical information about the product (mp, IR, ^{1}H NMR, and mass analysis)	• the equation used to calculate yield • disposables (gloves, pipettes) • names of common glassware (funnels, beakers) • steps taken in routine procedures (weighing, diluting, purifying) • mistakes made (e.g., spilling the product) (Note: If such mistakes are made, the synthesis must be repeated!)

How authors convey the order of events in a synthesis is different in lab manuals and journal articles. In lab manuals, order is normally conveyed through numbered lists, but lists are not used in journal articles. In lieu of lists, novice writers tend to resort to ordinal language; they overuse words such as *first*, *second*, *next*, and *then* to convey order. More experienced writers learn to avoid such language. Consider these examples:

Inappropriate	Next, the extracts were combined, then they were reduced to 1 mL, and then they were frozen.
Inappropriate	First, the extracts were combined; second, they were reduced to 1 mL; and third, they were frozen.
Appropriate	The extracts were combined, reduced to 1 mL, and frozen.

In the appropriate version, the sequencing of events is implied through the order in which the steps are presented. Words such as *first* or *next* are redundant and unnecessary; by omitting these words, the writing is more concise. Similarly, take care not to overuse the word *then*, which should be used only sparingly. The use of ordinal language, although grammatically correct, detracts from the flow of the text.

Use Ordinal Language Sparingly

Ordinal language indicates order or position in a series. Examples include the following:

- first, second, third
- next, then

Ordinal language should be used sparingly when describing synthetic procedures in journal articles. Instead, convey order through the arrangement of your sentences.

Submove 2.2: Provide Compound Characterization Data

We now call your attention to submove 2.2 in which authors report data regarding the physical properties and identity of any new compound synthesized. These data must also be included for known compounds prepared by a new method. Manuscripts submitted without this information will most likely be rejected. As shown in excerpts 3S and 3T, submove 2.2 often begins after a colon in the sentence that ends submove 2.1. Information regarding the product's yield, physical appearance (e.g., a pale yellow oil), melting point (mp), and diastereomeric ratio (dr) are reported first. These results are followed by data that confirm the compound's identity (e.g., results from IR, ^{1}H NMR, ^{13}C NMR, and elemental analyses).

Perhaps you are surprised to see such data included in the Experimental Section. Most students view NMR and IR data as "results" and expect to see the actual spectra in the Results section. But spectra take up a great deal of space and are rarely included in published articles. Moreover, specific formats have been developed to present data from each analytical method (e.g., excerpt 3T reports data from IR, ^{1}H NMR, ^{13}C NMR, and elemental analyses.) To write like a chemist, you must adhere to these formats.

Reporting NMR Results

In synthesis articles, NMR results and other data used to confirm product purity or identity are reported in the Experimental Section (not the Results section) and formatted as text (not as spectra). Excerpts 3S and 3T illustrate conventional ways to report IR, NMR, and elemental analysis data.

Exercise 3.18.

Using excerpt 3T as a guide, rewrite the following ^{1}H NMR data in the appropriate format for an Experimental Section of a synthesis article. (Hint: it is customary to arrange NMR signals from smallest to largest chemical shift (**δ**)).

The ^{1}H NMR (400 MHz) solvent was $CDCl_3$. Three signals were detected. The first had a chemical shift of 1.2 ppm (a triplet, integrated area equal to 3H; J = 3.6 Hz). The second had a chemical shift of 2.4 ppm (a quartet, integrated area equal to 2H, J = 3.6 Hz). The third had a chemical shift of 2.2 ppm (singlet, integrated area equal to 1H).

Perfecting Your Writing

Here we introduce two writing conventions that apply to Methods sections in both analysis and synthesis articles: the use of verb tense (past and present) and voice (passive and active).

Verb Tense and Voice

In Methods sections, past tense and passive voice predominate; however, in some cases, present tense and/or active voice are used. Like other writing conventions, the proper use of tense and voice reveals authors' familiarity with disciplinary expectations, objectivity, and more expert-like writing abilities.

Past and Present Tenses

The Methods section is largely written in the past tense. In general, the Methods section describes work that was done in the past, making the past tense the appropriate choice. This is different from a lab manual, which gives a set of instructions in the present tense.

Lab manual	Set the flow rate to 1 mL/min.
Journal article	The flow rate was 1 mL/min.
Lab manual	Stir the mixture. Heat to reflux.
Journal article	The mixture was stirred and heated to reflux.

Although the Methods section is overwhelmingly written in the past tense, there are a few correct instances of present tense. The general rule of thumb for deciding when to use past or present tense in the Methods section (and elsewhere in the journal article) is as follows:

Work was done in the past; knowledge exists in the present.

Work done in the past is described using past-tense verbs (e.g., analyzed, built, heated, synthesized). Knowledge that exists in the present (and presumably into the future) is described using present-tense verbs (e.g., contains, implies, is expected to, suggests). Present tense is also used to describe fixed features of an apparatus (e.g., the flange *is* made of stainless steel 465) or to refer readers to information presented elsewhere in the text (e.g., Figure 2 *shows* . . .). Consider these additional examples:

Past tense	The water *was* triply distilled.
Present tense	Triply distilled water *contains* less than 1 ppb of the impurity.
Past tense	The probe *was* modeled after work described elsewhere (4).
Present tense	The probe *projects* through the tee into the main chamber.

Exercise 3.19.

For each italicized pair, select the correct verb tense:

a. Phase data, which *measure/measured* the phase shift in the cantilever oscillation, *are/were* taken with an AFM operated in tapping mode. (Adapted from Clancy et al., 2000)
b. This area *is/was* also part of the Klondike Plateau Ecoregion, which *is/was* unglaciated in the past. (Adapted from Bandara et al., 2019)
c. The tests in this study *are/were* conducted in TCE-contaminated groundwater. The A-zone *is/was* an unconfined shallow layer composed mainly of placed fill over Bay Mud. (Adapted from Hageman et al., 2001)
d. Protease activities *are/were* measured by zymography. (Adapted from Koga et al., 2019)
e. Metabolites *are/were* identified using their UV spectral properties and molecular mass, and whenever possible by comparison with authentic standards. (Adapted from García-Villalba et al., 2019)
f. The samples *are/were* hereafter referred to as *Fg*-infected wheat/grain samples. (Adapted from Koga et al., 2019)

Passive and Active Voice

The Methods section is also written largely in passive voice. Passive voice is most often combined with past tense:

Inappropriate	We heat the mixture to 80 °C.	[present tense–active voice]
	We heated the mixture to 80 °C.	[past tense–active voice]
	The mixture is heated to 80 °C.	[present tense–passive voice]
Appropriate	The mixture was heated to 80 °C.	[past tense–passive voice]

You may have been taught in other writing courses not to use passive voice because it is considered "weak." However, passive voice, when used appropriately, strengthens writing in chemistry journal articles (and other scientific genres). Passive voice is often preferred because it sounds more objective. Passive voice essentially removes the human subject (i.e., the scientist) from the sentence so that the focus of the sentence is on the science. As a result, the word *we* and other personal pronouns (e.g., *I* and *our*) are seldom used in the Methods section.

Active (avoid)	We added solid Se (0.030 g) to the pale orange solution.
Passive (preferred)	To the pale orange solution was added solid Se (0.030 g).
Active (avoid)	We stirred the mixture for 10 min at room temperature.
Passive (preferred)	The mixture was stirred for 10 min at room temperature.

A list of common past tense–passive voice combinations used in Methods sections and a summary of the functions of various verb tense–voice combinations are shown in Tables 3.3 and 3.4, respectively.

Table 3.3. Past Tense–Passive Voice Combinations Commonly Used in the Methods Section

was added	was determined	was maintained	was refluxed
was allowed	was dissolved	was performed	was removed
was assigned	was dried	was poured	was separated
was carried out	was evaporated	was prepared	was stirred
was collected	was extracted	was purified	was treated
was concentrated	was filtered	was quenched	was used
was cooled	was heated	was recorded	was washed

Table 3.4. Common Verb Tense–Voice Combinations for Key Functions in the Methods Section of a Journal Article (other combinations are possible)

function	tense–voice combination	example
a. to describe materials	past–passive	Methanol (99.8%) *was purchased* from Aldrich.
	past–active	All other chemicals *were* reagent grade.
b. to describe procedures and numerical methods	past–passive	The samples *were centrifuged* at . . .
		To the solution *was added* . . .
		Each sample *was diluted* with . . .
		Statistical analyses *were performed* using . . .
c. to describe fixed features of instrumentation	present–active	The length of the tube *measures* 10 cm.
d. to refer to a graphic	present–active	Figure 1 *shows* . . .
	present–passive	The apparatus *is shown* in Figure 1.

Joining Sentences in Passive Voice

When you join two sentences that are in past tense and passive voice, each subject must have a verb that is preceded by *was* or *were*.

Incorrect	The eluant *was added* to the column, and the samples *collected* in 10 mL increments.
Correct	The eluant *was added* to the column, and the samples *were collected* in 10 mL increments.

Passive Voice

See Chapter 16, "Language Tips," Part 3: Writing Conventions, for tips on using active and passive voice.

Exercise 3.20.

Rewrite these sentences so that they are more appropriate for the Methods section of a journal article using passive voice and past tense.

a. We recrystallized the product from ethanol in a fume hood.
b. Filter the precipitate. Wash three times with 10 mL of ethanol each time.
c. Add chlorosulfonic acid (0.350 mL) dropwise to the mixture.
d. We used a Nicolet model 590 FTIR spectrometer to analyze the water-ice films.

Chapter Review

Check your understanding of what you've learned in this chapter by defining each of these terms, in the context of this chapter, for a friend or colleague who is new to the field:

abbreviations	numerical methods	passive voice
ab initio	operational parameters	QA/QC
acronym	ordinal language	semiempirical
active voice		

Explain the following to a friend or colleague who has not yet tried to write a Methods section for a journal article:

- Main purpose of a Methods section
- Different moves in the Methods sections of analysis and synthesis journal articles
- Guidelines for what details to include in a Methods section (revisit exercise 3.1)
- Appropriate use of capitalization, numbers, and units in a Methods section
- Guidelines for conveying the order of events in a Methods section
- Ways to describe a synthesis in a journal article versus a lab manual
- Guidelines for presenting ^{1}H NMR data
- Use of tense (past/present) and voice (active/passive) in a Methods section

4 *Writing the Results Section*

All sections of a journal article lead up to or away from the results section, and the results section may retain its value long after the methods and conclusions have become obsolete.

—Paradis and Zimmerman (2002)

This chapter focuses on the Results section of the journal article. The Results section makes use of text and **graphics** (figures and tables) to highlight the essential findings of a study and to tell a story of scientific discovery. In this chapter, we focus on writing text; the preparation of figures and tables is covered in Chapter 14. After reading this chapter (and Chapter 14), you should be able to do the following:

- Distinguish between description and interpretation of data
- Organize and present your results in a clear, logical manner
- Refer appropriately to tables and figures in the text
- Use appropriate tense, voice, and word choice
- Prepare properly formatted figures and tables

Graphics

We use the term *graphics* to refer collectively to figures and tables. Figures and tables are used to display, clarify, and summarize results, helping readers comprehend data more quickly.

Writing on Your Own: Results Section

Visit the *Write Like a Chemist* (2nd ed.) website for step-by-step guidance in writing your own Results section.

Write Like a Chemist. Marin S. Robinson and Fredricka L. Stoller, Oxford University Press. © Oxford University Press 2022.
DOI: 10.1093/oso/9780190098940.003.0004

The purpose of a Results section is to present the most essential data collected during a research project. A well-written Results section guides the reader's attention back and forth between text and graphics while highlighting important features of the data and telling the story of scientific discovery. Months (possibly years) of accumulated knowledge and wisdom, and countless pages of data, are distilled into only a few pages; hence, only the essential threads of the story are described in the results.

Traditionally, Results sections stood alone in most journals, separated from Discussion sections. This manner of presentation gave readers the opportunity to analyze data on their own, without being influenced by the authors' theories and interpretations. Today, however, most journals combine these sections. Combined Results and Discussion (R&D) sections are preferred because they are more concise. Combined R&D sections also provide a more continuous, unbroken chain of thought, making the science easier to follow. Many ACS journals use combined R&D sections, including *Analytical Chemistry*, *Biochemistry*, *Environmental Science & Technology*, *The Journal of Organic Chemistry*, and the *Journal of Physical Chemistry*. However, as of this printing, *Inorganic Chemistry* still prefers separate sections. Because conventions vary, it is important to check the Author Guidelines of your targeted journal before beginning to write.

Combined R&D sections are not all alike. If viewed on a continuum, three patterns emerge: blocked R&D, alternating R&D, and integrated R&D. In the **blocked R&D** pattern, a block of results is followed by a block of discussion. For example, for three sets of results, the pattern would be [results 1, results 2, results 3] [discussion 1, discussion 2, discussion 3]. This approach is reminiscent of fully separated sections, but they are merged under a single "Results & Discussion" heading. More often, authors use an **alternating R&D** pattern, first presenting a result and then discussing it. For three results, the pattern would be as follows: [results 1, discussion 1] [results 2, discussion 2], [results 3, discussion 3]. The story of scientific discovery is often easier to tell (and understand) if each finding is presented and discussed before moving on to the next. With the **integrated R&D** pattern, results are presented and discussed together, often in the same paragraph or even the same sentence. The text is organized in a way that best conveys the story of scientific discovery, with no obvious delineation between results and discussion. This pattern is less common but can be quite effective when done properly. These patterns are intended to serve as guiding constructs only. In practice, most authors combine features of several R&D patterns, making a pure example difficult to find.

Combined R&D Patterns

Combined R&D sections are not all alike. They fall on a continuum that can be represented by three configurations:

Blocked R&D	[R1, R2, R3] [D1, D2, D3]
Alternating R&D	[R1, D1] [R2, D2] [R3, D3]
Integrated R&D	no obvious pattern

Despite the frequency with which combined R&D sections now appear in chemistry journals, we address the sections separately in this book. The different purposes of Results and Discussion sections are important to understand and distinguish, even if you ultimately write a combined R&D section. In this chapter, we focus on the Results section. The Discussion section and the integrated R&D pattern are examined in Chapter 5.

Reading and Analyzing Writing

In Results and Discussion sections, the reader should be led step-by-step through the subject, showing how conclusions unfold logically as the results accumulate.

—Charles H. DePuy, University of Colorado Boulder

We begin by asking you to read and analyze a Results section on your own. Excerpt 4A is a continuation of excerpt 3A (in Chapter 3) regarding the analysis of proanthocyanidins (PAs) in pomegranate fruit. The excerpt includes the first two (of five) subsections that describe (1) the qualitative identification of PAs by LC–ESI–MS and (2) their phloroglucinol adducts after phloroglucinolysis degradation. The subsections that describe PA quantification and method validation are omitted.

Exercise 4.1.

As you browse through excerpt 4A, consider these questions:

a. What writing and formatting conventions do you notice in the subheadings, text, table, and figures (e.g., capitalization, use of italic or bold font, use of parentheses)?
b. Identify one sentence that is only results. Identify one sentence that is only discussion. How does the language help you differentiate between the two?
c. Are the graphics (Figure 2, Table 1, and Figure 3) referred to at the beginning or end of the paragraph that describes them? To what extent are the data in the graphics repeated in the text?
d. What did the authors do to make their writing concise?

Excerpt 4A (adapted from Díaz-Mula et al., 2019)

RESULTS AND DISCUSSION

Qualitative Identification of PAs by LC–ESI–MS. The PA profile in pomegranate samples was studied using the general method for polyphenol analysis. Masses of PA monomers, oligomers, and polymers were extracted from the MS chromatogram. Monomers (*m/z* 289 and 305) and dimers (*m/z* 577, 609, and 593) of catechins and/or gallocatechins were detected with high intensity (Figure 2), and the presence of different isomers was observed. The same monomers and dimers were previously detected in pomegranate samples.[18] Some trimers composed of catechins and gallocatechins (*m/z* 881 and 913) with masses that could also correspond to dimers of ECG and EGCG were identified but with much lower intensities. Masses of other trimers, tetramers, and higher polymers were not detected. This could be due to (1) the absence of these polymers in pomegranate samples, (2) their presence at undetectable concentrations, or (3) significant amounts of polymeric PAs remaining in the residue after extraction. Quantification was difficult due to the lack of available standards and the possible co-elution of the oligomers as a large unresolved peak. Acid-catalyzed cleavage in the presence of phloroglucinol was applied for an accurate quantification of PAs.

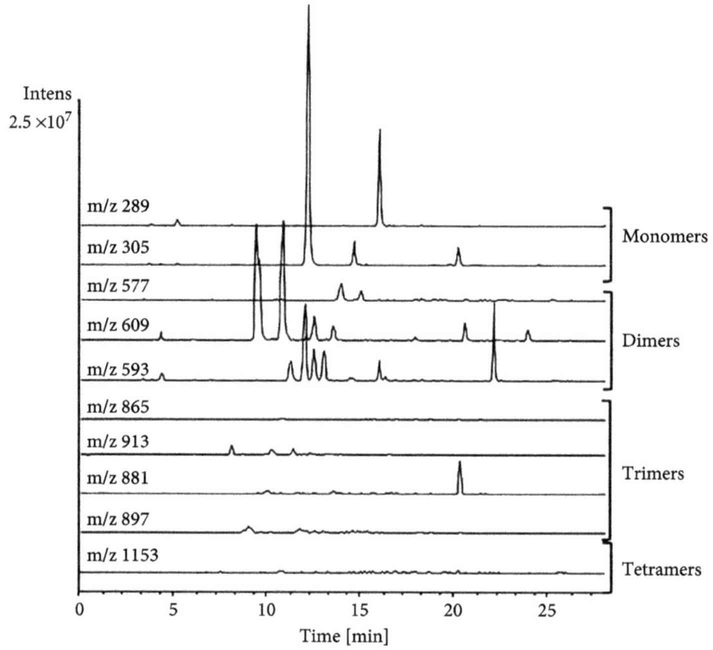

Figure 2. EIC of the main monomers, dimers, trimers, and tetramers searched in the pomegranate samples: *m/z* 289, (epi)catechin; *m/z* 305, (epi)gallocatechin; *m/z* 577, (epi) catechin–(epi)catechin; *m/z* 609, (epi)gallocatechin–(epi)gallocatechin; *m/z* 593, (epi) gallocatechin–(epi)catechin; *m/z* 865, (epi)catechin–(epi)catechin–(epi)catechin; *m/z* 913, (epi)gallocatechin–(epi)gallocatechin–(epi)gallocatechin; *m/z* 881, (epi)gallocatechin–(epi) catechin–(epi)catechin; *m/z* 897, (epi)gallocatechin–(epi)gallocatechin–(epi)catechin; and *m/z* 1153, (epi)catechin–(epi)catechin–(epi)catechin–(epi)catechin.

Phloroglucinolysis Reaction. A variety of flavan-3-ol monomers and their phloroglucinol adducts were detected in pomegranate samples after the phloroglucinolysis degradation (Table 1 and Figure 3). All showed similar spectra with maxima around 275 nm. The fragmentation profile was similar for all adducts, with losses of *m/z* 126 and fragments with *m/z* 125 corresponding to a phloroglucinol residue. Other fragments at *m/z* 261 and 177 or 175 were also common in adducts. The monomers, C, EC, GC, and EGC, showed similar fragmentation profiles with losses of *m/z* 18 and 44 and a common fragment at *m/z* 179. EGCG and ECG showed losses of *m/z* 44 and 152 and fragments at *m/z* 169, corresponding to the gallic acid residue.

Table 1. Proanthocyanidin Cleavage Products Found in Pomegranate Samples after Acid Catalysis in the Presence of Phloroglucinol

compound	no.[a]		rate(min)	$[M{-}H]^-$	γ_{max}	MS/MS
(−)-gallocatechin	1	adduct	5.6	429	278	**303, 261, 177, and 125**
	3	monomer	7.9	305	278	**287, 261, 219, and 179**
(−)-epigallocatechin	2	adduct	6.2	429	271	**303, 261, 177, and 125**
	7	monomer	11.6	305	271	**287, 261, 219, and 179**
(+)-catechin	4	adduct	9.1	413	278	**287, 261, 175, and 125**
	8	monomer	13.0	289	278	**271, 245, 205, and 179**
(−)-epigallocatechin gallate	5	adduct	9.4	581	275	**537, 455, 429, 261, and 175**
	10	monomer	16.8	457	277	**413, 331, 305, and 169**
(−)-epicatechin	6	adduct	9.8	413	278	**287, 261, 175, and 125**
	11	monomer	17.1	289	278	**271, 245, 205, and 179**
(−)-epicatechin gallate	9	adduct	15.2	565	277	**439, 413, 395, 261, and 175**
	12	monomer	23.2	441	273	**397, 331, 289, 193, and 169**

[a] Numbers correspond to peaks in Figure 4.

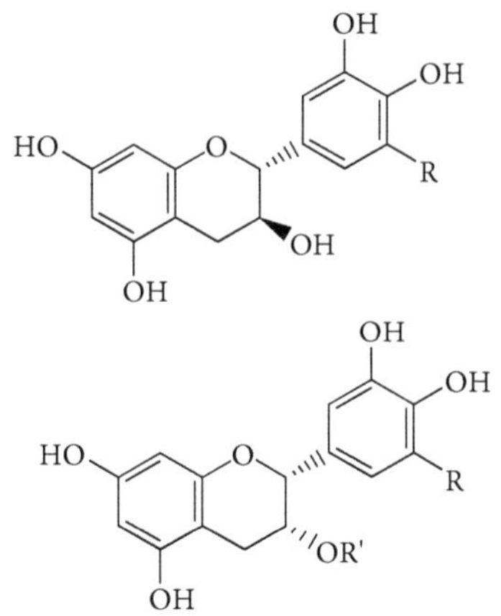

R = H Catechin
R = OH Gallocatechin

R = H R' = H Epicatechin
R = H R' = Gallate Epicatechin Gallate
R = OH R' = H Epigallocatechin
R = OH R' = Gallate Epigallocatechin Gallate

Figure 3. Basic structures of flavan 3-ols in pomegranate.

Exercise 4.2.

Compare the following graph (a bar graph) with Figure 2 in excerpt 4A (an extracted ion chromatogram or EIC). Although the graphs are of different types, use the EIC in excerpt 4A as a guide to identify at least five formatting errors in the bar graph.

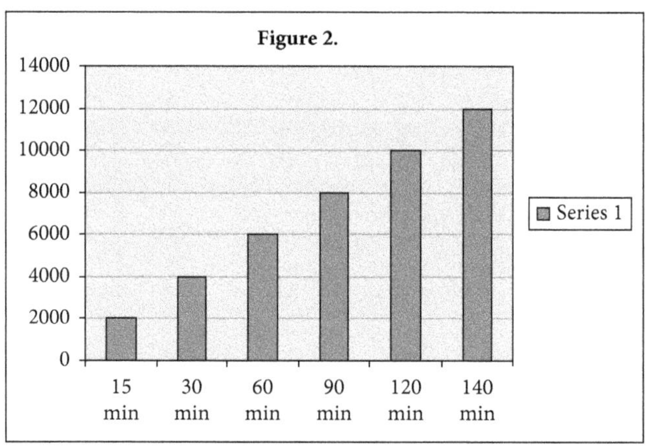

Exercise 4.3.

Compare the table here with Table 1 in excerpt 4A. Although the tables contain different types of information, use Table 1 in excerpt 4A as a guide to identify at least five formatting errors in this table.

Table 1. Correlation coefficient (R^2), coefficient of variations (CV), and relative recovery (RR) of analyzed aldehydes.

	R^2	CV	RR
2-Methylpropanal	.963	4.7%	110%
2-Methylbutanal	.972	4.6%	104%
3-Methylbutanal	.970	4.0%	109%
Pentanal	.995	3.9%	114%
Hexanal	.992	4.3%	103%
Furfural	.989	5.1%	99%
Methional	.998	2.4%	90%
Phenylacetaldehyde	.983	10.3%	98%
(*E*)-2-Nonenal	.994	12.0%	89%

Analyzing Audience and Purpose

An author's central obligation is to present an accurate and complete account of the research performed, absolutely avoiding deception, including the data collected or used, as well as an objective discussion of the significance of the research.

—American Chemical Society, Ethical Guidelines to Publication in Chemical Research

The central purpose of the Results section is to describe your research findings to other scientists (an expert audience) in a clear and concise manner. As you will see in Chapter 5, the central purpose of the Discussion section is to interpret those findings. The distinction between description and interpretation is not always clear-cut. The following rule of thumb helps to distinguish between the two:

Description (Results) answers the question *What did you find?*
Interpretation (Discussion) answers the question *What do your findings mean?*

Truth?

Words such as *truth* and *prove* seldom appear in scientific writing. In a computer-based analysis of 180 journal articles, *prove* was found only twice, and *truth* never occurred. Hedging words are used instead. For example, data *suggest* (not prove), results offer *evidence* (not proof), and *findings* (not truths) are reported. Hedging is described in more detail in Chapter 5 and in Chapter 16, "Language Tips," Part 2: Word Choice.

An objective description of results allows readers to examine the data unbiased by interpretation. Results are sometimes viewed as a glimpse at the "truth"; alternatively, interpretations are educated opinions that are likely to change over time. As a writer, you must learn to distinguish between description and interpretation, especially if you write a combined R&D section. The following sentences from excerpt 4A (referring to Figure 2 within that excerpt) help to clarify the difference:

Description	Masses of other trimers, tetramers, and higher polymers were not detected.
Interpretation	This could be due to (1) the absence of these polymers in pomegranate samples, (2) their presence at undetectable concentrations, or (3) significant amounts of polymeric PAs remaining in the residue after extraction.

That masses of other trimers, tetramers, and higher polymers were not detected conveys a "truth" or "fact" that will not change; thus, the statement is descriptive. However, the reasons they were not detected are based on speculation; hence, the statement is interpretive.

Exercise 4.4.

Sentences a–d (adapted from Fabjan et al., 2003) are taken from the R&D section of an article that investigated tartary buckwheat as a source of dietary rutin. For each sentence, decide whether its primary purpose is to describe, interpret, or both:

a. The highest content of rutin, 2.5–3% dry weight, was observed as sampling started.

b. Trends in rutin content were rather similar in all of the buckwheat varieties, indicating a more important influence of environment than genotype on the rutin content of the buckwheat herb.

c. Later sowing had essentially no impact on rutin content in herb, as shown in Figure 2.

d. Based on this study, it is clear that buckwheat herb production is feasible and that it could readily be produced as a nutritionally rich food, a rutin-rich herb tea, or food additive.

Analyzing Organization

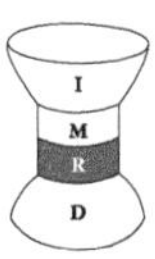

The purpose of the Results section is to present—without interpretation—the results of the study. Two moves are suggested to accomplish this task (Figure 4.1). The first move, Set the Stage, serves to transition the reader from Methods to the Results section. Two submoves are involved: in submove 1.1, the reader is briefly reminded how a particular set of results was obtained; in submove 1.2, the reader is referred to a graphic (a table or figure) that displays those results. These complementary submoves are often accomplished in a single sentence (a poignant reminder of conciseness in chemistry writing). After the graphic has been introduced, the authors shift to the second move, Tell the Story of Scientific Discovery, where important findings and discoveries (as well as obstacles and limitations) are identified, trends are highlighted, and unexpected results are underscored. Importantly, the story is rarely told in chronological order; rather, it is told in a way that logically leads the reader to the conclusions of the paper. These two moves are repeated, as needed, for each set of results. Because these moves describe quite specific information, the Results section is in the narrowest part of the hourglass.

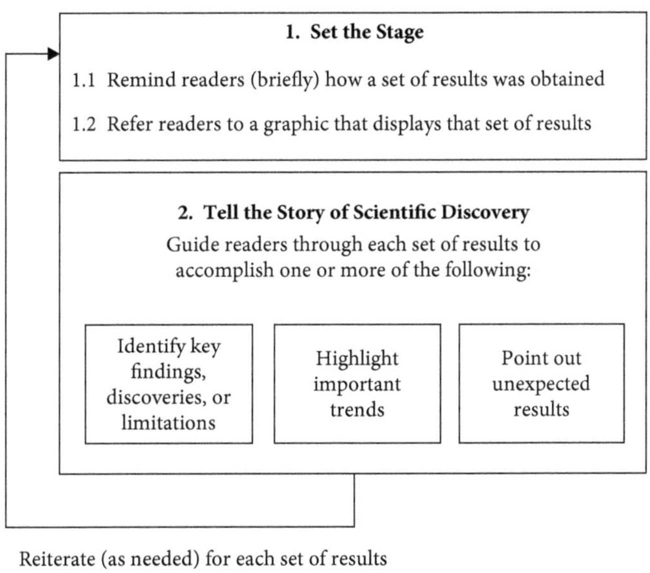

Figure 4.1. A visual representation of the move structure for a typical Results section.

Sets of Results

Results sections often include multiple sets of results. Each set presents a different piece of evidence or a different part of the project. The various sets are linked to lead logically to the conclusions of the paper.

Exercise 4.5.

The authors of excerpt 4A present the first two sets of results using both text and graphics. Each set is denoted with a bolded subheading. How well did the authors adhere to the move structure in Figure 4.1? Are submoves 1.1 and 1.2 addressed in each set? For move 2, what part of the story do the authors share (e.g., findings, trends, unexpected results)?

Exercise 4.6.

Browse through three articles in one of these journals: *The Journal of Organic Chemistry*, *Analytical Chemistry*, *Environmental Science & Technology*, or *Chemical Research in Toxicology*. How well do the results (in the combined Results and Discussion sections) adhere to the move structure in Figure 4.1?

Analyzing Excerpts Move by Move

Because we do not work in a vacuum in the academic world, we must learn to communicate with others. The written word remains the foundation of this communication, and I would hope that some of the capstones in my career have come from being able to communicate in a clear and controlled manner.

—Richard Malkin, University of California–Berkeley

We are now ready to read and analyze excerpts of Results sections in more depth, one move at a time. After examining the individual moves, we look holistically at writing conventions that run across the Results section.

Move 1: Set the Stage

The goal of move 1 is to transition the reader from the Methods section to the Results section. The move begins with a brief reminder of how a set of results was obtained and then refers the reader to a graphic that displays these results. Consider passage P1:

P1 The EPR spectra of samples of $PM_{2.5}$ from five different sites in the U.S. are shown in traces A–E of Figure 1. (From Dellinger et al., 2001)

In one sentence, the authors remind readers that $PM_{2.5}$ (particulate matter with effective diameters <2.5 µm) was collected from five sites and analyzed with EPR, and they refer us to Figure 1, which displays the EPR spectra. (Details about $PM_{2.5}$ and EPR were included in the Methods section.)

Referring to Figures and Tables

Refer to a figure or table at first mention of the data in the text (not at the end of the paragraph). Also, every figure and table must be "called out" in the text; do not include figures or tables in a paper that are not first mentioned in the text.

Note that the reference to the graphic (move 1, submove 1.2) comes *before* guiding the reader through the graphic (move 2). In this way, the reader can view the data before reading the associated prose. Often, the graphic is introduced in parentheses, as shown in the next two passages (P2 and P3). Note, too, the use of subheadings in these examples; subheadings are a particularly effective way to direct the reader's attention to each set of results.

P2 **Chromium Accumulation in *E. coli* Cells.** The cellular uptake of chromium is presented as milligrams of chromium per dry weight of treated *E. coli* cells (Table 1). (From Plaper et al., 2002)

P3 **Experiment 1.** When fed at a concentration of 3 µg of Se/g of diet, high-Se broccoli significantly reduced the incidence and total number of mammary tumors as compared to rats fed 0.1 µg of Se/g of diet as either selenite alone or 0.1 µg of Se/g of diet in combination with low-Se broccoli (Table 1). (From Finley et al., 2001)

Another common way to refer to a graphic in the text is with the phrase "as shown in Figure" or "are shown in Figure." Consider this example:

P4 The spectra of pure hazelnut oil were compared with the spectra of seven other oil types, *as shown in* Figure 1. (Adapted from Ozen and Mauer, 2002)

"As shown in Figure #"

Based on a computer-based analysis of 200 chemistry journal articles, the most common four-word combination in chemistry writing is "*as shown in Figure* __."

Exercise 4.7.

Reread and analyze passages P1–P4 to answer these questions.

a. How well do these passages set the stage (move 1)? For each passage, predict what the results will be about, even though you have not read the Introduction or Methods section.

b. Both past tense and present tense are used in these passages. Find instances of both and explain when each is used. Based on your answer, select the right tense for this sentence: The data *are/were* shown in Table 1.

c. Based on these passages, describe how the words "table" and "figure" should be formatted in the text. Identify three ways in which you can refer to a table or figure in the body of your text.

Move 2: Tell the Story of Scientific Discovery

Like a picture, a well-crafted graphic is worth a thousand words. (Of course, authors must avoid the temptation to include both the graphic and the thousand words.)

—Sharon R. Neufeldt, Montana State University

After the stage is set, you are ready to tell the story revealed by the data. The story is told in both text and graphics (typically tables and figures). Most authors determine the order of their graphics first (Table 1, Figure 1, Figure 2, Table 2, etc.), as well as the order of the data within each graphic (entry 1, entry 2, etc.), and then write the prose to complement the graphics.

Ultimately, the graphics and prose should work together, reinforcing but not duplicating one another; the reader's attention should naturally be shifted back and forth between the two. That said, we point out that graphics are not required in all instances, and on rare occasions an article is published with no figures or tables. Indeed, graphics should only be included if there are sufficient data (see Chapter 14 for guidelines on the amount of data required for a table or figure) and if the graphics make the data more accessible to readers. Graphics that are superfluous or that repeat content that is easily described in the text detract from, rather than enhance, the Results section.

The best way to learn how to write a Results section is to read and analyze Results sections from the literature. To this end, we examine excerpts (prose and graphics) from six published Results sections. We guide you through the content of these articles, elucidate what the authors have "discovered," and analyze how the authors have organized their results. (In each case, the original articles had multiple sets of results, but for brevity, we include only a few.) As you read these excerpts, pay attention to how the authors use text, in combination with graphics, to describe their data (i.e., to identify important findings, highlight trends, and underscore unexpected results). In subsequent chapters, we read additional excerpts from these same articles, taken from Discussions (Chapter 5), Introductions (Chapter 6), and abstracts (Chapter 7).

We begin with an excerpt from *Analytical Chemistry* (excerpt 4B). In a combined R&D section, the authors explore what happens when Fuji apples are treated with the fungicide imazalil. Imazalil is commonly applied to fruits postharvest to control fungi growth and prevent spoilage. However, little is known about its toxicity and whether it can diffuse into the fruit after application. To address these concerns, the authors submerged 12 apples for 2 min in an aqueous solution of imazalil (1000 ppm). This concentration mimics the concentration applied to fruits after harvest (between 1000 and 2000 ppm). The distribution of the fungicide in the apples over time (7 days) is monitored using three techniques: (1) matrix-assisted laser desorption ionization (MALDI) mass spectrometry; (2) laser ablation of the apple slices from the skin (0 mm) to the inner region (7 mm); and (3) off-line analysis of the ablated samples using electrospray ionization-mass spectroscopy (ESI–MS).

With that background, let's look at excerpt 4B. The section begins with experimental confirmation that MALDI mass spectrometry can detect protonated imazalil (*m/z* 297), thereby reminding readers of the technique used (move 1). The readers are referred to Figure 1, which shows the MALDI spectrum of an apple slice treated with imazalil; an intense base peak is observed at *m/z* 297, with the expected cluster peaks for chlorine isotopes. Moreover, peaks associated with the matrix are small and will not interfere with the protonated imazalil peak. (Hooray—the technique works!) The limit of detection (LOD) for imazalil (0.05 ppm) is determined by spiking a standard solution of imazalil on an apple slice and obtaining the corresponding MALDI spectrum. The authors do not include this spectrum; they realize it is enough to report only **representative data**. Instead they state that it was obtained "in the manner described above."

Representative Data

Avoid the temptation to report all data. Instead, present only representative data—the data needed to illustrate your key findings.

Laser-ablation sampling combined with images of the intensities of protonated imazalil from the MALDI spectrum (shown as a heat map) were used to monitor the penetration of the fungicide into the apple slices over 7 days. The results at days 0, 1, 4, and 7 are shown in Figure 2 (of the excerpt) along with a control (untreated) sample. It is possible that the authors collected data every day (0–7) but realized that only four days (0, 1, 4, and 7) were needed to effectively tell the story. In the corresponding text, the authors highlight important trends observed over these days. We end the excerpt here, but the article goes on to examine possible effects of vacuum drying apple sections prior to analysis and quantifying the concentrations of imazalil from the skin (0 mm) to the inner region of the apple (7 mm).

Exercise 4.8.

Read and analyze excerpt 4B (text and figures) to answer these questions:

a. How quickly do the authors refer to Figure 1? Figure 2?

b. Explain why the authors include data from only 4 days in Figure 2.

c. What data in Figure 2 do the authors choose to emphasize in the text?

d. Excerpt 4B comes from a combined R&D section, but only two of nine paragraphs are shown. Would you characterize these two paragraphs as results alone or results and discussion combined? Explain.

e. Are the captions for Figures 1 and 2 placed above or below the figures? What other writing conventions do you notice in the formatting of the captions?

Excerpt 4B (adapted from Pereira et al., 2019)

RESULTS AND DISCUSSION

An initial experiment was performed to determine whether the signal from protonated imazalil can be used for MALDI imaging. Figure 1 shows a MALDI spectrum obtained after spiking 1 μL of a 1000 ppm imazalil solution (in ACN) with 1.4 μL of a 10 mg/mL solution of CHCA (in 1:1 H_2O/ACN with 0.1% formic acid) on a 0.2 mm thick apple slice. The base peak in the mass spectrum is m/z 297, which corresponds to the nominal mass of the protonated molecule. The set of characteristic isotopic cluster peaks at m/z 297, 299, and 301 is consistent with the presence of two chlorine atoms. Peaks associated with the CHCA matrix are the dominant remaining peaks in the mass spectrum and do not interfere with the protonated imazalil peak. The limit of detection (LOD) for imazalil was evaluated by spiking imazalil standard with CHCA on an apple slice in the manner described above. Based on a value of three times the noise level of the blank, the LOD was determined to be 0.05 ppm.

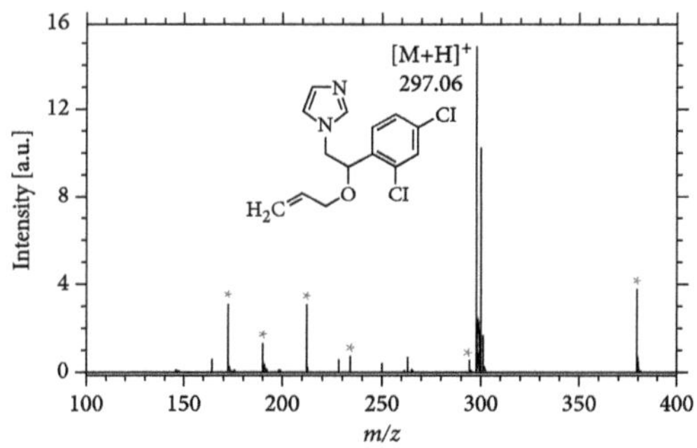

Figure 1. MALDI mass spectrum obtained from the application of 1 μL of a 1000 ppm imazalil (*m/z* $[M + H]^+$) solution on an apple slice; matrix peaks are labeled with asterisks.

Figure 2 shows MALDI and optical images of sections of imazalil treated apples containing skin (top) and endocarp. These apples were treated with imazalil and stored for the indicated number of days. Figure 2a depicts the optical images of apple slices, Figure 2b shows the MALDI imaging of imazalil, and Figure 2c shows the merged optical and MALDI images. The heat map of Figure 2b was obtained by plotting the intensity of *m/z* 297.06 ± 0.2. The control section was treated as indicated above, and imazalil was not detected. On day 0 (30 min after imazalil application), imazalil was detected in the skin but was not detected in the inner region. The same was found on day 1, with the compound slightly more distributed. The images of apple sections 4 days after fungicide application showed that a small quantity of imazalil migrated approximately 1 mm toward the inner region, even though the largest quantity remained in the skin. On day 7, the imaging results show the migration of imazalil approximately 3 mm inside the apple.

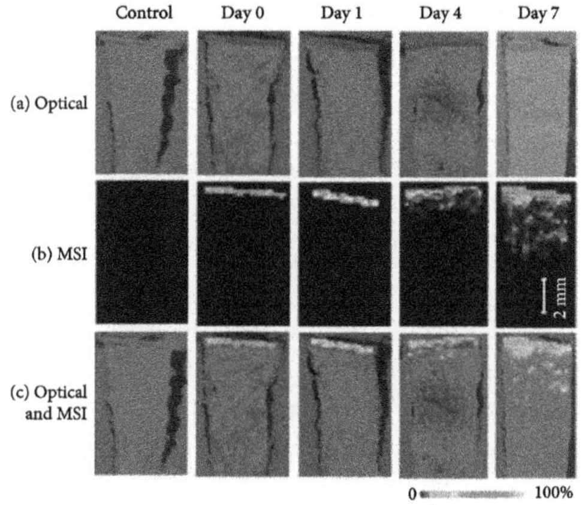

Figure 2. Penetration analysis of imazalil (*m/z* 297.06 $[M + H]^+$) in apples by MALDI imaging on days 0, 1, 4, and 7 after dipping application of imazalil: (a) optical image of apple slices, (b) MALDI imaging of imazalil, and (c) merged optical and MALDI image. (To view Figure 2 in color, consult the original article.)

Excerpt 4C is taken from an article in *The Journal of Physical Chemistry A*. Using high-level ab initio quantum chemistry calculations, the authors propose a new hybrid model to identify environmentally sustainable dielectric gases to replace SF_6. Alternatives to SF_6 are highly desirable because SF_6 is a potent greenhouse gas with a global warming potential (GWP) 23,500 times greater than naturally occurring greenhouse gases such as CO_2 (Yu et al., 2018).

This except illustrates a common and effective way to organize the "story of scientific discovery." Novice writers often think that results should be presented in chronological order, but this is often the least effective way to share data. Instead, authors should purposefully sequence their results to make the story easy to follow. Excerpt 4C does just that. In the first paragraph, the authors present in Figure 1 (not included) optimized geometrical parameters for 11 hybrid molecules, molecules that have not yet been synthesized in the laboratory. In the next statement, beginning with the key word *however*, the authors offer a logical framework for presenting the rest of their results: *However, before the synthesis of these novel compounds is called on, the following three questions should be answered unambiguously*: . . . The three questions are stated (excerpt 4C) and then answered to complete the Results section. Subheadings are used to clearly identify each question. For brevity, we include only a brief excerpt from each subsection. Together, the results unfold in a way that logically leads to the authors' conclusion: new hybrid molecules to replace SF_6 have been successfully identified.

Chronological Order?

Results may appear to be presented in chronological order, but usually they are not. Actual chronological order is often quite messy because of false starts, dead ends, and "wrong turns." True chronological order only confuses the reader. Instead, effective writers use hindsight to intentionally rearrange their results in a logical sequence of events.

Exercise 4.9.

As you read excerpt 4C, complete these tasks:

a. What are the three questions the authors pose to organize their results? How well do the subheadings correspond to these questions?

b. In each set of results, the authors refer to at least one table or figure. Find one sentence in each section that highlights results displayed in these graphics.

c. Were the authors able to answer the three questions posed? Were their findings encouraging or disappointing with respect to finding a replacement for SF_6?

d. In subsection 3.1, both present and past tense are used. Give two examples of each tense. Can you discern the convention that guided their selection?

Excerpt 4C (adapted from Yu et al., 2018)

RESULTS AND DISCUSSION

Two types of hybrid schemes, namely integrated and fractional chimeras, can be realized according to the specific molecular structures of the parent gases to be used for construction. For instance, in comparison with the 50% SF_6 + 50% N_2 gaseous mixture, the hybridization between SF_6 and N_2 can occur in two manners: first, the N_2 molecule is integrated as a whole into one of the S–F bonds of SF_6, forming SF_5N_2F; second, half SF_6 and half N_2 structures are combined together to generate SF_3N, or SF_4NF by two-thirds SF_6 with half N_2.

A series of hybrid molecules has been predicted accordingly (Figure 1), making use of SF_6, C_2F_4, C_2F_2, NF_3, c-C_4F_8, CF_4, C_2F_6, SO_2, and CO_2 as the parents. All new molecular structures involve the unsaturated chemical bonds. For example, the S≡C triple bond exists in SF_3CF (½ SF_6 + ½ C_2F_2), consisting of 67% C^{2p} and 33% ($0.62S^{3p} + 0.37S^{3d}$) with the bond order 2.37. A similar bonding mechanism occurs in the S=C double bond of SF_4CF_2. Neither fully fluorinated S=C nor S≡C molecules have ever been known in the laboratory. However, before the synthesis of these novel compounds is called on, the following three questions should be answered unambiguously: Are the compounds environmentally friendly, namely, with low GWP? Is the dielectric strength comparable to or higher than SF_6 and, simultaneously, is the boiling temperature low? Are they thermally stable enough to exist under ambient conditions?

3.1. GWP Assessments. The unsaturated bonds in the hybrid structures are very helpful for reducing GWP. Using the theoretical anharmonic IR spectra (Figure 2), the radiative efficiencies for the hybrid molecules were calculated to be in the range 0.2–0.4 (RE, Table 1) with the largest value of 0.68 for SF_5N_2F, which are comparable to those of SF_6 (0.52) and perfluorocarbons (0.1–0.5). Apparently, all of the hybrid molecules absorb IR energy in the window at 8–12 μm, which is largely transparent in the natural atmosphere. However, the high reactivity toward the atmospheric radicals (e.g., hydroxyl, OH) allows them to be removed rapidly from the atmosphere. The lifetimes were estimated to be at most 30 years in view of the typical atmospheric concentration of OH radicals (106–107 cm^{-3}).[40] Therefore, the GWP for these hybrid molecules is in the range 0–2000, which is considerably lower than that of SF_6 or perfluorocarbons. . . .

3.2. Dielectric and Boiling Point Assessments. The dielectric performance of the hybrid molecules can outperform SF_6 significantly. Dielectric strengths (Er) relative to SF_6 and boiling point (Tb) are listed in Table 1. It is very encouraging that the dielectric strength of the hybrid molecules is generally higher than or at least comparable to that of SF_6, as determined by the unique molecular electrostatic potential surface (Figure 4). [*section omitted*]

3.3. Thermal Stability Assessments. Thermal stability of the hybrid molecules has been enhanced considerably, as shown by the greatly negative enthalpies of formation at standard state, $\Delta H_{f,\,298.15}$, and relatively high electron affinities and ionization potentials (Table 2). At first sight, the species with multiple bonds seem to be less stable in view of the values of $\Delta H_{f,\,298.15}$. For example, the SF_3N molecule involves the S≡N triple bond and its $\Delta H_{f,\,298.15}$ is −83.3 kcal/mol, which is considerably higher than that of SF_6 (−291.7 kcal/mol). However, ab initio calculations confirmed that the thermal unimolecular decomposition or isomerization of SF_3N is negligible even at extremely high temperatures (e.g., ~3000 K whenever partial discharge or arc occurs) because the significant barriers (e.g., >70 kcal/mol) have to be surmounted for all of the possible reaction routes (Figure 2). . . .

Excerpt 4D from *Analytical Chemistry* also illustrates how subheadings can be used to organize the presentation of results. Each subheading signals a new set of results; cumulatively, the results build to tell the authors' story. Again, we include only brief selections from each subheading to illustrate this approach. The authors also included a scheme, a table, and five figures to present their results, skillfully shifting the readers' attention back and forth between text and graphics.

The excerpt focuses on the development of an assay that used nanospheres to rapidly detect Ebola virus (EBOV) glycoprotein. You may recall the West Africa Ebola epidemic in 2014–2016, one of the largest epidemics in history. Rapid detection of this highly contagious pathogen helped to improve survival rates and reduce the spread of infection. The authors use several abbreviations in their Results section that were defined earlier in their paper. We define those terms here:

NPs	nanoparticles
QDs	quantum dots
RNs	red fluorescent nanospheres
RNs@Au	the assay, fabricated by adhering gold nanoparticles onto the surface of red fluorescent nanospheres. Each assay contains hundreds of quantum dots and dozens of gold nanoparticles to enhance fluorescence and colorimetric signals.

Exercise 4.10.

As you read excerpt 4D, answer these questions:

a. Based on only the subheadings in excerpt 4D, what steps did the researchers take to determine the effectiveness of their assay?

b. What evidence do the authors present to convince readers that their assay was effective?

c. Compare the formatting of the subheadings in excerpts 4C and 4D. What differences and similarities do you notice? (Note: Formatting preferences are generally determined by the journal, not the authors.)

d. Novice writers often place in-text references to graphics (e.g., Figure 1) at the end of sentences. Browse through excerpts 4C and 4D again. What do you observe?

Excerpt 4D (adapted from Hu et al., 2017)

RESULTS AND DISCUSSION

Characterization of RNs@Au. RNs@Au was obtained from RNs and Au NPs (Figure 1A). Transmission electron microscopy (TEM) (Figure 1B) showed that there were some 20 nm Au NPs (red arrows) on the nanosphere besides numerous QDs (blue arrows). We calculated that dozens of Au NPs were adhered onto the surface of one hydrophilic carboxylterminated poly(styrene/acrylamide) (PSt-AAm) (for 100 RNs@Au). Additionally, there were approximately hundreds of QDs in one PSt-AAm.[30,33,34] From the scanning electron microscopic (SEM) image, we could also see the Au NPs (red arrows) on the surface of the nanosphere clearly (Figure 1C).

Principle of Ebola Virus Glycoprotein and Whole Virion Detection Using Enhanced Strip Biosensor. As illustrated in Scheme 2, Ab–RNs@Au–SA was used to detect EBOV glycoprotein or EBOV whole virion via two steps. In the first step, the target was labeled with Ab–RNs@Au–SA and captured by the test line of antibody to glycoprotein via antigen–antibody reaction. In the second step, biotin–RNs@Au solution was applied to the sample pad. . . .

Proof-of-Concept Assay for Detection of Ebola Virus Glycoprotein. This work serves as a proof of concept for an Ebola glycoprotein detection assay. In the absence of glycoprotein, no color change was observed (negative, Figure 2A) because no RNs@Au/glycoprotein complexes can be captured on the test line. In this case, neither colorimetric signal nor fluorescence signal was observed on the test line. From Figure 2B, the red signal can be read from the positive sample with the naked eye due to the localized surface plasmon resonance effect of Au NPs. . . .

Detection of Ebola Virus Glycoprotein. To evaluate the sensitivity and dynamic range of the enhanced lateral flow test biosensor for EBOV glycoprotein detection, under optimized detection conditions (Figure S5), different concentrations of glycoprotein were loaded onto the strip biosensor. As shown in Figure 3A, in the absence of glycoprotein, there was no red band observed on the test line. For the positive samples, the intensity of red lines on the test line increased with increasing concentrations of glycoprotein. . . .

Superiority of Enhanced Strategy over Conventional Test Strips and Unenhanced Strategy. Because each RNs@Au contains dozens of Au NPs, and biotin–RNs@Au can further amplify the signal of Au NPs, more Au NPs can be captured on the test line.

Therefore, the assay has advantages in sensitivity over the conventional test strip, which is based on Au NPs. As shown in Figure 4A, the visualization limit of the Au-based lateral flow assay is 100 ng/mL glycoprotein, which is 50-fold higher than the enhanced assay in this work. . . .

Robustness of Enhanced Lateral Flow Assay for Clinical Application. After we demonstrated that the enhanced RNs@Au-based lateral flow assay performs better than the unenhanced RNs@Au-based and conventional Au-based lateral flow assays, the specificity of this strategy was evaluated. . . .

Because vs. Since

Note the use of *because* rather than *since* in excerpt 4D. In more formal writing, *since* is used to connote passage of time (e.g., *ACS Nano* has been published since 2007). See Chapter 16, "Language Tips," Part 6: Word Usage, for additional tips on *because* and *since*.

Another technique commonly used in Results sections is what we call the "broad-to-narrow" approach. At the start of a research project, scientists often test out multiple possibilities; however, over time, as data and clues accumulate, they narrow their focus to just those areas that proved most fruitful. Scientists then want to describe their results in a way that captures this discovery process, illustrating (concisely, of course) how initial tests led them down the path to their ultimate findings.

Broad-to-Narrow Approach

Scientists often limit the number of variables in later experiments, based on information acquired in earlier work. This approach also applies to writing. Authors begin by sharing results for many variables, then limit their focus to only those that advance their story.

Excerpt 4E from *Chemical Research in Toxicology* illustrates this broad-to-narrow approach. The authors begin by presenting genotoxicity results from three different Cr^{3+} compounds (chromium chloride, chromium nitrate, and chromium oxalate). In the first set of results (not included in excerpt 4E), they share data from a Pro-Tox (C) assay. Chromium chloride ($CrCl_3$) and chromium nitrate ($Cr(NO_3)_3$) showed results indicative of DNA damage, but chromium oxalate ($KCr(C_2O_4)_2$) did not. As the authors state (our italics): *Interestingly, none of the 13 stress promoters was induced when bacteria were treated with chromium oxalate*, and they refer readers to a figure that supports these findings. The authors use the word *interestingly* to call the reader's attention to the observation that chromium oxalate is somehow different from chromium chloride and chromium nitrate. The word also foreshadows a new direction in their research: they must now explain why chromium oxalate is different.

Interestingly

Authors sometimes use the word *interestingly* to foreshadow a result that is explained in more detail elsewhere in the paper. The word is most effective when used only once in a paper.

This difference is explored in their second set of results, which is where excerpt 4E begins. Presumably, for DNA damage to occur, chromium must get inside the *E. coli* cells. Thus, the researchers decide to measure intracellular chromium concentrations from each source of chromium. Flame atomic absorption spectroscopy (FAAS) is used to measure the uptake of chromium inside the cells, and the results are reported in Table 1 (included in excerpt 4E).

More information on *E. coli* and FAAS are provided here. Read the definitions of key terms, excerpt 4E, and complete exercise 4.11 to discern what happens next in their story of scientific discovery.

E. coli — *Escherichia coli*, bacteria that live in the human intestinal tract. Some strains of *E. coli* are harmless; others cause diarrhea-like symptoms. The Pro-Tox (C) assay and the Cr^{3+} tests use *E. coli* K-12 strains. (Note, the genus and species names are italicized, and the *E.* is capitalized.)

FAAS — Flame atomic absorption spectroscopy; the flame atomizes metals in solutions. Once in the gas phase, the atoms absorb UV–vis light, exciting electrons to higher energy levels. The amount of light absorbed is used to determine the metal concentration.

Genus and Species Names

Genus names, as formal names, should be capitalized and italicized. Species should not be capitalized, but they should be italicized. Genus names are spelled out in full at first mention and abbreviated thereafter (e.g., *Escherichia coli* then *E. coli*).

Exercise 4.11.

Complete these tasks as you read excerpt 4E:

a. What data in the table do the authors choose to emphasize in the text?

b. Using the data shown in Table 1, fill in the blanks: Each Cr compound was tested in concentrations of ___, ___, and ___ mM.

c. What Cr compound was not detected intracellularly, even at the highest concentration (1.25 mM)?

d. How do these results build on the authors' story of scientific discovery and explain the results of the first Pro-Tox (C) assay?

e. How well do the authors adhere to table-formatting conventions (see Chapter 14)? How are concentrations aligned (centered or by decimal)? Are column headings capitalized?

Excerpt 4E (adapted from Plaper et al., 2002)

Chromium Accumulation in *E. coli* Cells. The cellular uptake of chromium is presented as milligrams of chromium per dry weight of treated *E. coli* cells (Table 1). FAAS measurements of total chromium concentrations in *E. coli* cells showed that only chromium chloride and chromium nitrate accumulate intracellularly but not chromium oxalate. As shown in Table 1, at all concentrations of chromium oxalate used, the quantity of chromium was below the detection limit.

Table 1. Uptake of Chromium by *E. coli* Cells Treated with Different Cr^{3+} Compounds

Cr^{3+} compound	concentration of added compound (mM)	intracellular content of Cr (mg of Cr/g of dry weight)
$CrCl_3 \times 6H_2O$	0	<0.021
	0.63	1.8
	1.25	6.7
$Cr(NO_3)_3 \times 9H_2O$	0	<0.021
	0.63	2.4
	1.25	5.7
$KCr(C_2O_4)_2 \times 3H_2O$	0	<0.025
	0.63	<0.022
	1.25	<0.02

As shown in excerpt 4E, Cr^{3+} enters the cells when either chromium chloride or chromium nitrate is the source but not with chromium oxalate, explaining why this form of chromium is different than the other two. With this discovery, the authors now present their third set of results (not shown), to explore a possible mechanism for DNA damage once Cr^{3+} is inside the cell. They examine how Cr^{3+} affects gyrase, an enzyme that regulates the ability of supercoiled DNA to relax. Results, however, are reported only for chromium chloride, not for chromium oxalate or chromium nitrate. Chromium oxalate can be omitted because it does not release Cr^{3+}, in accord with the broad-to-narrow approach. Chromium nitrate can be omitted because it produced results very similar to chromium chloride. Recall in excerpt 4B, we learned that authors need only share representative data, and this is what the authors of excerpt 4E did. At the start of the project, there were three possibilities; however, as knowledge was gained, the final test focused on only one compound, saving both time and money.

It is possible that the authors conducted the gyrase test on all three chromium compounds. Even so, the Results section is best written without this information. Authors risk diluting relevant results by including extraneous data that fail to advance their story. As in the previous excerpts, only a small fraction of the data is reported. Readers benefit from the researchers' hindsight. Learning to tell your story of scientific discovery in retrospect, by reorganizing your data and highlighting only the most illustrative pieces, is an essential skill in effective writing.

Before moving on to excerpt 4F, we call your attention to two ways in which the concept of zero is addressed in excerpt 4E. First, we consider the concept of zero in measured concentrations (i.e., the concentrations reported in the far-right column of Table 1 in excerpt 4E). Recall that no chromium oxalate was detected in the cells; however, the authors do not report this with a zero. Rather, they use the phrase "below the detection limit" in the text and the less-than symbol (e.g., <0.025 mg/g) in Table 1, which puts an upper limit on the amount of chromium oxalate present. Novice writers might (incorrectly) report that "no chromium was present" in the text and use a zero in the table (0 mg/g). Such uses of zero, however, are incorrect because (for measured concentrations) zero varies with the sensitivity of the detecting instrument. For example, on one instrument, zero will be less than one part per million; on a more sensitive instrument, zero will be less than one part per billion. Instead of reporting zero, authors report that the measurement was below the detection limit for that instrument. Some common ways to express this concept in the text and table are as follows:

In the text	In a table
x was not detected	<0.02
x could not be detected	bdl (for below detection limits)
We did not detect X	ND (for not detected)
x was below detection limits	

It is correct to use "0" when referring to compounds *added*, where the "0" indicates that none was added. For example, you may say that "no chromium oxalate was added" and use a "0" to indicate this in a table. Thus, the second column of Table 1 (in excerpt 4E) includes three zeroes, indicating that no chromium compounds were added to the cells in these three experiments.

Zero?

Measured concentrations should not be reported as zero, even if the substance is not detected. A more sensitive instrument (one with a lower detection limit) might be able to detect the substance. In other words, even if you don't "see" the substance, it might be there!

The final two excerpts (excerpts 4F and 4G) illustrate a few ways in which results from organic syntheses are described. The results often include a hint of what was tried and failed, as well as what ultimately succeeded. Reading these excerpts will give you a feel for the trial-and-error process of science and the "aha" experience that sometimes occurs when a successful approach is discovered. These excerpts also illustrate how organic chemists condense a great deal of data into a few carefully prepared tables.

Excerpt 4F is taken from an article written by Demko and Sharpless. (Barry Sharpless was a co-recipient of the Nobel Prize in Chemistry in 2001 for his work on chirally catalyzed oxidation reactions.) In this article, the authors propose a way to synthesize aromatic tetrazoles from nitriles in water, using only sodium azide and a zinc salt. Water, despite its obvious advantages (i.e., safe and inexpensive), rarely succeeds as a solvent in organic reactions. Thus, a synthesis that uses water successfully is an important scientific accomplishment.

We begin our analysis of excerpt 4F by examining Demko and Sharpless's eq 1 (our eq 4.1), which uses an "R" group, to represent unspecified organic radicals or residues, and the compound labels **1** and **2**, to represent a nitrile and a tetrazole, respectively:

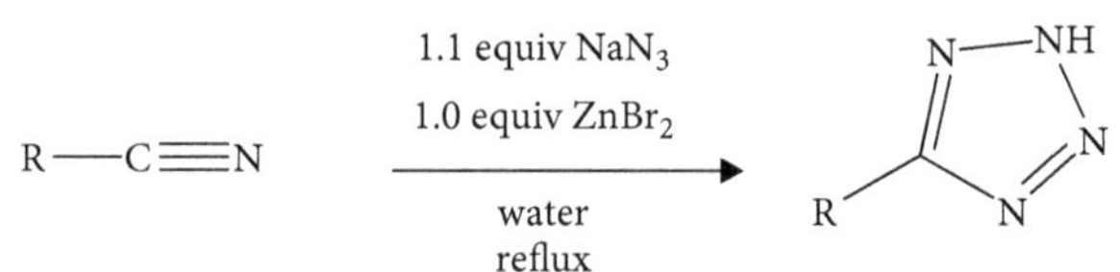

1 **2** (4.1)

You are likely already familiar with R groups from organic chemistry; you may be less familiar with **compound labels**. Compound labels are bolded numbers (or numbers and letters) that are used to represent compounds. After being introduced, the label can be used instead of the compound's name or structure, thereby saving considerable space. Compound labels are bolded so that they can be easily differentiated from other numbers in the sentence, such as citation numbers, which are never bolded.

Compound Labels

Compound labels, in addition to being concise, focus attention on important features of a reaction. They allow authors to communicate the versatility and generality of a reaction, mechanism, or scheme efficiently. See Chapter 16, "Language Tips," Part 3: Writing Conventions, for more details on Compound Labels.

In Table 2 of excerpt 4F, the structures of specific tetrazoles are shown, and each is assigned a unique compound label (**2a–2j**). These structures also provide the information needed to assign R groups in **1** and **2**. Consider, for example, the tetrazole labeled **2a** in excerpt 4F (Table 2). The R group on the tetrazole is phenyl; hence, phenyl is also the R group on the nitrile, and the authors' eq 1 can be rewritten with **1a** and **2a** specified (eq 4.2):

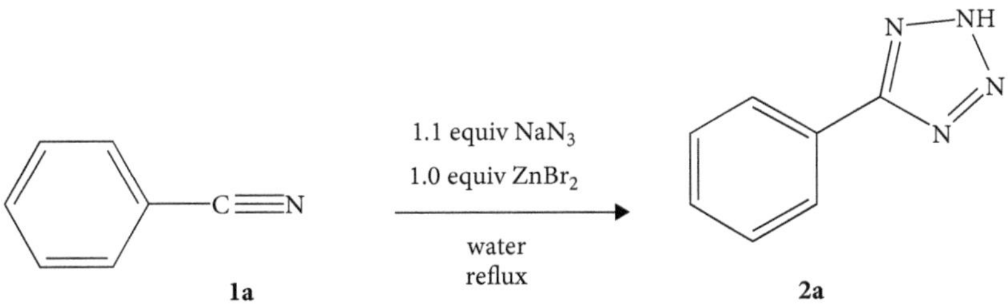

1a **2a** (4.2)

After introducing Table 2 in excerpt 4F, the authors do a good job of moving the reader's attention back and forth between the table and text, highlighting important results, but not repeating the data. As you read the text and table together, it becomes clear that the authors have organized their table not in an arbitrary way, but according to nitrile type (aromatic, electron-poor aromatic, electron-rich aromatic, etc.). This organization helps the reader detect trends in the data.

Exercise 4.12.

Examine the connections between the compound labels used in eq 1 and Table 2 in excerpt 4F. Use these connections to replace R in eq 1 when **2c** is the product formed.

Excerpt 4F (adapted from Demko and Sharpless, 2001)

[*From Introduction*]
We report here a safer and exceptionally efficient process for transforming nitriles **1** into tetrazoles **2** in water; the only other reagents are sodium azide and a zinc salt (eq 1).

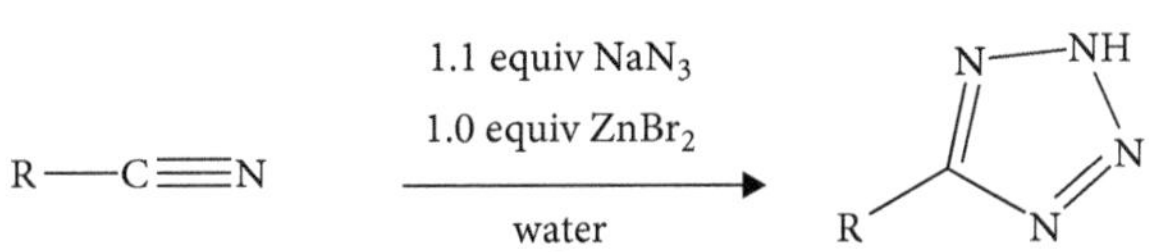

1 **2** (1)

[*From a combined R&D section, although no heading was included*]
A wide variety of nitriles were converted to tetrazoles on a 20 mmol scale. Other things being equal, the more electron-poor a nitrile, the faster it reacts. Aromatic nitriles (see Table 2) with a variety of substituents (**2a,b,c,i**) reach completion within several days at reflux. Electron-poor aromatic and heteroaromatic nitriles, such as 2-cyanopyridine and cyanopyrazine (**2d,e**), are complete within a few hours.

Table 2. Aromatic Tetrazoles

entry	tetrazoles[a]	temperature/time	yield (%)	mp (°C)
2a	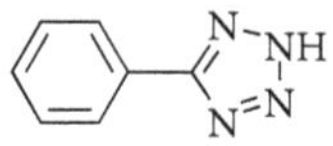	reflux/24 h	76	215–236
2b	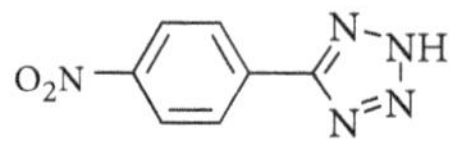	reflux/24 h	94	220
2c	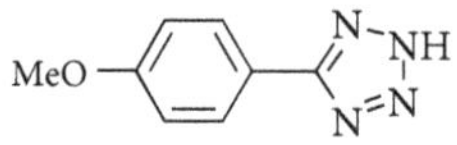	reflux/48 h	86	231–232
2d	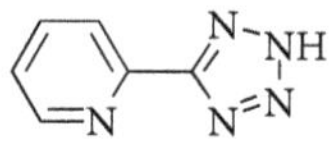	reflux/6 h	79	211
2e	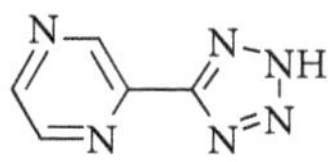	reflux/2 h	83	193–195
2f	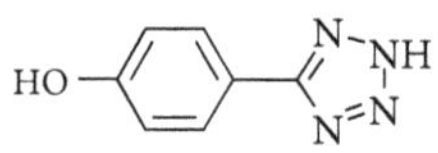	140 °C/24 h	96	234–236
2g	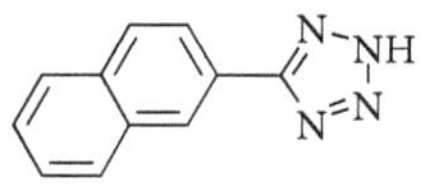	140 °C/48 h	73	205–207
2h	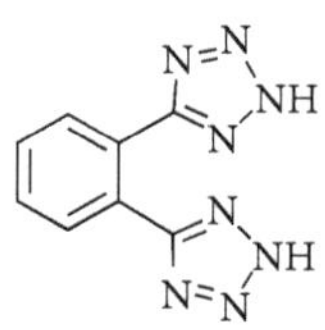	reflux/48 h	64	228–230
2i	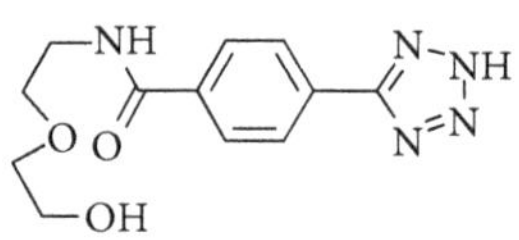	reflux/12 h	67	158–160
2j	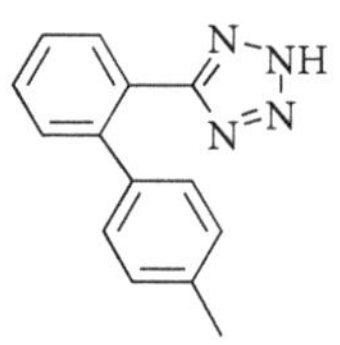	170 °C/48 h	67	150

[a] These reactions were run on 20 mmol scale.

In journals such as *The Journal of Organic Chemistry* and *Organic Letters*, reactions such as eq 1 in excerpt 4F are often included in the table itself, just under the table title. The reaction equation and table entries are linked by compound labels. Exercises 4.13 and 4.14 illustrate this common convention.

Exercise 4.13.

Consider Table 1, adapted from Usugi et al. (2004). The table includes an equation depicting a general chemical reaction (1 + 2 → 3). The table entries describe five specific reactions by defining R and R' in each case. Using Table 1, write out the specific reactions for entries 2–5. Entry 1 has been completed for you (R = *n*-Bu and R' = Ph):

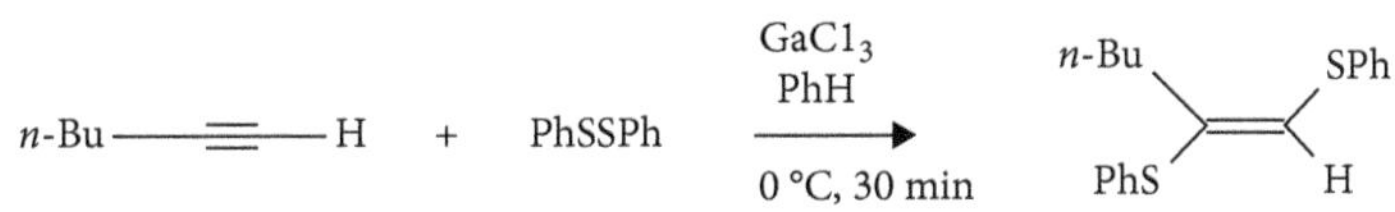

1 **2** **3**

Table 1. $GaCl_3$-Mediated Reaction of Various Alkynes with Disulfide[a]

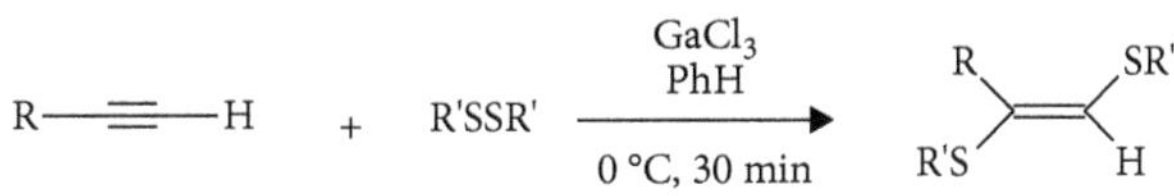

1 **2** **3**

entry	R	1	R′	2	3	yield (%)
1	*n*-Bu	**1a**	Ph	**2a**	**3a**	83
2	*i*-Pr	**1b**	Ph	**2a**	**3b**	58
3	*t*-Bu	**1c**	Ph	**2a**	**3c**	50
4	Ph	**1d**	Ph	**2a**	**3d**	87
5	Ph	**1d**	*p*-Tol	**2b**	**3e**	84

[a] Disulfide (0.50 mmol), alkyne (0.75 mmol), $GaCl_3$ (0.50 mmol), and PhH (2 mL) were employed.

Exercise 4.14.

Imagine that you are in a research group that is measuring relative rates of nucleophilic substitution (S_N2) reactions. The reactions that you have investigated are listed here. Prepare a table, like Table 1 in exercise 4.13, to summarize these reactions. Include the following equation at the top of the table: R–Br + Nu^- → R-Nu + Br^-.

reaction							relative rate
t-BuBr	+	Cl^-	→	*t*-BuCl	+	Br^-	<1
*neopentyl*Br	+	Cl^-	→	*neopentyl*Cl	+	Br^-	1
i-PrBr	+	Cl^-	→	*i*-PrCl	+	Br^-	500
EtBr	+	Cl^-	→	EtCl	+	Br^-	40,000
MeBr	+	Cl^-	→	MeCl	+	Br^-	2,000,000
MeBr	+	$CH_3CO_2^-$	→	$MeOOCH_3$	+	Br^-	1,000,000
MeBr	+	HO^-	→	MeOH	+	Br^-	32,000,000
MeBr	+	I^-	→	MeI	+	Br^-	200,000,000

Excerpt 4G involves a reaction that you most likely remember from organic chemistry, the Grignard reaction. Recall Grignard reagents (RMgX) add to carbonyl compounds to form alcohols. Excerpt 4G provides insights into how *allylic* Grignard reagents add to carbonyl compounds. The allylic Grignard used in this work is prenylmagnesium chloride, shown as reagent **5** in eq 2 (adapted from Bartolo and Woerpel, 2018). When **5** is reacted with a carbonyl, two alcohols are possible. Alcohol **6** will form if the α-carbon of **5** adds to the carbonyl carbon. However, if allylic transposition occurs (resulting in migration of the carbon-carbon double bond within the 3-carbon system), the γ-carbon of **5** will add instead, producing alcohol **7**. The work shows that the γ-addition is preferred when the carbonyl compound is unhindered.

With that background in mind, let's examine two sets of results that the authors present in their Results and Discussion section. They begin with an unhindered carbonyl, the ketone propiophenone. The ketone is allowed to react competitively with prenylmagesium chloride (**5**) and two different *alkyl* Grignard reagents (RMgX, where R = methyl or isopentyl). The product ratios tell us if the alkyl or allyl Grignard reagent reacted faster (i.e., the fastest pathway will give the product in greatest yield). The authors also compare the reactivity of **5** to allylmagnesium chloride ($H_2C{=}\,CHCH_2MgCl$), also allylic but with hydrogens rather than methyl groups on the terminal vinylic carbon. Again, the product ratios tell us which Grignard reagent reacted faster. Read excerpt 4G (part 1) and complete exercise 4.15 to learn what they discovered.

Exercise 4.15.

Refer to excerpt 4G (part 1) as you answer these questions.

a. Draw the structures for RMgX when R is represented by **a**, **b**, and **c**. Which one is also allylic?

b. Examine the three products (**10**, **11**, and **12**). Which results from α-addition of **5**? Which results from γ-addition of **5**? Which results from α-addition of RMgX?

c. The product ratios for the three products are shown. Based on these ratios, do allylic or alkyl Grignard reagents react faster with an unhindered ketone? Was the α- or γ-addition pathway of **5** preferred?

Excerpt 4G (part 1) (from Bartolo and Woerpel, 2018)

RESULTS AND DISCUSSION

Initial studies focused on establishing that prenylmagnesium chloride reacts at a comparable rate to allylmagnesium halides (i.e., at the diffusion rate limit) with unhindered ketones.[8] Just like allylmagnesium halides, prenylmagnesium chloride was found to be more reactive with propiophenone (**9**) than other alkylmagnesium reagents were (eq 2). The addition products of the other Grignard reagents were not observed when prenylmagnesium chloride was competed against both methylmagnesium chloride and isopentylmagnesium bromide. When prenylmagnesium chloride was competed against allylmagnesium chloride, however, approximately equal amounts of prenylated and allylated products were formed. These observations indicate that the addition of allylmagnesium chloride and prenylmagnesium chloride occur at comparable rates. Considering that previous studies indicate that allylmagnesium chloride reacts at rates approaching the diffusion limit, it may be concluded that prenylmagnesium chloride also reacts at that rate with relatively unhindered ketones.[8]

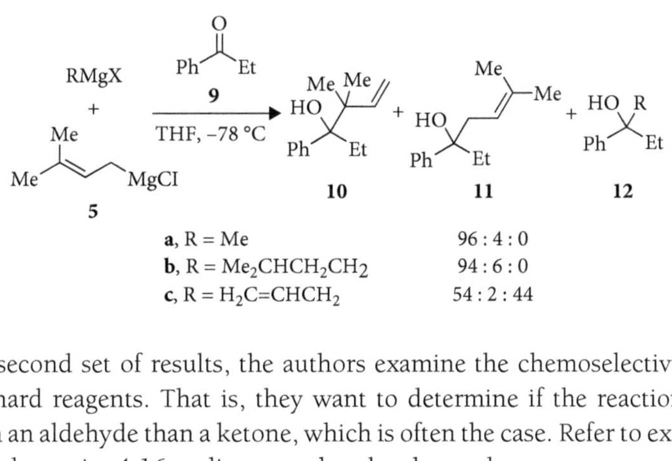

(2)

a, R = Me — 96 : 4 : 0

b, R = $Me_2CHCH_2CH_2$ — 94 : 6 : 0

c, R = $H_2C{=}CHCH_2$ — 54 : 2 : 44

In the second set of results, the authors examine the chemoselectivity of allylic Grignard reagents. That is, they want to determine if the reaction occurs faster with an aldehyde than a ketone, which is often the case. Refer to excerpt 4G (part 2) and exercise 4.16 to discover what they learned.

Exercise 4.16.

Refer to excerpt 4G (part 2) as you answer these questions.

a. Consider the two products (**15** and **16**) in the reaction in Table 1. Which product is formed from the aldehyde? Which from the ketone?

b. Entries 1 and 2 in Table 1 show product ratios when alkyl Grignard reagents react competitively with an aldehyde and ketone. Which carbonyl clearly reacts faster?

c. Entries 3 and 4 in Table 1 report product ratios when allylic Grignard reagents are reacted with aldehydes and ketones. Do they show the same preference in reactivity (chemoselectivity) as alkyl Grignard reagents? Explain your answer.

Excerpt 4G (part 2) (from Bartolo and Woerpel, 2018)

[*Continuation of excerpt 4G (part 1)*]

Intermolecular competition experiments demonstrate that prenylmagnesium chloride, like allylmagnesium chloride, reacts with low chemoselectivity between an aldehyde and a ketone. This lack of selectivity observed with allylic magnesium reagents is consistent with diffusion-controlled addition. Competition experiments with Grignard reagents and an excess of benzaldehyde (**13**) and acetophenone (**14**) showed that both allylmagnesium chloride and prenylmagnesium chloride do not exhibit chemoselectivity in instances where other alkylmagnesium reagents exhibit high chemoselectivity (Table 1).[8] These experiments provide further evidence that prenylmagnesium chloride and allylmagnesium chloride react at similar rates.

Table 1. Intermolecular Competition Experiments to Evaluate Chemoselectivity in the Additions of Grignard Reagents to Carbonyl Compounds

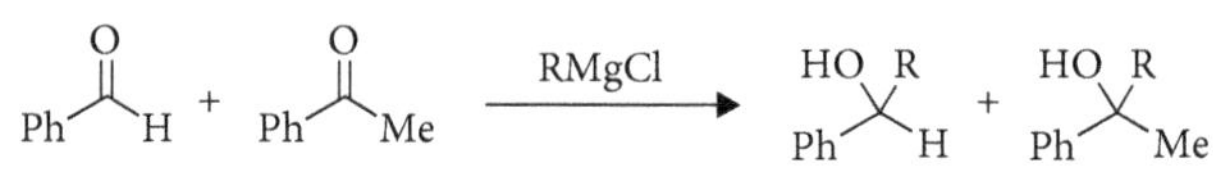

13 **14** **15** **16**

entry	R	product	15:16[a]
1	Me	a	221:1[8]
2	*n*-Pr	b	143:1[8]
3	$H_2C{=}CHCH_2$	c	51:49[8]
4	$Me_2C{=}CHCH_2$	d[b]	60:40[8]

[a] Ratios determined by GC analysis. [b] Regioisomers were inseparable by GC.

We hope that excerpts 4F and 4G and the accompanying exercises have increased your awareness of how organic chemists use tables and equations to pack in a lot of results. When you understand how to read these graphics, stories unfold about what the chemists discovered in their work. Each set of results shares new insights, and collectively the series of results lead logically to new insights about chemical reactions.

Perfecting Your Writing

Here we spotlight a few writing features that are characteristic of the entire Results section, including verb tense–voice combinations and word choice (i.e., use of *we*, *respectively*, *quantitative language*, *very*, and *scientific plurals*).

Verb Tense and Voice

Past and present tense are both used in the Results section as are active and passive voice. Past tense is generally used to remind readers how results were obtained (e.g., concentrations *were measured* . . .) and to highlight results (e.g., results *indicated* . . .). A list of commonly used past-tense verbs in Results sections is shown in Table 4.1. Present tense is generally used to refer to a table or figure (e.g., Figure 1 *shows* . . .) or to state a result that is expected to be true over time (e.g., Water *is* more soluble than the alcohol.). Table 4.2 highlights some tense–voice combinations and their functions that are common in Results sections.

Table 4.1. Commonly Used Past-Tense Verbs in Results Sections

catalyzed	observed	resulted
formed	obtained	revealed
found	prepared	showed
gave	produced	studied
led	reported	used

Table 4.2. Common Verb Tense–Voice Combinations for Key Functions in the Results Section of a Journal Article (other combinations are possible)

function	tense–voice combination	example
a. to remind readers how results were obtained	past–passive	Concentrations *were measured* using . . .
		The PA profile *was studied* by . . .
b. to refer to a graphic	present–active	Figure 1 *shows* . . .
	present–passive	Concentrations *are reported* in Table 1.
c. to identify key results	past–active	Other combinations *failed* to produce higher yields.
		Kinetic measurements *confirmed* this trend.
	past–passive	Higher mass compounds *were not detected*.
		Only values below detection limits *were observed*.
d. to share results expected to be true over time	present–active	PCBs *are* more strongly retained in the sample matrix as the fat content *increases*. (From Llompart et al., 2001)

Exercise 4.17.

Sentences 1–5 are from a Results section (adapted from Weston et al., 2004). Using Table 4.2 as a guide, do the following:

a. Determine the function of each sentence.
b. Select the appropriate verb for that function.

1. The tailwater ponds (are/were) the most contaminated of all sites studied (Table 2).
2. A toxicity unit (TU) approach (is/was) used to identify pesticides potentially responsible for the observed toxicity.
3. Esfenvalerate concentrations (are/were) at least 0.5 TU in five samples.
4. This approach assumes that the other pyrethroids (are/were) comparable to cypermethrin in the bioavailability of particle-adsorbed residues.
5. TU calculations for samples not toxic to *C. tentans* (are/were) shown in Table 3.

Use of *We*

In the Results section, the word *we* (and other personal pronouns, e.g., *I* and *our*) is used only sparingly. This is because results should be reported objectively, and the use of *we* can make the writing sound less objective. *We* is typically avoided entirely in move 1, where the purpose is to remind readers of the research methods:

inappropriate	We measured X.
appropriate	X was measured.

Alternatively, *we* is sometimes used in move 2, as authors tell their story of scientific discovery. Specifically, *we* may be used to highlight a decision, action, or choice made by the authors during their work. For example, *we* was used three times in excerpt 4D (*we* calculated, *we* could also see, and *we* demonstrated). Here are some additional examples:

- In MALDI–MS, this challenge is overcome by coadding individual spectra. *We* have adopted a similar approach to achieve reproducible SERS spectra. (Adapted from Jarvis and Goodacre, 2004)
- After injection of the methanol/water extract of kelp powder and kelp powder spiked with 0.5 μg mL^{-1} of As(III), DMA, MMA, and As(V), *we* observed an overlap of DMA and phosphate ribose. *We* therefore decided to change the mobile phase. (From Almela et al., 2005)

- *We* reasoned that the host-guest interactions at the nanoparticle-solution interface investigated in this work could be used for similar purposes. (From Liu et al., 2001)
- *We* have defined our sets of compounds for cross-comparison more broadly than in previous studies. (From Vieth et al., 2004)

_We_ in the R&D Section

In a computer-based analysis of 60 Results and Discussion sections, the word *we* occurred only 3 times per 1000 words. Thus, its use is uncommon in this section. The most compelling reason to use *we* in the Results section is to highlight a decision, action, or choice made while conducting your work.

Exercise 4.18.

Consider the most compelling reasons for using *we* in a Results section. For each passage (a–e), decide whether the use of *we* is appropriate. Explain each decision.

a. The results for the As and Pb concentrations *we* obtained for the 83 samples are reported in Table 3. *We* note from Table 3 that the range of As and Pb concentrations in the two populations is quite distinct.
b. *We* use the term *KD*, the distribution coefficient, in the following discussion, although equilibrium may not have been achieved in all cases.
c. To reduce the problems of ligand-specific bias, *we* developed a modified rating for each molecule. *We* have called this corrected score the multiple active site correction rating (MASC).
d. Within the reference group, *we* found that the mean oxidative damage for smokers was significantly higher than that for nonsmokers.
e. At the end of each experiment, *we* measured the release of adsorbed alkenes from air–water interfaces.

Use of *Respectively*

The word *respectively* (meaning "separately, in the order specified") often appears in science writing and can be used to make your writing more concise. Most often, *respectively* is used at the end of the sentence, but it can also be used within the sentence. Note, too, that when two or more items have the same unit, the unit is stated only once.

- The concentrations of 2-ABP, 3-ABP, and 4-ABP in PPD were estimated at 70, 310, and 500 ppb, *respectively*.
- The turnover of 2-propanol and tosylate must exclusively take place via equilibria 5 and 6, *respectively*.
- With a recovery of 70%, the ^{239}Pu and ^{240}Pu concentrations would be approximately 10.5 and 1.9 fg/mL, *respectively*.
- Curves 1–3 correspond to film thicknesses of 10, 25, and 50 nm, *respectively*, on a water–ice substrate that is 100 nm thick.

Compare the following sentences with and without the word *respectively*.

Less concise	A was measured at X °C, B was measured at Y °C, and C was measured at Z °C.
More concise	A, B, and C were measured at X, Y, and Z °C, respectively.

In these examples, you can see how the word *respectively* helps achieve conciseness. More important, it aids clarity. By grouping values together, trends in the data are easier to discern. To ensure that the correct meaning is conveyed when using *respectively*, it is crucial that the order of the first set of items (e.g., A, B, C) parallels the order of the second (e.g., X, Y, Z).

Respectively

See Chapter 16, "Language Tips," Part 1: Conciseness, for tips on using *respectively* in your written work.

Exercise 4.19.

Practice using *respectively* by rewriting each of these sentences to include the word correctly.

a. Coefficients A are plotted in Figure 7, whereas coefficients B are plotted in Figure 8.
b. The conductivity increased by 0.5% in trial 1, 5.6% in trial 2, and 10.1% in trial 3.
c. Two irrigation canals, sites MA and TL, contained substantial amounts of bifenthrin: MA contained 8.8 ng/g and TL contained 10.4 ng/g. (Adapted from Weston et al., 2004)
d. The markers are shown at 600 nm in **1**, 650 nm in **2**, and 700 nm in **3**.

Use of Quantitative Language

Results are best described using quantitative rather than qualitative language. Qualitative words that convey an opinion (e.g., *excellent*, *good*, *poor*) or even neutral terms (e.g., *high* and *low*) should be replaced with quantitative values whenever possible, for example:

Vague	Heating the mixture to 93 °C gave very good yields.
Still Vague	Heating the mixture to 93 °C gave high yields.
Better	Heating the mixture to 93 °C gave a 98% yield.
Vague	The solution was very acidic.
Better	The pH of the solution was 1.2.

Exercise 4.20.

Each of these sentences can be made more quantitative by including a numerical value in the parentheses. Select the correct value (on the right) for each set of parentheses.

a. The correlation was weak ( ).	$R^2 = 0.999$
b. The signal was strong ( ).	>99%
c. The acidity was low ( ).	200 °C
d. The calibration curve was linear ( ).	pH > 20
e. The temperature was high ( ).	<1 ng/g
f. The yields were excellent ( ).	S/N = 15
g. The concentration was below the detection limit ( ).	r = 0.2

Use of *Very*

One of the most overused words by inexperienced writers is the word *very*. Some scientists would argue that all instances of the word *very* should be eliminated from journal articles because its use contributes to wordiness, minimizes objectivity, and indicates a lack of precision on the part of the writer. Nevertheless, *very* is observed in the chemical literature, although infrequently. In excerpts 4A–4G, it was used only twice (both in excerpt 4C) (and, in our opinion, both uses were unnecessary):

- The unsaturated bonds in the hybrid structures are *very* helpful for reducing GWP.
- It is *very* encouraging that the dielectric strength of the hybrid molecules is generally higher than or at least comparable to that of SF_6.

There are a few cases where *very* is appropriate. For example, in infrared spectroscopy, absorbance intensities can be referred to as *very strong* or *very weak*. The word is also used to describe solubility: substances can be referred to as *very soluble* or *very poorly* soluble. If *very* is used to refer to a *very slow* reaction or a *very poor* yield, it is best to also include a quantitative value (e.g., The yield was very low (<1%)).

Very

Be careful not to overuse the word *very* or similar adjectives (e.g., *exceedingly, extremely*) in your writing.

Exercise 4.21.

Consider the use of *very* in these sentences. For each sentence, indicate if you think its usage is appropriate, inappropriate, or borderline. Which sentences could be improved by including a quantitative value?

a. The spectrum of maltohexaose is *very* similar to that of . . .
b. Additional experiments were carried out, some of which yielded *very* surprising results.
c. Low-molecular-weight hydrocarbons are volatile and *very* poorly soluble in water . . .
d. The gel filtration experiments confirmed that *very* little Zn^{2+} was released . . .
e. The rate constants for the slowly and *very* slowly desorbing fractions were . . .
f. When BINAP or DPPF as a ligand was used, the yields were *very* low.

Use of Scientific Plurals

The word *data* is commonly misused by writers; the mistake involves using *data* as a singular noun. The singular form of *data* is *datum*, but the word *datum* is rarely used; a single point of data is typically referred to as a *data point* (rather than a datum). If the term *data set* is used instead of *data*, the singular verb form is correct (e.g., "The second *data set* confirms this trend."). In nearly all instances, the word *data* is plural and should be used with a plural verb:

Incorrect	Data is . . .
Correct	Data are . . .

Some correct uses of data, with plural and singular verb forms, are shown in Table 4.3.

Table 4.3. Common Uses of *Data* in Chemistry Writing

Data with a plural verb	
These data show . . .	These data are supported by . . .
The data imply . . .	The data suggest . . .
The data indicate . . .	The data reveal . . .
Data with a singular verb	
The profile suggests that the data (*set*) is well converged.	
Data followed by a singular verb that agrees with a different noun	
Inspection of the data *reveals* . . .	
A *copy* of the data is available.	
A key *feature* of the bond length data is . . .	

Data (a plural noun)

In nearly all instances, the word *data* is plural. The singular form, *datum*, is rarely used. The word *data* is almost always used with a plural verb: data are (~~data is~~).

In addition to the word *data*, other scientific plurals are often problematic for novice writers. Table 4.4 includes a list of confusing singular and plural word forms. See also Chapter 16, "Language Tips," Part 4: Grammar, for more details on plural and singular scientific words.

Table 4.4. Singular and Plural Forms of Common Scientific Words

singular form	plural form (preferred form is given first)
apparatus	apparatus, apparatuses
appendix	appendixes, appendices
bacterium	bacteria
basis	bases
criterion	criteria
formula	formulas, formulae
fungus	fungi, funguses
hypothesis	hypotheses
index	indexes (indices if mathematical)
matrix	matrixes (matrices if mathematical)
medium	media, mediums
spectrum	spectra

Exercise 4.22.

Select the correct singular or plural verb form in these sentences.

a. The data in Table 1 (illustrates/illustrate) agreement between theory and experiment.
b. The entropy data in Table 2 (is/are) helpful for predicting their usefulness as cancer drugs.
c. Analysis of the data (yields/yield) the expected result.
d. The three IR (spectrum/spectra) in Figure 2 (was/were) obtained before crystallization.
e. The (basis/bases) for our decision was the demonstrated inertness of the PTFE toward the solvent.

Chapter Review

As a self-test of what you've learned in Chapter 4, define each of the following terms, in the context of this chapter, for a friend or colleague who is new to the field:

alternating R&D	compound labels	integrated R&D
blocked R&D	detection limit	representative data
broad-to-narrow approach	graphics	sets of results

Also explain the following to a friend who hasn't yet given much thought to writing a Results section for a journal article:

- Main purpose of a Results section
- Moves of a Results section
- Relationship between text and graphics in a Results section
- Differences between description (in Results sections) and interpretation (in Discussion sections)
- Circumstances in which the use of *we* could be acceptable in a Results section
- Uses of the past tense and present tense in a Results section
- Common plural words used in scientific writing (e.g., data, spectra)
- Uses of the word *respectively*
- Use of *very* in a Results section

5 *Writing the Discussion Section*

A Discussion section should be as satisfying to read as the last chapter of a mystery novel. The groundwork is laid in the Introduction, technological tools are described in the Methods section, evidence is revealed in the Results section, but it is in the Discussion section where the mystery is solved.

This chapter focuses on the Discussion section, the last part of the standard IMRD structure for a journal article. The Discussion section, as mentioned in Chapter 4, can stand alone or can be part of a combined Results & Discussion (R&D) section. In either case, it serves the same major purpose: to interpret the results of the study. In this chapter, we analyze excerpts from various Discussion sections, including those that accompany results presented in Chapter 4. Upon completion of this chapter, you should be able to do the following:

- Organize a Discussion section following the major moves
- Interpret your results (but avoid overinterpretation)
- Describe the broader implications and/or applications of your findings
- Follow appropriate writing conventions

Writing on Your Own: Discussion Section

Visit the *Write Like a Chemist* (2nd ed.) website for step-by-step guidance in writing your own Discussion section.

In the Discussion section, authors interpret their data, address *why* and *how* questions (e.g., Why was the reaction faster? How did the mechanism proceed?), and, ultimately, extend their findings to a larger context (e.g., What value will these findings have to the scientific community?). Ideally, the Discussion section explains

Write Like a Chemist. Marin S. Robinson and Fredricka L. Stoller, Oxford University Press. © Oxford University Press 2022.
DOI: 10.1093/oso/9780190098940.003.0005

the story revealed by the data, postulates reasons for the observed behaviors, and furthers our fundamental understanding of the underlying science.

Although interpretation is the primary goal of this section, authors must be careful not to overinterpret their data, misinterpret their results, overstate their assumptions, or stray too far from scientific evidence. The excerpts selected for this chapter illustrate ways to avoid these pitfalls. As you will see, the excerpts illustrate that the language of the Discussion section is typified by restraint and understatement. Words such as *fact*, *truth*, and *prove* are rarely used in Discussion sections. Words such as *theory* and *evidence* are much more common, as are such **hedging words** as *appear*, *indicate*, *seem*, and *suggest* (Tables 5.1 and 5.2). By hedging, writers acknowledge the limitations of their knowledge, the tentativeness of their claims, and the recognition that their knowledge will be subjected to scientific scrutiny over time.

Hedging Words

Words that soften interpretations and suggest that interpretations are not absolute facts. See Table 5.1, Table 5.2, and Chapter 16, "Language Tips," Part 2: Word Choice, for more details on hedging.

Table 5.1. Common Hedges in Chemistry Writing

verbs	helping verbs	adjectives	adverbs	
appear	can	possible	apparently	possibly
imply	could	probable	generally	potentially
indicate	may		largely	presumably
seem	might		likely	probably
suggest	should		mainly	typically
support	would		partially	

Table 5.2. Verbs Commonly Used with Helping Verbs (see Table 5.1) to Hedge

assign	conclude	expect	make	rationalize
attribute	deduce	explain	observe	relate
cause	determine	form	obtain	use

Chemical professionals should seek to advance chemical science while striving for the highest standards of scientific integrity. This includes sharing ideas and information, keeping accurate and complete records, and giving due credit to the contributions of others.

—The Chemical Professional's Code of Conduct

To begin our analysis of the Discussion section, we ask you to consider excerpt 5A, which continues the Results and Discussion section shown in Chapter 4 (excerpt 4A) on proanthocyanidins (PAs) in pomegranate fruit. The excerpt comprises four paragraphs. The first three were taken from a section with the subheading **Proanthocyanidin Composition**, near the end of the R&D section. The fourth paragraph is the last paragraph in the paper. Each paragraph contains passages that are largely discussion.

Exercise 5.1.

Read excerpt 5A and complete these tasks:

a. The first paragraph begins by reminding readers of results, but its purpose shifts with the sentence that begins "Similar qualitative profiles _____." What is the purpose of the latter part of this paragraph, after the shift?
b. In addition to reminding readers of results, what purposes are accomplished in the second paragraph?
c. What is the major purpose of the third paragraph?
d. What is the major purpose of the last paragraph?
e. Based on your answers to a–d, suggest at least three possible purposes of a Discussion section.
f. What do you notice about how numbers are formatted in this excerpt? Consider citations and numbers greater than one thousand.

Excerpt 5A (adapted from Díaz-Mula et al., 2019)

[*From Results and Discussion section*]

Proanthocyanidin Composition. Flavan-3-ol monomers and their phloroglucinol adducts were quantified in different parts of the pomegranate fruit and juices (pressed aril juice and commercial juice) from the Mollar cultivar (Table 3). From a qualitative point of view, pomegranate samples showed a rich profile of PAs with a variety of monomeric flavan-3-ol units present in their molecules (C, EC, GC, EGC, ECG, and EGCG). Similar qualitative profiles have been described in green tea,[23] Quercus leaf infusions,[24] and red wine.[25] Previous studies reporting concentrations of flavan-3-ols in common Spanish[18] and American[26] food products showed that, in general, catechins (C and EC) are found in all of the flavanol-containing products; however, the presence of gallocatechins (GC and EGC) is only relevant in some products, including broad beans, lentils, grapes, wine, beer, coffee, tea, and most of the berries. Galloyled flavanols (ECG and EGCG) are still less common, and their presence was important only in green and black teas, strawberries, and grapes.

Pomegranate husk contained the largest PA content (1840 µg/g) compared to seeds (316.2 µg/g) and membranes (51.4 µg/g). The amount in the arils, the edible part of pomegranate fruit, was even lower (34.4 µg/g). The predominant flavan-3-ol units in husk and membrane were GC (51 and 41%, respectively) and C (15 and 43%, respectively), whereas in seeds and arils, C (55 and 38%, respectively) and EC (30 and 32%, respectively) were the most abundant units. The presence of gallocatechins and prodelphinidins in pomegranate peel was previously reported.[20] In the case of juices, the amount in commercial juice (146.9 mg/L) was much higher than in pressed aril juice (21.3 mg/L), probably because, in commercial juice, the whole fruit is pressed and the contact with husk and membranes could enrich the juice in these compounds. Different flavan-3-ol units were present in both juices with high percentages of EC and GC. In all pomegranate samples, EGCG and ECG were present but at low concentrations (less than 8%). It was difficult to properly compare these results to those found in the literature as a result of scarce information available about the PA content in pomegranate samples and the different methodologies used to quantify them.

[*one paragraph omitted*]

In comparison to commonly consumed foods that are rich in PAs (apples, green tea, cocoa, red wine, grapes, and berries), the amount found in edible pomegranate fruit was low. In some varieties of apples, PA concentrations were between 800 and 1410 µg/g;[26] after phloroglucynolusis, the PA concentrations were between 1703 and 5329 µg/g.[21] Chocolate showed concentrations around 5000 µg/g,[29] and grape seeds, also quantified after pholoroglucinolysis, showed concentrations of 2395 µg/g.[30] Plums and berries showed concentrations of 2159[26] and between 230 and 6637 µg/g,[31] respectively. In particular, different varieties of strawberry showed concentrations between 539 and 1632 µg/g, also applying the phloroglucinolysis method.[14]

[*several paragraphs omitted*]

In conclusion, this report shows that PAs are present in significant amounts in the edible parts of pomegranates and are particularly relevant in commercial juices, both from the Mollar cultivar. Future studies should be considered that analyze and compare the content of PAs in fruit and juices from other pomegranate cultivars. PAs complement the large amounts of ellagitannins and anthocyanins present in pomegranates and should also be considered as contributors to the health effects of pomegranate juice, particularly because they are mainly flavan-3-ol monomers and dimers that show higher bioavailability than larger oligomers and polymers.

Analyzing Audience and Purpose

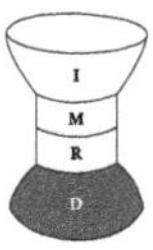

Two major purposes of the **Discussion section** are to interpret or explain results presented in the paper and to suggest broader implications of these findings. Not surprisingly, each purpose is associated with a slightly different audience. The interpretation of results, like the Results section itself, is typically written for an expert audience. Alternatively, the broader implications of the work are typically accessible to a less sophisticated audience. This shift in audience completes the hourglass structure of the IMRD format. The Discussion section forms the bottom of the hourglass; it begins with a specific focus but ultimately expands to offer a more general perspective.

Discussion Section

Two major purposes of a Discussion section are to *interpret* results and *suggest* broader implications of findings.

Exercise 5.2.

Re-read the first sentence of the concluding paragraph in excerpt 5A. Is it written for an expert audience or a more general scientific audience? Justify your choice.

Analyzing Organization

As shown in Figure 5.1, the Discussion section is organized around two major moves. The first move (Discuss Specific Results) is divided into two submoves. The first submove reminds readers about the results that are about to be discussed. Such a reminder may not be needed in a combined R&D section, but it is necessary in a stand-alone Discussion section. This submove is often accomplished in only a sentence or two.

Submove 1.2 is the heart of the Discussion section. Here is where you explain your results and offer interpretations of your findings, using caution not to overinterpret your data. There are countless ways to tell your story. How you choose to do this depends, in part, on your area of research; however, most authors address at least two of the bulleted items in submove 1.2. Relevant literature should be cited, highlighting works that offer **corroborating evidence** and/or conflicting points of view. In this way, your work connects to a larger body of evidence, moving toward the ultimate goal of scientific consensus.

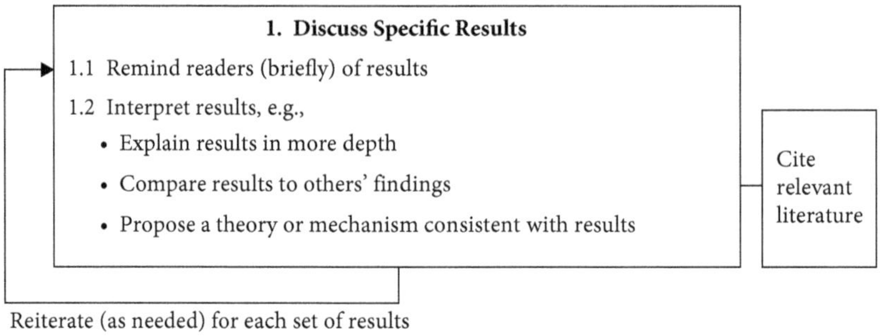

Figure 5.1. A visual representation of the move structure for a typical Discussion section.

Corroborating Evidence

You can strengthen and support your results and interpretations by citing works that validate and corroborate your findings.

Move 1 is repeated as needed for multiple sets of results, ideally paralleling the order used in the Results section. As in the Results section, you can use subheadings to help the reader locate the discussion for each result. If, while writing the discussion, you question the logic of your sequencing, you need to revise both the Results and Discussion sections to align them.

Align Results and Discussion

The order in which you present findings (in Results) and interpret findings (in Discussion) should be parallel.

Move 2 of the Discussion section entails concluding the work. Some journals (e.g., *Analytical Chemistry* and *The Journal of Organic Chemistry*) use an explicit heading to demark this section (e.g., **Conclusions**). Others (e.g., *The Journal of Agricultural and Food Chemistry* and *Environmental Science & Technology*) incorporate concluding remarks into the last paragraph of the Discussion section, often signaled by a phrase such as *In conclusion* or *In summary*. In either case, authors use this section to highlight their principal accomplishments, taking care to not repeat or summarize information presented elsewhere in the paper, including the abstract.

Concluding remarks are also used to suggest broader implications or applications of the work. Authors are asked to look beyond the specific details of their own work and focus instead on the broader goals of the research project by addressing questions such as these:

- What are the implications of the work?
- What new insights were gained?
- How has this work increased our fundamental understanding of the research area?
- What are the practical applications of this work?
- How will the work affect society (e.g., industry, medicine, technology, the environment)?

Attaining this broader outlook can be challenging, especially for students who spend most of their time focused on only a small part of a larger project. Over time, however, your grasp of the broader picture will improve as you continue to read the literature, attend seminars and conferences, and read and write research proposals in your research area. Finally, authors may elect to suggest ideas for future work in their concluding remarks (as in excerpt 5A). This is not required and, in fact, is somewhat controversial, particularly if your research group is already engaged in this work. You might run the risk of getting scooped!

Signaling the Conclusion

Some journals expect a separate section for conclusions, signaled by a **Conclusions** heading. Others integrate conclusions into the last part of the Discussion section, signaled with a phrase such as

In conclusion, In summary,

Note that these phrases are always followed by a comma.

Exercise 5.3.

Consider these passages taken from Discussion sections. Although they are presented out of context, specify if move 1 or move 2 (Figure 5.1) is being accomplished in each. Briefly justify your answer.

a. The stress promoters induced in the Pro-Tox (C) test indicated that Cr^{3+} could cause changes in DNA topology. (From Plaper et al., 2002)

b. Our findings suggest that the surface properties and pore structure of minerals change dramatically upon RAMEB addition. (From Jozefaciuk et al., 2003)

c. A possible explanation for this effect can be attributed to an increased density and viscosity of the milk–NaOH phase when NaOH concentration increases. (From Llompart et al., 2001)

d. Because it is known that humans exposed to different Cr^{3+} species accumulate high levels of Cr^{3+} intracellularly (*17*), presented results may have an impact on human intake of Cr^{3+} as a nutrition additive. (From Plaper et al., 2002)

e. In conclusion, we have shown that $PM_{2.5}$ contains stable radicals that can be detected by EPR. The EPR parameters, persistence in air, and DNA-damaging capacity of the $PM_{2.5}$ radicals are similar to those of the radicals in cigarette tar. (From Dellinger et al., 2001)

f. The data in this article are consistent with a wealth of evidence showing that dietary Se consumed in excess of the Recommended Dietary Allowance lowers the risk of several important cancers (*2*, *14*, *26*, *27*). (Adapted from Finley et al., 2001)

Analyzing Excerpts Move by Move

With the audience, purpose, and organization of the Discussion section in mind, we are ready to analyze excerpts of Discussion sections from the chemistry literature (i.e., journal articles introduced in Chapter 4, where we focused on the Results section). Our goal is to examine the different ways in which authors explain and interpret their results in accord with the move structure in Figure 5.1. We first consider moves 1 and 2 in non-synthesis journals (*Analytical Chemistry*, *The Journal of Physical Chemistry A*, and *Chemical Research in Toxicology*). We then examine two articles from *The Journal of Organic Chemistry*. We conclude the chapter by highlighting writing features that are common to the entire Discussion section.

Move 1 in Non-Synthesis Articles: Discuss Specific Results

Let's begin by analyzing the article on imazalil in apples. The authors use a blocked R&D approach where results are presented first, followed by discussion. In their Results section, the authors showed MALDI spectra to confirm that imazalil can be detected; they also shared images of heat maps that monitored the penetration of imazalil inside the apple (excerpt 4B). Concentrations (in ppm) of imazalil from the skin to the inner region of the apple (on days 0, 1, 4, and 7) were also reported but were omitted from excerpt 4B. We pick up their story here, where they begin to discuss their findings (excerpt 5B).

In just a few words, the authors remind readers of their results (the rate and extent of transport of imazalil) and, in the same sentence, begin to link their work to previous studies. Specifically, they examine results regarding imazalil's transport in oranges, lemons, and apples using either HPLC or GC–MS (in contrast to their use of MALDI imaging, IR laser ablation sampling, and ESI–MS). Indeed, the entire paragraph focuses largely on providing corroborating evidence. Knowledge is strengthened when similar results can be obtained through different methods (herein, different fruits and instrumentation). Collectively, these related works build toward a consensus view on the extent of imazalil transport in fruit. Because this paper was published in *Analytical Chemistry*, it includes a stand-alone Conclusion (not shown). In their conclusion (Move 2), the authors highlight their achievements, underscore the advantages of their approach, and suggest future applications for MALDI imaging and laser ablation.

Exercise 5.4.

Browse through excerpt 5B. How well do the authors accomplish move 1 (Figure 5.1)? Which bullet in submove 1.2 is addressed to the greatest extent?

Excerpt 5B (adapted from Pereira et al., 2019)

[*From near the end of the Results and Discussion section*]

The rate and extent of transport of imazalil are consistent with previous studies. For example, the distribution of imazalil in fruit after storage has been studied previously using gas and liquid chromatography. In a study of oranges using HPLC, it was found that 10–20% of the applied imazalil penetrated beyond the outer 1 mm of the fruit and into the interior of the fruit after 1 week.[6] The distribution of imazalil in lemons was probed by GC–MS, and it was found that the migration into the interior was slow and that approximately 10% of the pesticide was present in the albedo with the remainder in the flavedo after 60 days.[59] Some indications of slow degradation of the fungicide over time with a 10–20 week half-life were observed. A study of the distribution of imazalil in apples quantified by HPLC showed that 70–90% of the pesticide was observed in the peel after 20 days.[60] Approximately 20–25% of the pesticide remained in the fruit after 155 days.

We next consider an example in which the authors propose a theory to explain their results. In excerpt 5C, the authors discuss theoretically derived substitutes for SF_6. Recall, in excerpt 4C (in Chapter 4), the authors proposed a series of hybrid molecules and evaluated them sequentially for their global warming potential (GWP), dielectric strength and boiling point, and thermal stability. In excerpt 5C, we consider only their discussion of GWPs. As background for this excerpt, a molecule's GWP depends both on how long it survives in the atmosphere (its atmospheric lifetime) and how much outgoing infrared energy it absorbs (its absorbance cross section) near 10.5 μm (the "atmospheric window"—where infrared energy normally escapes to space). CO_2 serves as the standard and has a GWP of 1. In comparison, SF_6 has a GWP of 23,000 because of its extremely long atmospheric lifetime (3200 years) and large absorbance cross section in the atmospheric window. The new hybrid molecules have estimated GWPs between 1 and 2000, assuming an atmospheric lifetime of 30 years. The 30-year lifetime is based on reactivity with OH, a radical that reacts with, and destroys, many of the molecules in the atmosphere. The more reactive a molecule is with OH, the shorter its atmospheric lifetime.

In the next two paragraphs, the authors continue their story by discussing caveats to their GWP estimates. As it turns out, three of the hybrid molecules (SF_5NNF, CF_3SO_2F, and $CF_3OC(O)F$) do not react efficiently with OH. The reaction barrier is too high for that removal pathway to be important. Thus, in their discussion, the authors propose alternative theories for how these molecules might be removed from the atmosphere. For example, one of the molecules, SF_5NNF, also absorbs strongly in the UV and VUV regions of the electromagnetic spectrum; hence, the authors postulate that excited SF_5NNF—formed after absorbing VUV and UV radiation—could be photodissociated directly or become more reactive with OH. As you work through exercise 5.5, you will discover the alternative pathways proposed for the other two molecules. Based on the proposed loss mechanisms, the estimated GWPs of CF_3SO_2F and $CF_3OC(O)F$ are 3678 and 1739, respectively, both significantly lower than SF_6.

Exercise 5.5.

Skim the first paragraph of excerpt 5C. Then focus on the second paragraph to answer these questions:

a. What alternative mechanism do the authors propose for the losses of CH_3SO_2F and $CF_3OC(O)F$?

b. What evidence do they offer to support this mechanism?

c. What hedging language do they use to indicate that this is a proposed mechanism and not a fact? (If needed, consult Tables 5.1 and 5.2 for examples of hedging language.)

Excerpt 5C (adapted from Yu et al., 2018)

Note: To help you differentiate discussion from results in this integrated R&D section, we have italicized the discussion passages.

[*continuation of section 3.1 in excerpt 4C*]

However, three molecules including SF_5NNF, CF_3SO_2F, and $CF_3OC(O)F$ need further assessment. The addition of OH to the N=N double bond of SF_5NNF involves a barrier of 9.1 kcal/mol, *which is too high for its atmospheric depletion by OH radical to be important.* On the other hand, SF_5NNF was shown to have a broad and relatively strong UV absorption in the 250–350 nm range with cross sections between 10^{-19} and 10^{-18} cm^2 (Figure 3). Moreover, the absorption becomes much stronger in the VUV region. *Therefore, the excited molecule formed after VUV and UV radiation could be photodissociated directly or be more reactive with OH radicals. As a result, the atmospheric lifetime of* SF_5NNF *could be reduced significantly to a few days or weeks, leading to the negligible GWP.*

In contrast, neither CF_3SO_2F nor $CF_3OC(O)F$ has observable UV absorption above 250 nm (Figure 3). In view of the significant barriers for the reactions of CF_3SO_2F + OH and $CF_3OC(O)F$ + OH (Table S1), *it is conceivable that the dominant removal process for* CF_3SO_2F *and* $CF_3OC(O)F$ *could be heterogeneous hydrolysis in the upper mixed layer . . . where they are transported by diffusion. It has been shown experimentally that the hydrolysis of sulfuryl fluoride in basic ocean water occurs in minutes to hours and thus the transfer velocity across the air–sea boundary layer is the rate limiting step.*[30,31] The atmospheric lifetimes for CF_3SO_2F and $CF_3OC(O)F$ were estimated to be roughly 40 and 15 years, resulting in GWPs of 3678 and 1739, respectively, *which are much lower than that of* SF_6.

In excerpt 5C, the authors postulated new theories consistent with their results and existing knowledge. They could not "prove" their theories (hence their use of hedging language), but their theories are plausible, and they successfully avoided "hand-waving" arguments, that is, theories that lack supporting evidence. Although **hand-waving arguments** serve as useful constructs for scientists to think and talk about their results, they should be avoided in scientific papers.

Hand-Waving Argument

An expression used among scientists that implies that a researcher is advocating a theory or belief with little evidence to support it. Hand-waving arguments should be avoided. Interpretations included in Discussion sections should be supported by data.

We next consider an example from *Analytical Chemistry* (excerpt 5D) where the authors use the Discussion section to emphasize the significance of their findings. The excerpt continues excerpt 4D (in Chapter 4), which described the development of an assay to detect Ebola virus glycoprotein. Once again, an integrated R&D section is used (preferred by *Analytical Chemistry*).

Exercise 5.6.

In excerpt 5D, the authors suggest that their new assay may offer rapid and sensitive detection of EBOV. What support do they give? How do they hedge their claim?

Excerpt 5D (adapted from Hu et al., 2017)

Note: To help you differentiate discussion from results in this integrated R&D section, we have highlighted the discussion passages in italic font.

[from the last paragraph of the Results and Discussion section]

As shown in Figure 5, the results can be observed with the naked eye, and the fluorescence intensity can be collected with a smartphone camera and measured with Image-Pro Plus. Figure 5 also shows that the fluorescence intensity of glycoprotein was much higher than any of the negative controls. *These results indicate strong specificity for the glycoprotein and suggest that this strategy can be used for detection in complex samples.* The assay was then used to detect inactivated EBOV whole virion spiked in buffer, tap water, urine, and plasma (total protein concentration of EBOV is 400 ng/mL) to investigate the accuracy and practical applications of this method. As shown in Figure 5B, all positive samples presented fluorescence signals, whereas fluorescence was not detected in negative controls. In addition, the assay showed excellent performance in the detection of FBS samples (Figure S6). *These results suggest that the dual-readout enhanced lateral flow assay may meet requirements for rapid and sensitive detection in the diagnosis of EBOV infections.*

We next consider the Discussion section from Plaper et al. (2002), who investigated the genotoxicity of Cr^{3+} in *E. coli* cells. Recall from excerpt 4E that three sets of results were presented. The first set showed that chromium chloride and chromium nitrate induced similar stress promoters in *E. coli* test strains, but chromium oxalate did not induce any of the stress promoters tested. The second set indicated that Cr^{3+} from chromium chloride and chromium nitrate enters *E. coli* cells, but Cr^{3+} from chromium oxalate does not. The third set explored the effect of Cr^{3+} (from chromium chloride only) on gyrase, the enzyme that regulates the relaxation of supercoiled DNA.

This article has a stand-alone Discussion section. In accord with move 1, the Discussion Section begins by reminding readers of the results. There were three sets of results, and in passage P1, the authors remind the readers of the first set (that stress promoters were induced by chromium chloride and chromium nitrate but not chromium oxalate). They then hypothesize that the lack of induction by chromium oxalate *is probably due to* (nice hedging!) *the inability of that compound to enter the bacterial cells* (our italics). This provides a logical segue to the second set of results: that chromium oxalate does not enter the cells, confirming their hypothesis. Note that the authors do not speculate on how chromium chloride or chromium nitrate enters the cells, or on how chromium oxalate might be excluded. Again, to do so would be hand-waving. Their data show only that chromium concentrations are negligible in test cells when chromium oxalate is used; their analytical probe (FAAS) does not provide insights into the mechanisms of chromium transport.

P1 When *E. coli* test strains were treated with chromium nitrate, the induction profile obtained was very similar to the chromium chloride profile (data not shown). The promoters that were induced indicate that the action of Cr^{3+} was on DNA. However, none of the 13 stress promoters responded to chromium oxalate at any of the concentrations tested. These results indicate that it is the Cr^{3+} cations and not the anions that are responsible for the *lacZ* induction and that the lack of *lacZ* induction in the case of chromium oxalate is probably due to the inability of that compound to enter the bacterial cells. This was confirmed with FAAS measurements of total chromium concentrations in *E. coli* (Table 1). In cells treated with chromium oxalate, concentrations of chromium were negligible. (Adapted from Plaper et al., 2002)

In their next paragraph (excerpt 5E), Plaper et al. (2002) remind readers of the third set of results (the gyrase test). They also state why they conducted this test, motivated by the Pro-Tox(C) results, thereby continuing a logical progression in their story. Effective writers learn to arrange their results in the order that makes the most sense to their readers, often retrospectively, even if it is not the actual order in which they conducted the work. Complete exercise 5.7 as you read excerpt 5E to identify the types of information that the authors share as they discuss their findings.

Exercise 5.7.

As you read excerpt 5E, answer these questions:

a. Which of the three bullets listed in submove 1.2 (Figure 5.1) do the authors address as they discuss their third set of results?
b. What format do the authors use for citing references? What format was used in excerpt 5C? (See Chapter 15 for more information on formatting citations in your writing.)

Excerpt 5E (adapted from Plaper et al., 2002)

[from the Discussion section]

The stress promoters induced in the Pro-Tox (C) test indicated that Cr^{3+} could cause changes in DNA topology and in this way also affect proper DNA replication and transcription. For this reason, we examined the influence of Cr^{3+} on gyrase, an essential enzyme that relaxes and supercoils double stranded (ds) DNA in bacterial cells. Gyrase is type II topoisomerase that binds to ds DNA, then cleaves and opens one duplex, passes the second duplex through the opening, and then reseals the break (5). The enzyme binds through active site tyrosine at position 122 of the gyrase A-subunit. In a transesterification reaction, the phenolic oxygen on active site tyrosine attacks a phosphate in the DNA backbone, displacing a sugar hydroxyl group (38). In this way, gyrase becomes covalently linked to the DNA. Recently, Zhitkovich and co-workers (27) showed that Cr^{3+} caused the complexing of amino acids to DNA. Cysteine, histidine, tyrosine, methionine, and threonine were the most active amino acids in this binding to the double helix. Because Tyr122 is so important for proper gyrase binding to DNA, we expected that Cr^{3+} would have an influence on binding and, consequently, on gyrase activity. Indeed, we showed that Cr^{3+} bound to gyrase as well as to DNA. Additionally, we showed that Cr^{3+} may inhibit the enzyme's activity. A reason for the inhibition is likely the inability of gyrase to bind to ds DNA, because the quantity of gyrase DNA complexes decreased with increasing Cr^{3+} concentrations. As hydroxyl-containing amino acids are highly reactive with trivalent chromium (27), the inhibition of gyrase activity may well be a consequence of Cr^{3+} binding to the OH^- group of the Tyr in the active site, preventing the enzyme's interaction with DNA phosphate groups. It is also possible that Cr^{3+} binds to some other part of the enzyme and causes changes in the gyrase conformation so that Tyr122 is no longer available for interaction with DNA.

Affect vs. Effect

See Chapter 16, "Language Tips," Part 6: Word Usage, for tips on using these easily confused words.

Because vs. Since

See Chapter 16, "Language Tips," Part 6: Word Usage, for tips on using these easily confused words.

Move 2 in Non-Synthesis Articles: Conclude the Paper

To this point, our focus has been on move 1. In the upcoming excerpts, we illustrate move 2, where the authors conclude their work. We begin with excerpts from *The Journal of Physical Chemistry A* (excerpt 5F) and *Analytical Chemistry* (excerpt 5G). Both journals recommend a separate heading for concluding remarks. Consistent with move 2 (Figure 5.1), the authors highlight their principal findings, emphasize the importance of these findings, and suggest applications of their work. Neither mentions future work.

Excerpt 5F (adapted from Yu et al., 2018)

4. CONCLUDING REMARKS

In summary, a new hybrid model to identify the environmentally sustainable dielectric gases to replace SF_6 has been proposed successfully for the first time, in contrast to the conventional physical mixing of two or three individual gases. Both dielectric performance and thermodynamic stability of the hybrid molecules have been carried forward or even improved significantly from the parent molecules. Meanwhile, the atmospheric lifetimes are shortened considerably due to the enhanced reactivity toward chemical or photolytic degradations, and thus, the hybrid molecules are all environmentally friendly compounds with considerably less greenhouse emissions than SF_6. Except for a few molecules in the CAS registered reference databases (i.e., SF_3N, SF_4NF, CF_3SO_2F, CF_3OCOF, and CF_2NCF_3), the other fully fluorinated compounds involving multiple hybrid bonds (e.g., S=C and S≡C) are experimentally unknown. None has been considered as a potential gaseous dielectric before.

Excerpt 5G (adapted from Hu et al., 2017)

CONCLUSIONS

In summary, we designed a novel dual-signal readout nanosphere (RNs@Au) by the integration of many Au NPs and QDs in one nanosphere. The RNs@Au can provide both colorimetric signal and fluorescence signal to meet requirements of different clinical conditions. A dual-readout enhanced lateral flow assay for EBOV detection with high sensitivity is presented. The assay is based on RNs@Au and the introduction of a biotin–streptavidin system to amplify the dual signals. Unlike other reported lateral flow assays for EBOV glycoprotein, the dual-readout enhanced lateral flow assay is competitive as a rapid assay. It can increase sensitivity by two orders of magnitude and realize quantitative detection of EBOV glycoprotein. Moreover, this method is easy to operate and time saving, and its success in detecting EBOV-spiked samples validates its use for virus detection in real clinical samples. The merits of the novel reporter, dual-signal readout, ability for quantitation, high sensitivity, and anti-interference ability may substantially broaden the applications of lateral flow assays for early detection of disease biomarkers. These assays hold great potential in fields such as clinical diagnosis, food safety, and environmental monitoring.

Hyphenated Modifiers

Chemistry writing includes many two-word modifiers that describe nouns. Here are a few examples in which hyphens are (and are not) used. See Chapter 16, "Language Tips," Part 5: Punctuation, for more details on hyphenated two- and three-word modifiers.

- low-energy process (but "the process occurred at low energy")
- EBOV-spiked samples (but "samples were spiked with EBOV")
- dual-signal readout (but "the readout emitted dual signals")

- DNA-damaging capacity (but "had capacity to cause DNA damage")
- gas-phase process (but "the process occurred in the gas phase")

Exercise 5.8.

Glance through five articles in a current issue of *Analytical Chemistry*.

a. How many articles use a combined R&D section rather than separated Results and Discussion sections?
b. How many articles include a separate Conclusions section?
c. Select two Conclusions sections in *Analytical Chemistry*. How well do they adhere to move 2 in Figure 5.1?
d. Find at least five examples of hyphenated two-word modifiers in these articles.

The last example is from *Chemical Research in Toxicology*, where the final paragraph is used to conclude the work (P2). The conclusion is broad in scope, emphasizing the take-home message that Cr^{3+} can affect DNA. Implications of these findings are also suggested.

> P2 Our results suggest that Cr^{3+} has an impact on DNA, DNA topology, and consequently processes leading to cell growth and proliferation. This could ultimately lead to the mutagenic and carcinogenic potential of Cr^{3+}. Because it is known that humans exposed to different Cr^{3+} species accumulate high levels of Cr^{3+} intracellularly (*17*), presented results may have an impact on human intake of Cr^{3+} as a nutrition additive. (Adapted from Plaper et al., 2002)

Moves 1 and 2 in Synthesis Papers

We complete our analysis by examining Discussion sections of two organic synthesis articles. We first consider the Demko and Sharpless article (2001) on the synthesis of substituted tetrazoles from nitriles in water using a zinc catalyst (excerpt 5H). This excerpt is particularly useful because it illustrates several types of content that authors discuss in synthesis papers. After briefly reviewing the results of their kinetic studies (submove 1.1), the authors compare their findings to a reaction in the literature between N_3^- (prepared from hydrazoic acid) and RC≡N. The exact mechanism for this reaction is a matter of debate, as the literature supports both a two-step process and a concerted mechanism, both of which are illustrated in a scheme. Next, the authors build on these findings by exploring the role of the zinc catalyst used in their study. One possibility is that zinc serves as a Lewis acid, but they cannot support this hypothesis in tests using other Lewis acids. Rather than overinterpret their results, they simply state that the exact role of zinc is not yet clear.

Schemes and Charts

Schemes are used to show a sequence of reactions or a mechanism. Charts are used for groups of structures that do not show reactions.

Schemes have titles, not captions, as shown in excerpt 5H. Compound labels (e.g., **2**, **3a**) can be introduced in schemes for compounds that are referred to in the text; labels should not be assigned to compounds that are not mentioned in the text.

The authors go on to discuss several additional factors related to their synthesis, such as stoichiometric considerations (molar equivalents needed to ensure that the reaction goes to completion), chief competing reactions and how they might be minimized, and other reagents that might be successful substitutes in the reaction. Finally, they discuss their efforts to scale up the reaction, taking it from bench-scale to large-scale applications. Although some of these discussion points are not interpretive in a formal sense, they are representative of common topics in synthesis-paper Discussion sections. Following typical organizational conventions, the authors conclude with a brief summary and suggest broader implications and applications of their findings (following the move structure represented in Figure 5.1).

Discussion Points in Synthesis Papers

transition states and mechanisms
kinetic vs. thermodynamic considerations
possible roles of catalysts
stoichiometric considerations
competing reactions (and how to minimize them)
alternative reagents
efforts to scale up reactions

Exercise 5.9.

As you read through excerpt 5H, notice that the authors use a scheme to present two proposed mechanisms for the reactions. Using this excerpt as an example, propose a set of rules for formatting a scheme.

Excerpt 5H (adapted from Demko and Sharpless, 2001)

[*Continuation of excerpt 4F*]

Kinetic studies using the water-soluble nitrile **1i** revealed first-order dependence in both nitrile and azide and one-half order dependence for zinc bromide. The mechanism of the addition of hydrazoic acid/azide ion to a nitrile to give a tetrazole has been debated, with evidence supporting both a two-step mechanism[8b,21] (Scheme 1, eq 2) and a concerted [2 + 3] cycloaddition[22] (Scheme 1, eq 3). Our mechanistic studies to date imply that the role of zinc is not simply that of a Lewis acid; a number of other Lewis acids were tested and caused little to no acceleration of the reaction.[23] In contrast, Zn^{2+} exhibited a 10-fold rate acceleration at 0.03 M, which corresponds to a rate acceleration of approximately 300 at the concentrations typically used. The exact role of zinc is not yet clear.

Scheme 1

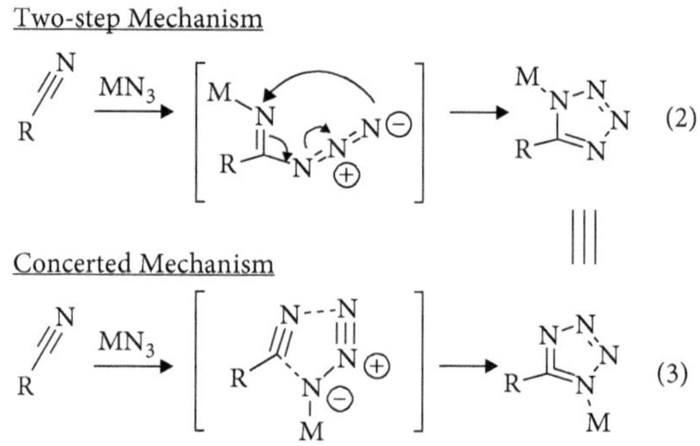

M = H, L_nZn, other metals

Empirically, we found that to ensure complete reaction one needs a 0.5 molar equiv of the zinc salt (ZnX_2); however, in many cases, lower loadings of zinc may be used.[24] The chief competing reaction is hydrolysis of the nitrile to the primary amide; therefore, in cases where the tetrazole-forming reaction is sufficiently fast, namely, with electron-poor nitriles, lower zinc loadings did not entail significant formation of the amide byproduct. Other zinc salts such as zinc perchlorate and zinc triflate also work; zinc chloride, while less expensive, led to more of the amide byproduct. Zinc bromide was chosen as the best compromise among cost, selectivity, and reactivity (see Table 3).

The process became even more attractive for large-scale applications when we found that it could be run at higher concentration without sacrificing yield and without the use of organic solvents in the workup or isolation phases. The resulting products were spectroscopically identical by ^1H NMR and ^{13}C NMR to those synthesized by the general method outlined below; however, the melting points were slightly lower.

In summary, we have demonstrated an exceedingly simple protocol for transforming a wide variety of nitriles into the corresponding 1*H*-tetrazoles. By using zinc salts as catalysts, we showed that water can be used as the solvent despite the relative insolubility of the starting materials. This discovery should facilitate the preparation of tetrazoles in the laboratory.

That vs. Which

See Chapter 16, "Language Tips," Part 6: Word Usage, for information on these easily confused words.

Lastly, we consider the article by Bartolo and Woerpel (2018), who investigated the reactivity of allylic Grignard reagents. Recall in excerpt 4G (part 2), their results suggested that allylic Grignard reagents lack chemoselectivity; that is, unlike their alkyl counterparts, they do not react faster with aldehydes than unhindered ketones. An explanation for this lack of chemoselectivity is offered in their discussion section (excerpt 5I). Citing supporting evidence from the literature, the authors postulate that the reaction first proceeds through a rapidly forming Lewis acid-base complex (**17**), which they show in a scheme (Scheme 3). They then propose two possible transition states for the addition: one that forms a six-membered ring (**18**) and one that forms a four-membered ring (**19**). They present arguments that with unhindered carbonyls, the six-membered transition state (which involves carbon-carbon bond formation and allylic transposition) forms faster than the four-membered transition state; hence, the major product is the γ-addition product (with allylic transposition). In subsequent paragraphs (not shown), the authors go on to provide additional corroborating evidence in support of the six-membered transition state with unhindered ketones. Alternatively, with sterically hindered ketones, they provide evidence that the α-addition product (via the four-membered transition state) becomes the major product. They conclude their work in a final paragraph, also shown in excerpt 5I.

Exercise 5.10.

Use excerpt 5I and Scheme 3 to complete these tasks:

a. Which rate is fastest for an allylic Grignard reagent: k_{-1}, k_2, or k_3?
b. What differences do you notice between Scheme 3 in excerpt 5I and Scheme 1 in excerpt 5H?
c. Give three examples of hedging in the concluding remarks.

Excerpt 5I (adapted from Bartolo and Woerpel, 2018)

[*Continuation of excerpt 4G (part 2)*]

The lack of chemoselectivity for the allylic Grignard reagents can be explained by consideration of the possible transition states for the addition. Complexation of the magnesium atom to the carbonyl compound (k_1) to form a Lewis acid–base complex (**17**) should occur rapidly (Scheme 3).[30] The rate of carbon–carbon bond formation with allylic transposition

(**18**) is faster than dissociation of the complex ($k_2 \gg k_{-1}$), resulting in no difference in the rates of addition to different types of carbonyl compounds. In the addition to unhindered carbonyl compounds, the rate of carbon–carbon bond formation through the transition state with allylic transposition (**18**) is much faster than the rate of carbon–carbon bond formation through the four-membered transition state (**19**, $k_2 \gg k_3$), leading to almost exclusive formation of the γ-addition product.

Scheme 3

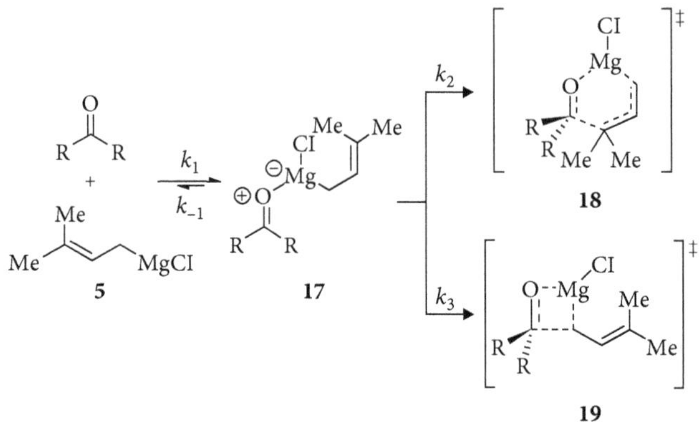

Taken together, these results are most consistent with a mechanism in which allylic Grignard reagents react through a six-membered concerted transition state with allylic transposition. This addition appears to be governed by diffusion with unhindered ketones, leading to low chemoselectivity and low chelation-controlled diastereoselectivity. When the products of the other transition states are observed in highly hindered ketones, they are formed with high diastereoselectivity, indicating that addition through these transition states occurs more slowly, likely through transition states similar to those of other Grignard reagents.

Perfecting Your Writing

Here, we examine writing conventions that are common throughout Discussion sections. We focus on verb tense and voice, in addition to the use of *we* and hedging words.

Verb Tense and Voice

Both past tense and present tense are common in the Discussion section. The same rule of thumb introduced in the Methods section applies to the Discussion section.

Work was done in the past; knowledge exists in the present.

For example, Plaper et al. (2002) state that "three Cr^{3+} compounds *were examined*" (work done in the past) but that the "hydroxyl radical *is* a known mediator of DNA damage" (knowledge that exists in the present). Similarly, the present tense–active voice combination is used in the Discussion section to state scientific "truths" (knowledge expected to be true over time), just as it is in the Results section. Note that interpretations and/or mechanisms put forth in a Discussion section are often considered to be "truths" and therefore are stated in present tense–active voice. Table 5.3 summarizes common verb tense–voice combinations and their functions, with example sentences.

Table 5.3. Common Verb Tense–Voice Combinations for Key Functions in the Discussion Section of a Journal Article (other combinations are possible)

function	tense–voice combination	example
a. to remind readers of results	past-active	Samples *showed* a rich profile of . . .
	past–passive	Fluorescence *was* not *detected* . . .
b. to interpret or explain results	present-active	A reason for the inhibition *is* likely . . .
		These findings *suggest* that . . .
c. to compare results to others' work	present-active	The rate *is* consistent with previous studies.
	past-active	Jefferson et al.[4] also *reported* . . .
	past-passive	Similar results *were observed* by . . .
d. to propose a theory or mechanism	present-active	It *is* possible that the process *involves* . . .
		These data *suggest* an unexpected . . .
e. to highlight principal findings	present-active	This work *shows* that . . .
	prt pf[a]-active	In summary, we *have demonstrated* . . .
f. to suggest implications/ applications	present–active	These assays *hold* great potential for . . .
	future-active	The proposed model *will allow* for new developments in . . .

[a] prt pf = present perfect

Exercise 5.11.

Examine sentences a–e taken from a Discussion section (adapted from Chen et al., 2018). Identify the verb tense (past or present), voice (active or passive), and most likely function (a–f in Table 5.3) of the italicized part of each sentence.

a. *This study showed* that Σ_{16}PAH concentrations in W-Ash were significantly lower than B-Ash ($p = 0.001$).

b. *This suggests* that W-Ash formed from more severe burning (i.e., higher temperatures) does not contribute to soil PAHs compared to background soil levels.

c. *Zhang et al.*[41] also *found* lower Σ_{16}PAH emissions during combustion of agricultural crop residues at higher burn intensities.

d. Additionally, decreased PAH emissions to the atmosphere *were reported* for high temperature burning in laboratory combustion studies.[41,42]

e. By implication, . . . *we expect* much higher concentrations and potential toxicity in moderately burned soils compared with more severely burned soils.

Use of *We*

The word *we* (used to refer to the authors of the work) is commonly used to achieve these purposes in the Discussion section:

To indicate a decision or course of action

P3 To correct this fluctuation, *we* incorporated caffeine standard measurements. (From Yan et al., 2007)

P4 To further validate the suitability of using caffeine as a standard, *we* tested 15 additional compounds. (Adapted from Yan et al., 2007)

To compare findings with previous works

P5 However, like others,[28] *we* often observed quantification inconsistencies when compounds with adjacent nitrogen atoms were analyzed. (Adapted from Yan et al., 2007)

P6 Previously, *we* reported that oral intake of BCAs has many beneficial physiological effects (2–6). (From Matsumoto et al., 2006)

To offer an interpretation

P7 *We* speculate that, due to structural variations, some unavoidable signal loss may be structure dependent and an error range of 10–20% is possible. (From Yan et al., 2007)

P8 Therefore, *we* assumed the decrease of mitochondrial membrane potential and release of cytochrome *c* to cytosol did not happen in the PDTC-pretreated cells. (From Cheng et al., 2006)

To report or highlight findings

P9 *We* found that sulfite alters the pattern of expression of the main aldehyde dehydrogenase gene, reinforcing the link at the molecular level between both metabolites. (Adapted from Aranda et al., 2006)

P10 In conclusion, *we* have demonstrated that PDTC inhibits luteolin-induced apoptosis, which it might do by causing the phosphorylation of caspase-9 in human leukemia HL-60 cells. (From Cheng et al., 2006)

Exercise 5.12.

Consider sentences a–e, all taken from Discussion sections. Which of the following best describes what the authors are trying to achieve in each sentence: (1) indicate a decision or course of action, (2) compare findings with previous works, (3) offer an interpretation, or (4) report or highlight findings?

a. *We* used this value as a correction factor for caffeine calibration, and all compound analyses were adjusted accordingly. (Adapted from Yan et al., 2007)
b. *We* rationalize that this is likely the root cause for the structure dependence. (From Yan et al., 2007)
c. In conclusion, *we* have shown that $PM_{2.5}$ contains stable radicals that can be detected by EPR. (From Dellinger et al., 2001)
d. In our previous study, *we* observed that luteolin could trigger cytochrome *c* released to cytosol. (From Cheng et al., 2006)
e. *We* confirmed that, when compounds contain isolated nitrogen atoms, the response is close to quantitative with a variation about 10–20% depending on structures. (Adapted from Yan et al., 2007)

Hedging Words

Hedging is the mark of a professional scientist, one who acknowledges the caution with which he or she does science and writes on science.

—Crismore and Farnsworth (1990)

As mentioned at the beginning of the chapter, the language of a Discussion section is typified by restraint and understatement. The goal is to let the science speak for itself. Words should be used to clarify, not convince. By overselling a point, even one that is well substantiated, you will appear to be biased. For this reason, chemists soften their interpretations and claims to "truth" by using hedging words.

Exercise 5.13.

Consider sentences a–e. Identify a word or phrase in each sentence that indicates that the authors hedged their claims. (Consult Tables 5.1 and 5.2, if needed.)

a. This could ultimately lead to the mutagenic and carcinogenic potential of Cr^{3+}. (From Plaper et al., 2002)
b. Because it is known that humans exposed to different Cr^{3+} species accumulate high levels of Cr^{3+} intracellularly (*17*), presented results may have an impact on human intake of Cr^{3+} as a nutrition additive. (From Plaper et al., 2002)
c. In all cases, this factor has a positive effect and it appears to be of increasing importance as the degree of chlorination of the PCBs increases. (From Llompart et al., 2001)
d. Because a substantial fraction of the fine particles in the atmosphere arise from combustion sources (*9*), it is possible that the deleterious health effects associated with $PM_{2.5}$ can be at least partially ascribed to radicals associated with combustion-generated particulate matter. (From Dellinger et al., 2001)
e. Our mechanistic studies to date imply that the role of zinc is not simply that of a Lewis acid. (From Demko and Sharpless, 2001)

Chapter Review

As a self-assessment of what you've learned in this chapter, define each of these terms, in the context of this chapter, for a friend or colleague who is new to the field:

hand-waving argument	hyphenated modifier	interpretation
hedging	implications	scheme

Also, explain the following to a friend or colleague who has not yet given much thought to writing a journal article Discussion section:

- Main purpose(s) of a Discussion section
- Standard moves of a Discussion section
- Use of present tense–active voice in a Discussion section
- Use of *we* in a Discussion section
- Purpose(s) of hedging and common hedges

6 *Writing the Introduction*

In writing an article about my chemical research, I like to use as a model a travel or travel-adventure article that might appear in a magazine or in the travel section of a newspaper. Because, after all, what is research but a voyage to a completely unknown place where everything is new? So the first few paragraphs of the article [the Introduction] should tell the potential reader where we are going and why.

—Charles H. DePuy, University of Colorado Boulder

This chapter focuses on the Introduction, the first major section of the journal article. The Introduction is often the first section to be read (by readers) but the last section to be written (by writers). This is because the Introduction must tell readers "where the article is going and why," a mission that is most easily accomplished after the rest of the sections have been completed. By the end of this chapter, you should be able to do following:

- Write an Introduction following its conventional organizational structure
- Compose the all-important opening sentence of an Introduction
- Establish the importance of your research area
- Cite and summarize others' works in concise and appropriate ways

Writing on Your Own: Introduction

Visit the *Write Like a Chemist* (2nd ed.) website for step-by-step guidance in writing your own Introduction.

Write Like a Chemist. Marin S. Robinson and Fredricka L. Stoller, Oxford University Press. © Oxford University Press 2022.
DOI: 10.1093/oso/9780190098940.003.0006

The Introduction, as its name implies, sets the stage for the rest of the journal article by introducing the research area, describing its importance, and hinting at what new knowledge and insights the authors have gained. The Introduction is also where authors summarize others' works, requiring several important writing skills such as paraphrasing, writing concisely, and correctly citing the literature (addressed in this chapter and in Chapters 15 and 16).

Citing the Literature

See Chapter 15 for guidance on how to cite the primary literature.

Reading and Analyzing Writing

At long last, we ask you to read the Introduction to the article on proanthocyanidins in pomegranate fruit (excerpt 6A). If you have progressed through the chapters in this book sequentially, you have already read the Methods, Results, and Discussion sections. Admittedly, this order may seem a bit unusual. Remember, however, we want you to read the Introduction through the eyes of the writer, not the reader. As authors write their Introduction, they already know what unfolds in the rest of their paper; now you, too, have this perspective. (If necessary, refer to excerpts 3A, 4A, and 5A in Chapters 3–5 to refresh your memory.) As you read the Introduction, consider how the authors introduce their story of scientific discovery.

Exercise 6.1.

Complete these tasks as you read excerpt 6A.

a. The following bulleted list specifies the major purpose of each paragraph in excerpt 6A but in scrambled order. Indicate which paragraph (1, 2, 3, or 4) is associated with each purpose. (You might try to guess their order first, and then read excerpt 6A to see if your ordering was correct.)
   - Introduce readers to the current study, which uses phloroglucinolysis combined with LC–ESI–MS/MS to quantify PAs in pomegranate fruit
   - Introduce readers to PAs and describe methods that have been used to detect PAs in foods, including phloroglucinolysis
   - Point out that little is known about PAs in pomegranates
   - Introduce readers to pomegranates and their health benefits

b. In which paragraphs are previous works cited? How detailed are the descriptions of previous works?

c. Based on this Introduction, is the "serial comma" (the comma placed before "and" or "or" in a series of three or more items) used in ACS journals? Support your answer with two examples from the excerpt.

d. Based on this Introduction, where should citations be placed? Should they be placed just at the ends of sentences? Should they be placed before or after punctuation?

Excerpt 6A (adapted from Díaz-Mula et al., 2019)

INTRODUCTION

The pomegranate (*Punica granatum* L.) is one of the oldest known edible fruits widely consumed as fresh fruit and juice. This fruit has generated great interest because its consumption has been associated with positive effects on human health[1,2] that have been attributed to its high content in phytochemicals, mainly polyphenols.[3] Most studies on pomegranate polyphenols have focused on two of the major classes of compounds: hydrolyzable tannins (ellagitannins) and anthocyanins.[4–7] More recently, a large number of colored anthocyanin–flavanol and flavanol–anthocyanin adducts have been reported in pomegranates, although at very low concentrations.[5,8] However, few data are available on other phenolic compounds, such as proanthocyanidins (PAs).

PAs, also known as condensed tannins, are a group of oligomeric or polymeric flavan-3-ols widely distributed in the plant kingdom, mainly in fruit and tree nuts.[9] PAs containing catechin (C) or epicatechin (EC) are named procyanidins; those containing gallocatechin (GC) or epigallocatechin (EGC) are called prodelphinidins, and those with afzelechin or epiafzelechin are propelargonidins.[10] Their potent in vitro antioxidant capacity suggested a protective effect for the prevention of diseases related to oxidative stress, such as cardiovascular diseases, type 2 diabetes, and cancer.[10,11] Their separation and identification are often difficult due to their polymeric nature, large structural diversity, and presence in complex matrixes.[12] The molar absorbance of PAs is also lower than those of other phenolic compounds, and this complicates their quantification even more using ultraviolet (UV) detectors. Different methodologies have been developed for the analysis of PAs that include spectrophotometric methods, normal-phase high-performance liquid chromatography (HPLC) coupled to fluorescence detection, derivatization with different reagents, and depolymerization reactions using toluene-α-thiol and phloroglucinol.[12] Phloroglucinolysis is the depolymerization of PAs in an acid environment in the presence of phloroglucinol and gives access to important information regarding PA composition.[13] This methodology has been widely reported in the literature for the characterization of PAs in different foods, including wine, apples, grapes, tea, cocoa, legume seeds, and strawberries.[14–16]

In the case of pomegranates, little interest in PAs has been evidenced thus far. Total PAs have been estimated in pomegranates by spectrophotometric methods,[17] which are generally non-specific and do not allow for individual compounds to be identified. In other publications, the occurrence of flavanols of a low degree of polymerization (mainly monomers and dimers) was reported in pomegranate samples.[18,19] In particular, the presence of (epi)-catechin, (epi)-gallocatechin, and different PA dimers in pomegranate fruit has been described.[18] In other studies, gallocatechins and a range of prodelphinidins were purified by HPLC from pomegranate peel.[20] However, knowledge of the occurrence of PAs in pomegranates is still fairly poor, and detailed quantitative information on the PA profile is lacking. Available reports do not provide data on PA subunit composition or the degree of polymerization; information is also lacking on PA content in edible parts of the pomegranate and its juices.

In this work, PA profiles (subunit composition, quantitation, and degree of polymerization) of different parts of pomegranate fruit and juices were studied using phloroglucinolysis combined with analysis by liquid chromatography–electrospray ionization–tandem mass spectrometry (LC–ESI–MS/MS).

That vs. Which

See Chapter 16, "Language Tips," Part 6: Word Usage, for tips on these easily confused words.

Exercise 6.2.

Different verb tenses are used in excerpt 6A, each with its own purpose. Examine the verbs in excerpt 6A as you answer these questions:

a. Present tense is commonly used to state a "truth" or known fact (e.g., "The pomegranate . . . *is*" in the first sentence). Find two additional examples of present tense in paragraph 3.

b. Past tense is used to describe work done in the past (e.g., "*were studied*" in paragraph 4). Find two additional examples of past tense in paragraph 3.

c. Present perfect combines *has* or *have* with a past participle (e.g., "*has generated*" or "*have been attributed*" in paragraph 1) and is used frequently to cite others' works. Find two additional examples of present perfect in paragraph 3.

d. Both active and passive voices are used in this excerpt. Which voice is used in the first sentence (paragraph 1)? Which voice is used in the last sentence (paragraph 4)?

Exercise 6.3.

A key purpose of the Introduction is to tell readers why the targeted area of research is important. Re-read the first and last paragraphs of excerpt 6A and then restate the importance of the work in your own words.

Analyzing Audience and Purpose

You have probably noticed that the Introduction is one of the most easily understood sections of a journal article. This is because the Introduction is written for a more general audience than the rest of the paper. A wide range of readers, from students to experts, should be able to understand at least parts of the Introduction. The purposes of the Introduction are also more general than other sections of the paper:

- To introduce the area of research
- To explain the importance of the research area
- To highlight relevant, precedent works
- To justify the need for the current work
- To introduce the current work

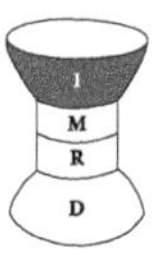

The Introduction begins with the most general information (the research area) and gradually shifts to a more specific focus (the current work), preparing the reader for the highly specific focus of the Methods section. This transition from general to specific is apparent in the now familiar hourglass shape of the journal article.

Exercise 6.4.

Each of these passages (a–e) is about a Grignard reaction. Four are from journal articles (two from Introductions and two from Discussion sections), and one is from a textbook. Based on your understanding of audience and purpose, assign each passage to its most likely genre (journal article or textbook). For passages from journal articles, assign the most probable section (Introduction or Discussion).

a. Entirely different regioselectivity to that observed with the BuMgX/CuCN reagents was obtained with $Bu_2Cu(CN)Li_2$ in THF[1] and BuLi/CuCN (cat.) in both THF and Et_2O. (Adapted from Ito et al., 2001)

b. The Grignard reagent has probably been the most widely used intermediate in organic chemistry since its introduction by Victor Grignard in 1900.[1] Despite this wide use, it is not possible to assign a specific structure for a particular reagent. This is because both RMgX and R_2Mg are formed during preparation of the reagent and are connected by the equilibrium described by the Schlenks.[2] (Adapted from Walter, 2000)

c. Organohalides (RX) react with magnesium metal in ether or tetrahydrofuran (THF) solvent to yield organomagnesium halides, RMgX. The products, called Grignard reagents after their discoverer, Victor Grignard, are examples of *organometallic* compounds because they contain a carbon–metal bond. (Adapted from McMurry, 2004)

d. The mechanism of the Grignard reaction with chlorosilanes is different from that with alkoxysilanes. An S_Ei mechanism, common for electrophilic substitution reaction with organometallic compounds,[1–3] can be assumed. (Adapted from Tuulmets et al., 2003)

e. One of the most important methods for forming carbon–carbon bonds is through the nucleophilic addition of an organometallic reagent to a carbonyl derivative. Such reactions are exemplified by Barbier–Grignard type reactions.[1–3] (Adapted from Li and Meng, 2000)

Analyzing Organization

The Introduction, like other journal-article sections, follows a conventional set of moves (Figure 6.1). The first move, composed of three submoves, has the broad purpose of introducing the general area of research. Submove 1.1 identifies the research topic, and submove 1.2 stresses its importance. These two submoves can be accomplished sequentially or together in the first few sentences of the paper. Note that only the general topic is mentioned at this point, *not* the **current work** (i.e., the specific work presented in the paper). Submove 1.3 is where authors summarize essential works in the field, thereby placing the current work in its appropriate context. This submove does not provide an exhaustive review of the literature, but rather includes "sound bites" that alert readers to works that critically influenced the current work or led to fundamental knowledge in the field. The entire first move is usually accomplished in a few paragraphs, with most attention devoted to submove 1.3. All three submoves are strengthened by citations to the literature. Works by other authors are cited as well as previous, relevant works by the current authors.

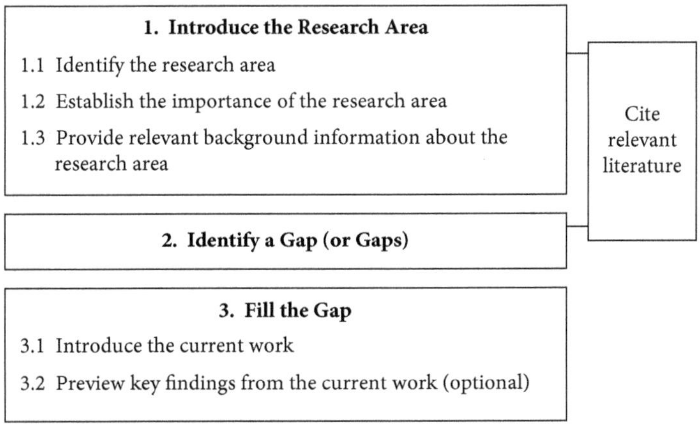

Figure 6.1. A visual representation of the move structure for a typical Introduction.

Current Work

The term *current work* is reserved for the specific work presented in the journal article. The phrase should not be used to refer to others' works or even past works by the authors.

The second move in the Introduction (Identify a Gap) shifts the reader's attention from what has been done (or learned or understood) to what still needs to be done (or learned or understood). The essence of this move is captured in the sentence, "Although much is known about X, little is known about Y." Gap statements vary in emphasis; a few possibilities are listed in Table 6.1. Of course, to correctly identify a gap, the authors must have thoroughly reviewed the literature; hence, citations to the literature are common in this move, too.

After a gap has been identified, the third and final move of the Introduction is to fill the gap. This move typically comprises a short paragraph at the end of the Introduction and begins with a phrase like "In this paper, we ___." At last, the authors can refer to the current work, and many use the personal pronoun *we* to accomplish this task. The authors give a short description of the current work (typically a few sentences), highlighting how the work fills the identified gap (submove 3.1). The Introduction can end here, or the authors can elect to preview their principal findings (submove 3.2). If the authors do the latter, care must be taken not to repeat sentences verbatim that appear elsewhere in the paper.

Table 6.1. Examples of Gap Statements (X represents what has been done, learned, or understood; Y represents what needs to be done, learned, or understood)

gap type	example
A question that remains unanswered	Numerous questions remain unanswered about Y.
A research area that remains poorly understood	Although much has been learned about X, Y remains poorly understood.
A next step that needs to be taken	The next step is to apply X to the study of Y.
An area that has yet to be studied	X has been the subject of several studies; however, to our knowledge, no studies on Y have been reported.
A procedure that needs to be improved (made less expensive, simpler, more efficient, etc.)	Although X achieves the desired detection limits, the method is costly and time-consuming.
A new hypothesis or observation that needs to be validated	Additional studies are needed to corroborate these findings.

Exercise 6.5.

Re-read excerpt 6A with the move structure of the Introduction in mind (Figure 6.1). Identify at least one sentence that accomplishes each of the following: submoves 1.1, 1.2, and 1.3; move 2; and submoves 3.1 and 3.2.

Exercise 6.6.

Read sentences 1–6, adapted from Motohashi and Mikami (2018). (Note: citation numbers have been removed to make this more challenging.)

a. Identify which sentences are from the Introduction and which are from the Discussion. (Hint: only two sentences are from the Discussion.)
b. For sentences from the Introduction, suggest what move or submove(s) are accomplished (Figure 6.1). Note that more than one submove can be accomplished in a single sentence.

1. Herein, we report the Ni-catalyzed cross-coupling reaction of the Grignard reagents under mild reaction conditions even at ambient temperature.

2. Organofluorine compounds have attracted explosive attention in pharmaceutical and agrochemical applications because fluorinated functional groups confer higher metabolic stability, mimic effect, and lipophilicity, based on their unique chemical, biological, and physical properties.[ref]
3. In summary, we have succeeded in the development of the cross-coupling difluoroiodomethylation of organomagnesium reagents with difluoroiodomethane in the presence of a Ni catalyst under mild reaction conditions at ambient temperature.
4. Synthetic methods to provide difluoromethyl arenes via selective benzylic C–H bonds[ref] and decarboxylative fluorination of α-fluoroarylacetic acids[ref] have also been reported.
5. However, the development of transition-metal catalyzed reactions has been limited.
6. Based on these results, a plausible reaction mechanism is proposed for construction of the Ni(0)/Ni(II) catalytic cycle . . . (Scheme 3).

Analyzing Excerpts Move by Move

If a chemical article is going to describe something new about a well-known reaction, the introduction might say, "Although the study of substitution reactions occupies a large portion of undergraduate chemistry and has been extensively studied for decades, no one has examined how the reactions change when the reagents are in the gas phase instead of in solution."

—Charles H. DePuy, University of Colorado Boulder

We now read and analyze excerpts, move by move, from Introductions, including excerpts from articles examined in Chapters 3–5. Later in the chapter, we highlight writing features that are characteristic of the Introduction section.

Move 1: Introduce the Research Area

As depicted in Figure 6.1, the research area is introduced in three submoves. In the first submove, the research area is described broadly. This step is initiated and often accomplished in the opening sentence of the paper. This all-important first sentence, in many genres, is used to set the tone for the work (see Figure 6.2); however, in chemistry journal articles, the first sentence tells the reader, with a broad stroke, what the story is about. Consider these examples, in which the general topic is mentioned early in each sentence (our bolding):

P1 Under the context of climate change,[1] **forest wildfires** are burning more acreage and occurring with greater frequency and intensity in the western United States[2] and throughout arid/semiarid regions of the world.[3] (Adapted from Chen et al., 2018)

P2 **Pollen loads** are produced by *Apis mellifera* L. from collecting hundreds of pollen grains and using their salivary secretions to agglutinate them. (Adapted from Bridi et al., 2019)

Inexperienced writers often want to begin the Introduction with their own work (e.g., In this paper, we investigated *X*); however, this approach is quite rare. More often, the current work is introduced at the *end* of the Introduction, and the general research area is presented first. For example, P1 mentions wildfires; the article itself discusses the effect of wildfire intensity on polycyclic aromatic hydrocarbon levels in soils. P2 mentions pollen loads; the article itself focuses on the antibacterial properties of honeybee pollen. In both cases, the opening sentences are general in focus. Other examples are shown in exercise 6.7.

Figure 6.2. Even Snoopy recognizes the importance of the opening sentence. In a novel, the first sentence sets the tone; in a journal article, it identifies the research area.

The All-Important Opening Sentence

The first sentence in the Introduction should convey the general research area, not the specific work described in the paper.

Exercise 6.7.

Consider the opening sentences in passages P3–P6. Deduce the general research area of each article based only on the first sentence.

P3 Sulfur hexafluoride (SF_6) is one of the most extensively studied polyatomic molecular gases because of its fundamental importance in chemistry, physics, and industry. (From Yu et al., 2018)

P4 Infections from the 2014–2016 Ebola virus (EBOV) resulted in over 28,500 cases and 11,300 deaths.[1] (Adapted from Hu et al., 2017)

P5 Chromium is a metal widely distributed in soil and plants (*1*). (From Plaper et al., 2002)

P6 The literature on tetrazoles is expanding rapidly.[1] (From Demko and Sharpless, 2001)

We call your attention to two writing features in these opening sentences. First, the research area is usually introduced in the present tense; only one passage (P4) is not in the present tense. Second, citations to the literature are quite common in the first sentence. Once again, all passages but one (P3) include a citation in the first sentence.

Equally important is what *not* to do in the opening sentence. We already cautioned against mentioning the current work in the opening sentence. We next caution against mentioning scientists by name in the opening sentence (and throughout the Introduction). Although the opening sentence often includes citations to the literature, names are normally not included. Lastly, we caution against using catchy language. Unlike many forms of writing (e.g., Snoopy's novel in Figure 6.2), when writing for an expert audience, catchy language is discouraged, not only in the opening sentence but throughout the work. Consider the following pair of opening sentences; the first is from the *Journal of the American Chemical Society*, written by scientists for an expert audience. The second is from the *Spotlights* section of the same journal, written by a science writer for a general audience. Catchy language is OK (and expected) in *Spotlights* but not in the actual journal article.

Written for an expert audience	Over the last two decades, research on monolayer-protected clusters (MPCs) has resulted in notable progress.[1-3] (Adapted from Kalenius et al., 2021)
Written for a general audience	Have you ever believed something to be true but felt that the evidence was just out of your grasp? That feeling is exactly how researchers studying monolayer-protected clusters (MPCs) felt about the existence of MPC isomers. (From Yardley, 2021)

Exercise 6.8.

These sentences (adapted from Burner et al., 2019) are from the Introduction of an article on the unimolecular reactions of small ionized polycyclic aromatic hydrocarbons. Based on content and language, which is likely the first sentence of the Introduction? Based on the move structure in Figure 6.1, predict the order of all three sentences:

a. This work centers on the unimolecular reaction dynamics of small ionized amino-substituted PAHs beyond aniline, with a focus on the 1-aminopyrene radical cation, as a means of characterizing the potential role of large amino-substituted PAH ions in the interstellar medium.

b. Polycyclic aromatic hydrocarbons (PAHs) are a class of organic molecules suggested to constitute roughly 20% of carbon in the interstellar medium.[1]

c. Based on aromatics identified in interplanetary dust particles, nitrogen-containing PAHs may be present as either amino- or cyano-substituted species.[3]

After the research area has been identified, the next step (submove 1.2) is to explain *why* the area is important. The reason is often hinted at in the opening sentence. For example, we learned earlier that SF_6 is of *fundamental importance* in chemistry, physics, and industry (P3); Ebola virus infections have resulted in over *28,500 cases and 11,300 deaths* (P4); chromium metal is *widely distributed in plants* (P5); and the literature on tetrazoles is *expanding rapidly* (P6). But a hint is not enough; the importance must be explained in more detail. To this end, four themes commonly emerge: (1) human health, (2) environmental impact, (3) product innovation, and (4) new knowledge. We examine each theme in turn.

We begin with human health. Concerns for human health are used to justify importance in three of the template articles in this book. Imazalil was studied in apples due to its possible toxicity, especially if it can diffuse into the fruit after application (P7). The Ebola assay was developed to rapidly detect this highly infectious disease and to prevent its spread (P8). Lastly, Cr^{3+} was studied to better ensure its safety as a food additive (P9).

P7 Concerns about the toxicity of imazalil derive from its potential effects on reproductive and neurobehavioral parameters[17] as well as possible genotoxic and teratogenic effects.[18] Of particular concern is whether imazalil can diffuse into the fruit after its application. The rate of diffusion can be influenced by the high polarity of imazalil, which promotes transport to the high water-content interior of the fruit.[19] (Adapted from Pereira et al., 2019)

P8 [The Ebola] virus is highly infectious, with a case fatality rate of 40–90%.[2,3] The high number of deaths was at least partly due to the lack of rapid diagnostic methods,[4] which can provide effective diagnosis for medical professionals and contribute to efforts to contain virus spread. It has been reported that if 60% of patients are diagnosed rapidly and isolated within 1 day of illness, the attack rate drops from 80% to nearly 0%.[5] This finding indicates the tremendous effect of a rapid diagnosis of Ebola virus disease. Rapid diagnostic tests are also crucial for infected patients early in the course of disease, at a time when they do not yet transmit widespread infection to others and have a greater chance of survival with clinical intervention. (From Hu et al., 2017)

P9 In numerous studies, Cr^{6+} compounds have been shown to be carcinogenic in vivo and mutagenic in vitro. Cr^{6+} can induce tumors in experimental animals and can neoplastically transform cells in culture (2). In cultured cells, Cr^{6+} induces DNA single-strand breaks, binding of amino acids and proteins to DNA, DNA–DNA cross-links, and Cr–DNA adducts (2, *4–8*).

In contrast to Cr^{6+}, trivalent chromium (Cr^{3+}) is an essential nutrient, needed for the expression of glucose tolerance (*1*). Because the normal human dietary intake of this element is less than 60% of the minimum suggested intake (*1*), many people, particularly in developed countries, supplement their diets with trivalent chromium. The use of chromium nutrition additives is widespread and unsupervised, and there is no awareness of possible side effects. It is therefore important that Cr^{3+} toxicity be investigated in further detail, to ensure safer application of this additive. (Adapted from Plaper et al., 2002)

In vivo and in vitro

In many chemistry journals, the Latin words in vivo (in the living cell) and in vitro (in glass, like a test tube) are no longer italicized.

Farther vs. Further

See Chapter 16, "Language Tips," Part 6: Word Usage, for tips on these easily confused words.

The second theme is environmental impact. Here, importance is established by focusing on areas such as pollution or the transport and processing of chemicals in air, land, or water. Environmental health disparities also fall into this category, as do long-term environmental challenges such as global warming and climate change. In this regard, the SF_6 article established importance by (1) focusing on greenhouse gases and global warming and (2) stating that substitutes for SF_6 are needed because SF_6 is an extremely potent greenhouse gas (P10):

> P10 SF_6 is well known because of its extreme chemical inertness and high dielectric strength for insulation in the transmission and distribution of electrical energy.[1,2] Recently, SF_6 has also been identified as a reactive fluorination reagent catalyzed by organometallic compounds in organic synthesis.[3,4] Moreover, SF_6 is the most potent greenhouse gas due to its extremely long atmospheric lifetime of 3200 years and efficient absorption of infrared radiation at wavelengths near 10.5 μm.[5–7] Consequently, SF_6 has the largest global warming potential (GWP), which is 23,500 times greater than that of naturally occurring greenhouse gases (e.g., CO_2). More seriously, the contribution of SF_6 to global warming is expected to be cumulative and virtually permanent because all of the SF_6 gases will be eventually released into the atmosphere by the electric power industry without disposal or recoverable methods that actually destroy it. It has been observed that the amount of SF_6 in the atmosphere is increasing at a rate of 9% per year from barely measurable quantities two decades ago to the current level near 7.3 ppq.[8] Because SF_6 has massive applications in the electrical industry, seeking alternatives has been taking on greater and greater urgency since the 1980s.[9,10] (Adapted from Yu et al., 2018)

Because vs. Since

See Chapter 16, "Language Tips," Part 6: Word Usage, for tips on these easily confused words.

The third theme, product innovation, embraces emerging, cross-disciplinary technologies in areas such as materials science, drug development, or green chemistry. The article on tetrazoles capitalizes on this theme (P11). In their Introduction, the authors stress the importance of tetrazoles in coordination chemistry, medicinal chemistry, and various materials science applications:

P11 The literature on tetrazoles is expanding rapidly.[1] This functional group has roles in coordination chemistry as a ligand, in medicinal chemistry as a metabolically stable surrogate for a carboxylic acid group,[2] and in various materials science applications, including specialty explosives.[3] Less appreciated, but of enormous potential, are the many useful transformations that make tetrazoles versatile intermediates en route to substituted tetrazoles, and especially to other 5-ring heterocycles via the Huisgen rearrangement.[4] (From Demko and Sharpless, 2001)

The final theme involves the acquisition of new knowledge. Scientists want to test new theories, understand current theories in more depth, and probe areas where current knowledge is limited. They want to look inside the "black box" and explain *why* something is happening. This theme was used to justify the importance of studying allylic Grignard reagents (P12). A puzzle existed: allylic Grignard reagents reacted faster with sterically hindered carbonyls than other reagents. Why was that? Kinetic data offered some explanation, but the mechanism was not known. This work is important because it sheds light on the underlying mechanism.

P12 Commercially available allylic magnesium halides are commonly used to incorporate alkene functional groups into a molecule.[1–3] Their high reactivity enables them to react with sterically hindered carbonyl compounds in cases where other allylmetal reagents do not add.[4–7] . . . The low selectivity observed in reactions of allylic magnesium halides arises because reaction rates approach the diffusion limit.[8–10] However, the mechanism responsible for this unique behavior has not been established. (From Bartolo and Woerpel, 2018)

These four themes are by no means exhaustive; indeed, numerous themes exist to explain the importance of a research area. Nonetheless, we hope that they serve as useful guides as you explain the importance of your work.

Exercise 6.9.

Select two or three chemistry journals in your own area of interest. Browse through several Introductions in these journals. How many address one of the four themes mentioned in the preceding section? What new themes did you discover?

Before leaving submove 1.2, we note that it is not enough to simply say that an area of research is important; importance must be substantiated with citations to the literature. As evidence, 38 different articles are cited in passages P7–P12. Let's take a closer look at these citations. In most instances, the authors used present tense when citing others' works. You might find this surprising because, after all, the cited works were written in the past; yet, the importance of the work is expected to be true today and into the future. Consider this example:

> P13 This functional group *has* roles in coordination chemistry as a ligand, in medicinal chemistry as a metabolically stable surrogate for a carboxylic acid group,[2] and in various materials science applications, including specialty explosives.[3] (From Demko and Sharpless, 2001)

It would sound funny to say, in the past tense, that the functional group "had" roles in coordination chemistry, implying that those roles are no longer important. In addition to present tense, another verb construction is commonly used when citing others' works. Consider this sentence:

> P14 In numerous studies, Cr^{6+} compounds *have been shown* to be carcinogenic in vivo and mutagenic in vitro. (From Plaper et al., 2002)

The "have been shown" construction in P14 is an example of the present perfect (see Table 6.2). There are two forms of the present perfect:

Present perfect–active	has shown, have shown
Present perfect–passive	has been shown, have been shown

Present perfect signals that knowledge gained from work completed in the past is still true in the present. When citing others' works, present perfect–passive is (by far) the conventional choice:

Present perfect–passive (more common)	Arsenic has been shown . . . (*1*)
Present perfect–active (less common)	Nagy et al. (*1*) have shown that arsenic . . .

Passive voice, as seen in these examples, places the emphasis on the science rather than the scientist. The preference for present perfect–passive is also observed in the Discussion section and other sections in which others' works are cited. As a result, the names of scientists seldom appear in most journal articles.

Exercise 6.10.

Browse through the 38 citations in passages P7–P12. How often were the names of the scientists who conducted the work mentioned in these citations? (You might be surprised by the answer!)

Present Perfect–Active and Present Perfect–Passive

Present perfect–passive voice is preferred when citing others' works. Passive voice emphasizes the science rather than the scientist. Present perfect–active voice is less common.

Table 6.2. Examples of Active- and Passive-Voice Constructions in Present Perfect[a]

active voice: has/have + past tense verb form (or past participle)	**passive voice**: has/have + been + past tense verb form (or past participle)
has demonstrated	has been demonstrated
has recognized	has been recognized
has shown	has been shown
have discovered	have been discovered
have observed	have been observed
have received	have been received

[a] Present perfect constructions require the use of the past participle when the past tense verb and past participle forms are different from one another. For most verbs, the past tense and the past participle verb forms are the same (e.g., for the verb *demonstrate*, both the past and past participle are *demonstrated*). With many irregular verbs, however, the past tense verb and past participle verb forms are different (e.g., for the verb *show*, the past tense is *showed*, but the past participle is *shown*). Common past participles include *become*, *chosen*, *given*, *grown*, *seen*, *shown*, *taken*, *undergone*, *undertaken*.

Exercise 6.11.

The following paragraph is from an Introduction in the *Journal of Agricultural and Food Chemistry*.

a. Identify each instance of present perfect–active and present perfect–passive.
b. Which construction is preferred?

> Progress in plant lectin biochemistry has exploded during the past decade (*1*). New lectins with interesting properties have been isolated and characterized; a great many lectins have been cloned, and homologies in their amino acid sequences and similarities in their molecular structures have been established. The X-ray crystallographic structure of a host of these lectins has been solved at high atomic resolution. The biosynthesis of many lectins has been elucidated. Many lectins have been expressed in bacteria and eukaryotic cells and their mutant forms studied to determine changes, if any, in their carbohydrate-binding specificity. Several lectins have also been transfected into food crops with the intention of conferring resistance to various insect vectors. (Adapted from Goldstein, 2002)

After the importance of the area has been established in the Introduction, the next step (submove 1.3) is to provide readers with relevant background information. Thus, citations are also abundant in submove 1.3. The goal here is to alert readers to essential works in the field, not to review the literature exhaustively. This point is often emphasized in Author Guidelines, as illustrated for three ACS journals:

- Background material should be brief and relevant to the research described. Detailed or lengthy reviews of the literature should be avoided. (*Chemical Research in Toxicology*, 2021)
- The introduction should state the purpose of the investigation and must include appropriate citations of relevant, precedent work but should not include an extensive review of marginally related literature. (*Analytical Chemistry*, 2021)
- Discuss relationships of the study to previously published work, but do not reiterate or provide a complete literature survey. (*Environmental Science & Technology*, 2021)

Despite the label assigned to submove 1.3, it is not the only submove (or move) that presents background information; indeed, relevant background information is (and should be) integrated throughout the Introduction, first to introduce the topic and establish its importance, and subsequently to identify the gap.

A few examples of how authors share background information are shown in P15–P19. As you read these passages, observe how concisely they are written. The authors do not summarize one work at a time, in whatever order they choose; instead, multiple citations are grouped together in a logical sequence, ultimately leading up to the current work. Also, note that no direct quotes are used. Although common in other genres, direct quotes are exceedingly rare in chemistry journal articles. Instead, chemistry authors must capture the essence of others' works in only a few words or sentences of their own. Consider P15; in a single sentence, the authors summarize and cite nine other works.

> P15 The most common cleanup methods are as follows: shaking the extract with concentrated sulfuric acid,[9–12] Florisil,[9–13] alumina,[14,15] and silica gel[8,13] and using size exclusion chromatography (SEC).[15,16] (From Llompart et al., 2001)

Avoid Direct Quotes

Although common in some written genres, chemists almost never use direct quotes in journal articles.

Next consider P16. The importance of chromium research has already been established (see P5, P9, P14); now the authors provide background information on what is already known about Cr^{3+} toxicity. To date, most cellular tests indicate that Cr^{3+} is nonmutagenic; however, recent studies suggest that Cr^{3+} can cause DNA adducts and crosslinks.

> P16 So far Cr^{3+} has been shown to be nonmutagenic in most of the studies performed on cellular test systems, probably due to the inability of the hydrated Cr^{3+} complexes to cross plasma membranes (*10*). However, in some recent studies, Cr^{3+} was shown to cause Cr–DNA adducts and DNA–DNA crosslinks (*13*, *14*). Cr^{3+} also increased DNA polymerase processivity and decreased its fidelity during DNA replication in vitro (*15–17*) besides causing formation of mutagenic adducts of amino acids to the DNA phosphate backbone (*18*). (From Plaper et al., 2002)

Numbering Citations

The citation numbers appear to be out of sequence in P16 because citations 11 and 12 are missing; their absence just means that 11 and 12 were already mentioned (in the correct order) earlier in the Introduction. (See Chapter 15 for more details on citations.)

In P17, the authors provide background information on imazalil in two areas; they first describe its use as a pesticide. Later, they describe how pesticide residues have traditionally been investigated in fruits and vegetables. This latter information provides a logical segue to their own study, which uses a new (and better) method to investigate pesticide distribution in fruits.

> P17 One of the most common uses of pesticides in fruits is fungicide.[9] Imazalil (1-{2-(2,4-dichlorophenyl)-2-[(prop-2-en-1-yl)oxy]ethyl}-1H-imidazole) is a postharvest fungicide that is widely used to prevent fruit degradation by controlling growth and sporulation of fungal strains.[10] The mode of action consists of inhibiting the biosynthesis of ergosterol, an essential component in the membrane of fungal cells.[11] Imazalil is efficient for the control of fungi of the *Penicillium* genus, such as *P. digitatum* and *P. italicum* in citrus fruits,[12] and *P. expansum* in pome fruits.[13] Usually, imazalil is applied to fruits after harvest by dipping, drenching, or spraying using concentrations between 1000 and 2000 ppm.[14–16] [. . .]

Traditionally, pesticide residue determination in fruits and vegetables has been performed using gas chromatography coupled to mass spectrometry (GC–MS)[20–23] and liquid chromatography coupled to mass spectrometry (LC–MS).[24–27] Both techniques demonstrate excellent analytical performance but provide limited information on the distribution of pesticides in the sample. (Adapted from Pereira et al., 2019)

Genus and Species Names

Genus names are capitalized and italicized (e.g., *Penicillium*); species names are italicized but not capitalized (e.g., *digitatum*). Genus names are spelled out in full at first mention and abbreviated thereafter (e.g., *Penicillium digitatum*, then *P. digitatum*).

En dashes

Did you notice that the dash used in GC–MS and LC–MS in P17 is slightly longer than the hyphen (-) used in two-word modifiers? It is called an en dash and is used to mean "through." Another example is UV–vis (i.e., UV through visible).

An en dash is also used in spans of three or more numbers in text (e.g., 10–15 mg), in citations (e.g., . . .[10–12]), and with page numbers in references (e.g., 1009–1020).

Exercise 6.12.

Based on this sentence and P15–P17, what general rules can you suggest for the proper placement of citations in a sentence?

> Many sensitive assays have been developed for EBOV detection, such as optofluidic biosensors based on plasmonic nanoholes,[12] interferometric measurement technique,[13] the luminescence resonance energy transfer strategy,[14] and the electroluminescence method.[15] (From Hu et al., 2017)

Lastly, we consider passages from two organic chemistry articles. In the substituted tetrazole article (P18), the authors succinctly summarize a "literature . . . replete" with previous methods by dividing these methods into three main categories. Each category has a single citation (superscripted numbers 6, 7, and 8, respectively), so it appears as if three publications are cited. However, these are group citations. In the Reference section, number 6 includes references 6(a)–6(c), number 7 includes references 7(a)–7(d), and number 8 includes references 8(a)–8(d). Thus, 11 articles are cited, not just three. This grouping method is discouraged in many journals but is still occasionally used in *The Journal of Organic Chemistry*. (See Chapter 15 for more information on formatting citations and references.)

P18 The most convenient route to 5-substituted 1*H*-tetrazoles **2** is the addition of azide ion to nitriles **1**.[5] The literature is replete with methods to perform this transformation; they fall into three main categories: those that make use of tin or silicon azides,[6] those that use strong Lewis acids,[7] and those that are run in acidic media.[8] (Adapted from Demko and Sharpless, 2001)

In the article on allylic Grignard reagents (P19), as background information, the authors share several mechanisms that have already been reported in the literature. Using only 89 words and one scheme, they highlight 11 relevant works, including as many as six citations in a single sentence. (Unlike P18, group citations are not used. *The Journal of Organic Chemistry* accepts either format.)

P19 Several mechanisms can be considered to explain the additions of allylic Grignard reagents to carbonyl compounds.[11–16] The additions of Grignard reagents to carbonyl compounds may proceed through monomeric, four-centered transition states resembling **1** or through dimeric, six-centered transition states such as **2** (Scheme 1).[14,16–18] Allylic Grignard reagents, however, have two other pathways available. They may add to carbonyl compounds through a monomeric, six-membered transition state with allylic transposition (**3**).[11,13] An open S_E2'-like transition state (**4**)[14] is also possible, although more recent mechanistic data suggest that this pathway is unlikely.[1,19,20] (From Bartolo and Woerpel, 2018)

Precede vs. Proceed

See Chapter 16, "Language Tips," Part 6: Word Usage, for tips on these easily confused words.

"Data suggest"

To many native English speakers, "data suggest" sounds wrong, and "data suggests" sounds right. However, the word "data" is plural. See Chapter 16, "Language Tips," Part 4: Grammar, for tips on "data" and other plural and singular scientific words.

Exercise 6.13.

Imagine you are writing an Introduction for a journal article about a new photovoltaic device. In your Introduction, you want to report the power efficiency and average visible transmittance previously achieved in TOPVs, as reported by Liu et al. (2019). Read through the abstract describing their work. Compose a single sentence for your Introduction to highlight this background information. Do not include the authors' names in your sentence, but do include a citation.

> **ABSTRACT.** Here, flexible transparent organic photovoltaics (TOPVs) based on IR-absorbing organic materials are presented, [offering] a simple, low-cost, and promising way to utilize the IR light for electricity generation and the penetrated visible light . . . for photosynthesis in plants. A power-conversion efficiency of ~10% with an average visible transmittance of 34% is achieved for TOPV devices. [Moreover], side-by-side comparison shows that plants grown under TOPV-filtered light and those [grown] under normal sunlight yield similar results. (Adapted from Liu et al., 2019)

Move 2: Identify a Gap

After the importance of the research area has been explained and relevant background information shared, the Introduction shifts from a focus on what has been done (or learned) to an emphasis on what remains to be done (or learned). This change in emphasis is signaled with a gap statement (Table 6.1). The gap statement points out what is lacking in the field and, in so doing, infers the next step that needs to be taken. Consider these examples:

P20 Currently, the only option is a trial-and-error method involving mixtures with buffer gases (N_2 or CO_2). However, operating gas mixtures in electric equipment has many perceived disadvantages including elusive optimal mixing ratios, difficult gas supply and recovery procedures, and problems with concentration control, leak handling, and disposal. Therefore, although much effort . . . has been invested, the current strategy to discover SF_6 alternatives has hit both experimental and theoretical bottlenecks. (Adapted from Yu et al., 2018) [*Next step: find a better strategy to discover SF_6 alternatives*]

P21 However, these methods are tedious to operate in the field and are not readily compatible with POC use without extensive, existing infrastructure. Hence, a POC assay allowing the rapid, sensitive, and quantitative detection of EBOV is vitally important for diagnosis in resource-poor areas. (From Hu et al., 2017) [*Next step: find a faster, more sensitive method to detect Ebola virus*]

P22 Each of these [methods for synthesizing tetrazoles] has one or more of the following drawbacks: expensive and toxic metals, severe water sensitivity, and the presence of hydrazoic acid, which is highly toxic and explosive as well as volatile. The few methods that seek to avoid hydrazoic acid liberation during the reaction, by avoiding acidic conditions, require a very large excess of sodium azide.[9] In addition, all of the known methods use organic solvents, in particular, dipolar aprotic solvents such as DMF. This is one of the solvent classes that process chemists would rather not use.[10] (From Demko and Sharpless, 2001) [*Next step: find a safer, less expensive, and water-based method*]

Although vs. While

See Chapter 16, "Language Tips," Part 6: Word Usage, for tips on these easily confused words.

Gap statements are not limited to move 2; they are also common in move 1. For example, the last sentence in P12 explains importance and identifies a gap: *However, the mechanism responsible for this unique behavior has not been established.* Similarly, the last sentence in P17 provides both background information (on techniques used previously) and identifies a gap: *Both techniques demonstrate excellent analytical performance but provide limited information on the distribution of pesticides in the sample.* In fact, gap statements are often observed multiple times in a single Introduction.

Exercise 6.14.

Re-read Excerpt 6A on proanthocyanidins in pomegranate fruit. Identify four gap statements in the Introduction and assign them to either move 1 or 2.

Exercise 6.15.

Select three different chemistry journals. In each journal, find two articles that include one or more gap statement(s) in their Introductions. Assign each gap statement to move 1 or move 2.

Move 3: Fill the Gap

After the gap has been established, the last move of the Introduction is to fill the gap (Figure 6.1). The move has two submoves. In the first submove (3.1), the authors introduce the current work. At long last, the authors get to focus on their own research, which presumably will fill the gap(s) identified previously. The submove typically starts with a new paragraph and a transitional sentence that begins something like "In this paper," or "In this study," (see Table 6.3 for other possibilities). The personal pronoun *we* often comes next (e.g., In this study, *we* . . .), which clearly signals a shift from describing others' works to the work at hand. As shown in Table 6.4, *we* is then followed by a verb in either present or past tense. Consider these examples, all in active voice:

P23 Here, *we propose* a one-step strategy for treatment of phosphite-laden wastewater using a single electrochemical reactor that integrates electrochemically induced oxidation (EO) and electrocoagulation (EC) processes. (From Liang et al., 2019) (*present tense–active voice*)

P24 *We report* here a safer and exceptionally efficient process for transforming nitriles into tetrazoles in water; the only other reagents are sodium azide and a zinc salt. (From Demko and Sharpless, 2001) (*present tense–active voice*)

P25 In this work, *we combined* MALDI imaging with laser ablation sampling and offline ESI–MS to analyze the distribution of imazalil in apples. (Adapted from Díaz–Mula et al., 2019) (*past tense–active voice*)

P26 Herein, *we designed* a novel reporter (RNs@Au) . . . (From Hu et al., 2017) (*past tense–active voice*)

Passive constructions (without *we*) can also be used. Here are three examples, one in present tense and two in past tense.

P27 Herein, pure VO_2 (M) powder with high crystallinity *is synthesized* through a two-step thermolysis and post-annealing process. (From Guan et al., 2018) (*present tense–passive voice*)

P28 In this work, PA profiles . . . of different parts of pomegranate fruit and juices *were studied* using phloroglucinolysis combined with analysis by liquid chromatography–electrospray ionization–tandem mass spectrometry (LC–ESI–MS/MS). (From Díaz-Mula et al., 2019) (*past tense–passive voice*)

P29 In this work, for the first time, a group of novel dielectric molecules superior to SF_6 *was constructed* theoretically by means of "hybrids" (Scheme 1), which is in distinct contrast to the commonly used binary or ternary mixtures. (From Yu et al., 2018) (*past tense–passive voice*)

Table 6.3. Common Phrases Used to Transition from the Second to the Third Move of the Introduction

In the present study,	In this context,	In this study,	In this paper,
In the present work,	In this investigation,	In this work,	Herein,

Table 6.4. Common Verbs that Follow *We* in the Fill-the-Gap Statement of the Introduction

In this work, we . . . (present-tense verbs)		In this work, we . . . (past-tense verbs)	
carry out	investigate	analyzed	found
demonstrate	present	calculated	initiated
describe	propose	chose	investigated
develop	provide	determined	measured
employ	report	examined	solved
examine	show	explored	studied
explore	use	focused on	synthesized

Move 3 (Fill the Gap) can conclude with submove 3.1 (as in excerpt 6A) or the authors can add a few more details about the current work, typically in a few sentences. Submove 3.2, preview key findings, is optional; however, many authors decide to include it, perhaps hoping to entice readers to read further. The preview is typically written in present tense, to describe "truths" gleaned from the work, as seen in P30–P32.

P30 The results *demonstrate* the feasibility of the RNs@Au-based lateral flow assay to visually and quantitatively detect EBOV-spiked samples (and potentially a broader range of diseases) with high sensitivity and specificity. (Adapted from Hu et al., 2017) (*present tense–active voice*)

P31 Importantly, we *show* that the choice of alkali-metal cation has a substantial effect on the structure and reactivity of these dimeric complexes. (From Broere et al., 2018) (*present tense–active voice*)

P32 In this article, we *provide* experimental evidence that is most consistent with the reactions of allylic Grignard reagents with unhindered carbonyl compounds reacting through a six-membered transition state with allylic transposition (i.e., transition state **3**). Furthermore, we *describe* experiments that suggest this transition state is responsible for the rapid and unselective reactions of allylic Grignard reagents. (From Bartolo and Woerpel, 2018) (*present tense–active voice*)

Exercise 6.16.

Select four or five chemistry articles in an ACS journal of your choice; browse through the last paragraph of the Introduction. What phrase (if any) is used to signal the transition to the current work? Do the authors preview their findings?

Perfecting Your Writing

I always take my writing seriously. Although I don't know my readers, they all meet me through my writing. I want my words to communicate my commitment to good science, my professionalism, and my desire to engage in a dialogue with the larger scientific community.

—David B. Knaff, Texas Tech University

Here we provide additional details about verb tense and voice across the Introduction. We also consider ways in which chemists achieve concise and fluid writing, not only in Introductions but also in other sections of the journal article and other professional genres.

Verb Tense and Voice

Various tense and voice combinations are used in the Introduction. We summarize common combinations, and their functions, in Table 6.5. In general, present tense and active voice can be used throughout, yet present perfect–passive constructions are commonly used when citing others' works.

Concise and Fluid Writing

It is not easy to write a clear and concise Introduction. Before pen is put to paper (or fingers to keyboard), authors must first find, read, and understand appropriate literature and consolidate key concepts, trends, and findings. Next, they must organize this information in a logical order, linking like concepts in fluent prose, using language that is neither repetitive nor choppy. These are not trivial writing skills, even for experienced writers. Here we focus on ways to help you develop these writing abilities, targeting conciseness and fluidity. These practices apply not only to the Introduction but also to other parts of the journal article, and to other genres as well.

Table 6.5. Common Verb Tense–Voice Combinations for Key Functions in the Introduction of a Journal Article (other combinations are possible)

function	tense–voice combination	example
a. to identify research area	present–active	Chromium *is* a metal widely distributed in soil and plants (*1*). (From Plaper et al., 2002)
b. to establish importance	present–active	Rapid diagnostic tests *are* also crucial for infected patients. (From Hu et al., 2017)
c. to provide background information	prt pf[a]-passive	Cr^{6+} compounds *have been shown* to be carcinogenic (*1*). (From Plaper et al., 2002)
d. to identify a gap	present–active	However, such methods *are* costly and tedious.
e. to introduce the current work	present–active	Here, *we propose* a one-step strategy to . . .
	past–active	In this work, *we combined* . . .
	present–passive	Herein, pure VO_2 powder *is synthesized* . . .
	past–passive	In this work, PA profiles *were studied* using . . .
f. to preview findings	present–active	The results *demonstrate* that Cr^{3+} causes DNA damage. (Adapted from Plaper et al., 2002)

[a] prt pf = present perfect

Conciseness in Writing

When newcomers to chemistry-specific writing first attempt to summarize others' works, they often use a wordy and repetitive writing style. For example, when multiple works are cited, novice writers often resort to a monotonous pattern:

> Garcia et al.[1] showed . . . Daloğlu et al.[2] showed . . . Manygoats et al.[3] showed . . .

Or worse (because it is wordier), inexperienced writers might compose the following:

> In a study by Garcia et al.,[1] it was shown that . . . . Another study was conducted by Daloğlu et al.[2] to show that . . . . In a more recent study conducted by the scientists Manygoats et al.,[3] additional evidence was provided to show that . . . . Finally, other researchers[4–6] have shown that . . . .

Although nothing is wrong grammatically with these sentences (even "et al." is used and punctuated correctly), they signal a novice writer because they are wordy and include unnecessary words such as *researchers* and *scientists*. To increase the conciseness and clarity of your sentences, eliminate references to *scientists* as well as their names and the titles of their works. The latter information is accessible from the references, so it is redundant to repeat it in the text. The authors of excerpt 6A and passages P1-P32 followed this advice; more than 180 articles were cited in these excerpts, but authors' names appeared only twice! Consider these examples:

Very wordy	In their article titled "Preparation of 5-Substituted 1*H*-Tetrazoles from Nitriles in Water," the researchers Demko and Sharpless[5] propose a way to synthesize 1*H*-tetrazoles using nitriles and sodium azide in water.
Wordy	The scientists Demko and Sharpless[5] recently proposed a way to synthesize 1*H*-tetrazoles using nitriles and sodium azide in water.
Wordy	Demko and Sharpless[5] recently proposed a way to synthesize 1*H*-tetrazoles using nitriles and sodium azide in water.
Concise	1*H*-tetrazoles have been prepared using nitriles and sodium azide in water.[5]

Researchers and Scientists

Avoid unnecessary words such as *researchers* or *scientists* in your writing. Your sentences should focus on the science, not the scientists who conducted the work.

Concise Writing

See Chapter 16, "Language Tips," Part 1: Conciseness, for additional tips on writing concisely.

Exercise 6.17.

Revise sentences a–c so that the names of all scientists, article titles, and book titles are removed. How does the sentence word count change with your revisions? Does the conciseness aid or detract from the readability of the sentences?

a. As explained in the well-known book *Fire and Climatic Change in Temperate Ecosystems of the Western Americas*, beehives are usually located in a "matorral," which is a shrubby sclerophyllous type of vegetation that covers the slopes of the coastal range.[19] (Adapted from Bridi et al., 2019) (41 words)

b. Both HNC and HCN have been reported to be ubiquitous in the interstellar medium according to researcher Loison et al.[5] (Adapted from Burner et al., 2019) (19 words)

c. The Diels–Alder reaction, as recently noted by García-Rodeja and co-workers,[3] predominantly leads to the formation of the corresponding endo cycloadduct. (Adapted from Wang et al., 2019) (22 words)

A second strategy for achieving conciseness requires that you group related ideas and use punctuation appropriately. For example, when summarizing the literature, introduce related ideas with the phrase *such as*. Consider these examples, juxtaposing a fabricated wordy passage with its more concise (authentic) counterpart:

Wordy	Several countries have established levels (recommended maximum limits, RMLs) for PCBs in dietary products. Levels of ~2000 ng/g have been established for fish. Levels ranging from 200 to 2000 ng/g have been established for meats. Levels between 100 and 300 ng/g have been established for eggs. (4 sentences, 46 words)
Concise	Several countries have established levels (recommended maximum limits, RMLs) for PCBs in dietary products, *such as* fish (~2000 ng/g), meats (200–2000 ng/g), *and* eggs (100–300 ng/g). (Adapted from Llompart et al., 2001) (1 sentence, 28 words)

An alternative way to achieve conciseness is to use the word *respectively*:

Concise	Several countries have established levels (recommended maximum limits, RMLs) for PCBs in dietary products, such as fish, meats, and eggs, with RMLs of ~2000, 200–2000, and 100–300 ng/g, *respectively*. (1 sentence, 31 words)

Respectively

See Chapter 16, "Language Tips," Part 1: Conciseness, for tips on how to use *respectively* correctly.

The use of colons, with grouped ideas, can also contribute to conciseness. In the following examples, the fabricated wordy passage uses no colon; the more concise (authentic) passage lists related items after a colon:

Wordy	The general analytical procedure for the determination of PCBs in full-fat milk includes four main steps. This first step involves extraction of the PCBs from the matrix. The second step involves preconcentration and cleanup. The third step uses gas chromatographic separation. Finally, the last step involves detection. (5 sentences, 47 words)
Concise	The general analytical procedure for the determination of PCBs in full-fat milk includes four main steps: extraction from the matrix, preconcentration and cleanup, gas chromatographic separation, and detection. (From Llompart et al., 2001) (1 sentence, 28 words)

Note that when you use a colon, it must be preceded by a complete sentence:

Incorrect	The procedure includes: extraction, cleanup, and detection.
Correct	The procedure includes three steps: extraction, cleanup, and detection.

Colons and Semicolons

See Chapter 16, "Language Tips," Part 5: Punctuation, for tips on using colons and semicolons correctly.

Items grouped in a series may also be numbered (with numbers enclosed in parentheses) to achieve conciseness.

Wordy	Over the past few years, we have encountered numerous examples of water as the "perfect" solvent. We observed this first in osmium-catalyzed dihydroxylation reactions.[12] We observed this next in nucleophilic ring-opening reactions of epoxides.[13] We also observed this in cycloaddition reactions.[13] Finally, we observed this in oxime ether, hydrazone, and aromatic heterocycle condensation processes.[14] (5 sentences, 54 words)
Concise	Over the past few years, we have encountered numerous examples of water as the "perfect" solvent in, for example, (1) osmium-catalyzed dihydroxylation reactions;[12] (2) nucleophilic ring-opening reactions of epoxides;[13] (3) cycloaddition reactions;[13] and (4) most oxime ether, hydrazone, and aromatic heterocycle condensation processes.[14] (Adapted from Demko and Sharpless, 2001) (1 sentence, 43 words)

When grouping related items to achieve conciseness, pay careful attention to parallelism. The preceding examples illustrate two types of parallelism:

Plural nouns	fruits, vegetables, and eggs
	reactions, mechanisms, and processes
Nominalizations	extraction, separation, and detection

Parallelism

See Chapter 16, "Language Tips," Part 4: Grammar, for tips on parallelism.

Exercise 6.18.

Imagine that you are writing a paper on lead concentrations in particulate matter in Houston, Texas. You want to summarize the results of Smith and Caine (2021) in your Introduction. Convert your notes (included here) into one or two concise sentences for your Introduction, grouping related items with one of the patterns described earlier. Check your sentence(s) for proper punctuation and parallelism.

> In 2021, researchers Dr. Sara Smith and Dr. Timothy Caine studied lead concentrations in particulate matter in Pittsburgh, PA. They wanted to see how much lead there was in the particulate and what the sources of the lead might be. They found about 65% of the lead was from diesel vehicles. About 20% came from emissions from non-diesel vehicles. The remaining 15% was from industrial emissions. The lead concentrations were reported to be somewhere between 2 and 20 parts per billion (ppb).

Fluidity in Writing

Experienced writers know how to make their words flow, logically linking thoughts and ideas. One way to achieve fluidity in your writing, particularly in the Introduction, is to use words or short phrases that create obvious linkages between sentences and/or add emphases to your writing. Consider the italicized examples in sentences a–f. Note the use of a comma after each italicized word or phrase.

a. *Furthermore*, the survival of some species is threatened by . . .

b. *Moreover*, the absorption becomes stronger in the IR region.

c. *Similarly*, **3** undergoes loss of N2 to generate . . .

d. *Alternatively*, the reaction could proceed through transition state **4**.

e. *To this end*, we systematically studied both isomers.

f. *Surprisingly*, only a single product was observed.

Words and phrases such as these are also used in the middle of sentences or between sentences connected by a semicolon. Consider these passages, paying special attention to the ways in which commas and semicolons are used.

a. Dairy products, and milk *in particular*, have received special interest due to their extensive and elevated consumption by the population.[6] (From Llompart et al., 2001)

b. Purification requires, *for example*, crystallization or chromatography, which may lead to losses. (From Boesten et al., 2001)

c. Although recently HSSPME has been applied to an enormous variety of matrixes,[22,24] *to date*, the number of publications that apply SPME technology to milk samples is relatively low.[25,26] (Adapted from Llompart et al., 2001)

d. Note also that all unimolecular dehydration reactions are favored by entropy; *however*, only the roaming reaction has a positive activation entropy. (From Asatryan et al., 2018)

e. This shallow minimum, *however*, is not reproduced by other DFT methods including B97D3. (From Asatryan et al., 2018)

Fluid Writing

See Chapter 16, "Language Tips," Part 7: Writing Effectiveness, for tips on how to make your writing fluid.

Of course, the use of these words and phrases is not confined to Introductions. When used appropriately, and in the right places, they can contribute to fluidity throughout the journal article. A list of useful phrases, organized by their common functions, is presented in Table 6.6. Many, but not all, of these phrases are conventionally followed by commas when they start a sentence.

Exercise 6.19.

Browse through the Introductions of three different journal articles and find at least five sentences that begin with a word or short phrase that serves one or more of the functions listed in Table 6.6.

Table 6.6. Common Phrases Used to Create Linkages and Their Functions

To show contrast	*To show cause and effect*	*To signal time*
Conversely,	Accordingly,	Afterward,
However,	As a consequence,	Initially,
In contrast,	As a result,	More recently,
Nevertheless,	Consequently,	Previously,
On the other hand,	Hence,	Recently,
Unfortunately,	Therefore,	Simultaneously,
To provide additional information	Thus,	Subsequently,
Additionally,	To this end,	To date,
Furthermore,	*To give examples*	Ultimately,
In addition,	For example,	*To refer to something previously stated*
Indeed,	For instance,	
Moreover,	*To add emphasis or clarify*	As mentioned previously,
Namely,	Indeed,	As shown in Figure 1,
To describe a typical case	In particular,	In the latter case,
In general,	More specifically,	In that/these case(s),
Typically,	Specifically,	In this context,
Usually,	Surprisingly,	In this respect,

Chapter Review

As a self-test of what you've learned in this chapter, define each of these terms, in the context of this chapter, for a friend or colleague who is new to the field:

background information	fill-the-gap statement	present perfect
current work	gap statement	review of the literature
direct quotes	parallelism	

Also explain the following to a friend who hasn't yet given much thought to writing an Introduction for a journal article:

- Main purposes of an Introduction
- Moves of an Introduction
- Role of a gap statement in an Introduction
- Types of gaps often identified in an Introduction
- Use of present, past, and present perfect in an Introduction
- Appropriateness of direct quotations and paraphrasing in an Introduction
- Common techniques for making writing more concise and fluid

7 Writing the Abstract and Title

The purpose of the abstract is to inform, to give away the punch line right at the start, and to let your readers decide whether they want to read the full document. Scientific writing is not like mystery writing in which the results are hidden until the end.
—Adapted from Alley (1996)

The title will be read by thousands of people. Perhaps few people, if any, will read the entire paper, but many people will read the title, either in the original journal, in one of the secondary (abstracting and indexing) databases, in a search engine's output, or otherwise. Therefore, all words in the title should be chosen with great care.
—Gastel and Day (2016)

This chapter addresses how to write abstracts and titles for journal articles. Both the abstract and title provide succinct, informative (not descriptive) summaries of the research. To this end, they are usually written in the final stages of the writing process. After completing this chapter, you should be able to do the following:

- Write a concise and informative abstract
- Write a concise and informative title

Writing on Your Own: Abstract and Title

Visit the *Write Like a Chemist* (2nd ed.) website for step-by-step guidance in writing your own abstract and title.

When compared to the Introduction, Methods, Results, and Discussion sections of a journal article, the title and abstract are quite short. The title usually

Write Like a Chemist. Marin S. Robinson and Fredricka L. Stoller, Oxford University Press. © Oxford University Press 2022.
DOI: 10.1093/oso/9780190098940.003.0007

has fewer than 20 words. Journals vary in how many words they allow in the abstract of a research article, but the maximum is typically between 150 words (e.g., *Journal of Agricultural and Food Chemistry*) and 500 words (e.g., *Environmental Science & Technology*). Despite their brevity (and perhaps because of it), the title and abstract are the most widely read sections of the journal article and thus are viewed by many as the most important sections of the journal article.

An Important 200 Words

Titles and abstracts are read by more readers than any other section of the journal article.

The Abstract

Concise and highly informative, the key purpose of the abstract is to inform readers about the principal findings of the work.

Reading and Analyzing Writing

We begin by asking you to read and analyze the title and abstract for the proanthocyanidins-in-pomegranates article (excerpt 7A). The questions in exercise 7.1 will guide your analysis of the excerpt.

Exercise 7.1.

As you read excerpt 7A, answer these questions.

a. Consider the title. Which of the following are included: research topic, importance, gap statement, procedures, instrumentation, results, interpretations, citations, conclusions?
b. The abstract contains nine sentences (numbered for your convenience). What purpose(s) is (are) accomplished in each sentence? Select your answer(s) from the following: describe what was studied, identify the methods used, share results, or offer a conclusion. (More than one purpose may be accomplished in a single sentence.)
c. Based only on the title and abstract, who are the intended audiences for this article (including subdisciplines of chemistry)? Give reasons for your choices.
d. What rules do the authors follow regarding the use of abbreviations in their abstract? What other writing conventions do you notice?

e. What verb tense(s) are used in the abstract?

f. Suggest why the authors include keywords at the end of the abstract. How many of these keywords also appear in the title and abstract?

Abstract Word Count

In this chapter, we conclude each abstract with a word count. We do this for instructional purposes; word counts do not appear in published abstracts.

Excerpt 7A (adapted from Díaz-Mula et al., 2019)

Pomegranate Fruit and Juice (cv. Mollar), Rich in Ellagitannins and Anthocyanins, Also Provide a Wide Range of Proanthocyanidins

[*Note: Sentences 1–9 are numbered for exercise 7.1b.*]

ABSTRACT: (1) Proanthocyanidins (PAs) were characterized in the husk, membranes, arils, and seeds of 'Mollar de Elche' pomegranate cultivar using liquid chromatography–electrospray ionization–ion-trap tandem mass spectrometry after acid catalysis in the presence of phloroglucinol (phloroglucinolysis). (2) PAs were also evaluated in commercial juice and in juice made in the laboratory by pressing fresh arils. (3) Pomegranate PAs were composed of a rich mixture of flavan-3-ols: catechin, epicatechin, gallocatechin, epigallocatechin, epigallocatechin gallate, and epicatechin gallate. (4) Catechins and gallocatechins were the most abundant PAs; galloylated flavanols were detected at lower concentrations. (5) The highest PA content was found in the husk (1840.5 µg/g) followed by seeds (316.2 µg/g) and membranes (51.4 µg/g). (6) Among the pomegranate's edible parts, arils and pressed aril juice showed concentrations of 34.4 and 21.3 mg/L, respectively, lower than those found in other traditional PA-containing foods and beverages (such as apples, grapes, chocolate, red wine, or tea). (7) Higher concentrations (similar to those found in red wine) were observed in commercial whole fruit pressed pomegranate juice (146.9 mg/L). (8) These PA concentrations represented between 0.1 and 7% of the total polyphenol content in the different pomegranate samples, which were richest in ellagitannins and anthocyanins. (9) The mean degree of polymerization (1.2–2.1) revealed that PAs in Mollar pomegranate samples are constituted mainly of monomers and dimers, which may influence the better absorption and bioavailability of these compounds. (*219 words*)

KEYWORDS: *pomegranate, proanthocyanidins, phloroglucinolysis, ellagitannins, anthocyanins*

Analyzing Audience and Purpose

Abstracts and titles are generally written for expert and scientific audiences; however, parts of each are also accessible to a more general audience. For example, the abstract regarding proanthocyanidins (PA) in pomegranate fruit (excerpt 7A) targets food and analytical chemists specifically, but most readers will understand that the article's purpose is to quantify the amounts of PAs in different parts of the fruit and to compare those values to other PA-containing foods and beverages.

The major purpose of the title is to inform readers about the specific content of the work, ideally identifying both what was studied and how it was studied. The major purpose of an abstract is to summarize, in one clear and concise paragraph, the purpose, experimental approach, principal results, and major conclusions of the work. Most journals also require that authors submit a graphic with their abstract. The graphic should be visually appealing and offer a quick overview of the article's content without sharing detailed results. Information about how to prepare the graphic is included in journals' Author Guidelines.

You likely noticed that excerpt 7A concluded with a list of **Keywords**. Some ACS journals ask authors to include a list of Keywords with their abstract, which appear in the PDF version of the article. Other ACS journals ask authors to provide a list of **Subjects**; unlike Keywords, Subjects do not appear in the PDF version of the article but only on the ACS website. Authors are asked to select Keywords or Subjects from a drop-down menu provided by the journal when they submit their manuscript. Both Keywords and Subjects facilitate electronic searches for articles. We have included these words with the abstracts presented in this chapter.

Keywords or Subjects

Searchable words that improve indexing and help readers find your work. Authors are asked to select words and phrases from a drop-down list that pertain to the subject matter of their work. Authors are encouraged to use these words in titles and abstracts.

Analyzing Organization

We consider the organizational structure for both the abstract and title. Because the abstract is generally written before the title, we begin with the abstract.

Organizing Abstracts

The Abstract comprises three moves (Figure 7.1), each one mirroring a section of the journal article: move 1, Introduction; move 2, Methods; and move 3, Results and Discussion. Each move is written to inform, not describe. In abstracts with at least 300 words, all three moves are typically present; however, in abstracts limited to 200 words (e.g., *The Journal of Organic Chemistry*), move 1 is often skipped. In such cases, the importance and purpose of the research are implied rather than stated explicitly.

In the first move, the research area is introduced. The first sentence typically identifies the research area (submove 1.1); as space permits, information about why this research area is important (submove 1.2) and what gaps exist in this area (submove 1.3) can also be shared. Move 2 states both *what* was done and *how* (i.e., the methods or instrumentation used), often in the same sentence. However, if the work focused on a novel technique or instrument, the *what* and *how* can be separated, allowing for more detail about the methods used.

Move 3 highlights the principal findings of the work and is typically the longest segment of the abstract. When appropriate, numerical results should be reported quantitatively along with estimates of error (e.g., "in concentrations of 5.4 ± 0.1 ppb"). Qualitative remarks (e.g., "the yields were good") should be avoided. That said, it is not appropriate to include a table in the abstract or to repeat all results reported in the paper. The abstract ends with an optional conclusion (submove 3.2) that underscores the major findings or impacts of the work.

Because the moves and submoves of the abstract directly parallel moves found elsewhere in the journal article, it is tempting for writers to repeat themselves and copy sentences from the journal article into the abstract. We caution you to resist this temptation. Effective writers avoid such repetition and find ways to reword the purpose and principal findings of their work.

1. Introduce Research Area

1.1 Identify research area
1.2 Describe why research area is important (optional)
1.3 Mention a gap in research area (optional)

2. State What Was Done and How

3. Report Principal Findings

3.1 Highlight major results (quantitatively when appropriate)
3.2 Offer a concluding remark (optional)

Figure 7.1. A visual representation of the move structure for a typical journal article abstract.

Organizing Titles

A document's title is the single most important phrase in a document. . . . In scientific writing, a title engages the audience by identifying the scope of the document. . . . An audience interested in that scope will continue to read.

—Alley (2018)

The title also has an organizational structure. After analyzing more than 300 titles of chemistry journal articles, we found that titles commonly follow an "X of Y by Z" pattern as illustrated in Table 7.1. In essence, X, Y, and Z are three moves linked together by common words (e.g., *of* or *by*). Y describes what was studied; X and Z modify or extend Y in some way. Y is required; X and Z are optional, but typically at least X or Z is present. Of course, this pattern illustrates only a conventional way to construct a journal article title; countless variations are possible.

Table 7.1. The "X of Y by Z" Pattern Common in Chemistry Journal Article Titles

X (*optional*)	(of)	Y	(by)	Z (*optional*)
A phrase that refers to, describes, or modifies Y; or a nominalization (e.g., *Determination, Investigation, Analysis, Measurement*)	of in for to . . .	What was studied	by at in via . . .	A target of Y, what was impacted by Y, or a method used to study Y
Effects of Refolding	on	Large-Scale RNA Structure[a]		
Dual-Signal Readout Nanospheres	for	Rapid Point-of-Care Detection	of	Ebola Virus Glycoprotein[b]
Mechanistic Insight	into	Additions of Allylic Grignard Reagents	to	Carbonyl Compounds[c]
Preparation	of	5-Substituted 1*H*-Tetrazoles	from	Nitriles in Water[d]

[a] Dethoff and Weeks (2019); [b] Hu et al. (2017); [c] Bartolo and Woerpel (2018); [d] Demko and Sharpless (2001).

Exercise 7.2.

Consider these five titles. Four are from ACS journal articles (each from a different ACS journal), and one is the title of a cover story in *Chemical & Engineering News* (*C&EN*). Which one is the *C&EN* title? Using Table 7.1, identify X, Y, and Z in the other four titles.

a. Degradation of Dicloran in Irradiated Water-Sediment Systems (from Vebrosky et al., 2019)

b. It's Time to Recycle Lithium-Ion Batteries (from Jacoby, 2019)
c. Single-Crystal Automated Refinement (SCAR): A Data-Driven Method for Determining Inorganic Structures (from Viswanathan et al., 2019)
d. Infrared Matrix-Isolation and Theoretical Study of the Reactions of Ruthenocene with Ozone (from Kugel and Ault, 2019)
e. Effects of Refolding on Large-Scale RNA Structure (from Dethoff and Weeks, 2019)

Analyzing Abstracts and Titles

In this section, we read and analyze abstracts and titles from journal articles published in several different ACS journals.

Analyzing Abstracts

We begin by considering how moves are accomplished in chemistry journal abstracts. Because abstracts are short, we consider the entire abstract at once (rather than move by move). Recall from Figure 7.1, the abstract comprises three moves (1, 2, and 3) and five submoves (1.1, 1.2, 1.3, 3.1, and 3.2). In the abstracts that follow, we use these numbers, placed inside brackets, to designate a specific move or submove within the abstract.

We first analyze an abstract that includes all three moves and submoves (excerpt 7B). The first sentence accomplishes submoves 1.1 and 1.2; the research area (field-based ecotoxicology studies) is introduced and its importance (to uncover the effects of emerging contaminants on aquatic organisms) is stressed. A gap is mentioned next (submove 1.3), which suggests that one way to study contaminant effects is through large-scale exposures; however, such exposures are rare and problematic. In move 2, the authors go on to state what they did (they evaluated a slow-release drug implant in fish) and how they did it (they measured drug uptake in fish tissues over time). Qualitative results are presented in submove 3.1, indicating that drug levels were highest in fish brain and liver. To conclude (submove 3.2), the authors emphasize that their proposed method offers a promising technique for monitoring pharmaceutical uptake in fish.

Excerpt 7B (adapted from McCallum et al., 2019)

ABSTRACT: [Submoves 1.1, 1.2] Field-based ecotoxicology studies are invaluable for uncovering the effects of contaminants of emerging concern (CECs) on aquatic organisms. [Submove 1.3] However, large-scale exposures are rare due to prohibitive costs, the availability of replicated habitats, and the potential for exposure to cause lasting damage to the environment. [Move 2] Here, we evaluated the viability of internal slow-release implants as

an alternative method for manipulating CEC exposures in aquatic wildlife using two fat-based carriers (coconut oil and vegetable shortening). We treated roach (*Rutilus rutilus*) with implants containing a high (50 µg/g), low (25 µg/g), or control (0 µg/g) concentration of the behavior-modifying pharmaceutical oxazepam. We measured oxazepam uptake in four tissues (plasma, muscle, liver, and the brain) over 1 month. [Submove 3.1] The two carriers released oxazepam differently: coconut oil was the superior implant type because it delivered a more consistent dose over time whereas vegetable shortening released oxazepam rapidly at the start of the exposure period. For both carriers and treatments, the brain and liver contained the most oxazepam. [Submove 3.2] Overall, the method is a promising technique for controlled manipulations of pharmaceuticals in fish, and we have provided some of the first data on the suitability and contaminant release kinetics from different implant types. (*196 words*)
SUBJECTS: *Plant-derived food, Anatomy, Lipids, Implants, Plasma*

The second example (excerpt 7C) includes all moves and submoves except submove 1.3. The research area (cryoconite, a granular aggregate) is introduced (submove 1.1), which is important because of its ability to accumulate atmospheric contaminants (submove 1.2); however, no gaps are identified (submove 1.3). The authors sampled cryoconite in high-glacier areas of China and tested it for mercury using atomic absorption spectroscopy (move 2). The authors report their findings quantitatively (submove 3.1); the total amounts of mercury found in cryoconite from various glaciers are reported numerically (with error terms). Only key findings are reported, not all results. The authors include units and error terms with each set of values (e.g., 66.0 ± 29.3 ng g^{-1}). (Note that units and error terms should *not* be omitted in abstracts to conserve space.) Notice, too, that the units appear only once, after the error term. This is also true if multiple values are listed in a series (e.g., 34.3 ± 17.4 and 0.65 ± 0.28 kg), where "kg" appears only once at the end of the series. The authors conclude by emphasizing the important role glacier cryoconite may play in mercury storage and transformation (submove 3.2).

Excerpt 7C (adapted from Huang et al., 2019)

ABSTRACT: [Submoves 1.1, 1.2] Cryoconite is a granular aggregate, composed of both mineral and biological material, and known to accumulate atmospheric contaminants. [Move 2] In this study, cryoconite was sampled from seven high-elevation glaciers in Western China to investigate the spatial and altitudinal patterns of atmospheric mercury accumulation in the cryoconite. Mercury concentrations were measured using atomic absorption spectroscopy. [Submove 3.1] The results show that total Hg (HgT) concentrations in cryoconite were significant with relatively higher Hg accumulation in the southern glaciers (66.0 ± 29.3 ng g^{-1}), monsoon-influenced regions, than those in the northern glaciers (42.5 ± 20.7 ng g^{-1}), westerlies-influenced regions. The altitudinal profile indicates that HgT concentrations in the northern glaciers decrease significantly with altitude, while those in the southern glaciers generally increase toward higher elevations. Unexpectedly high accumulation of methyl–Hg (MeHg) with an average of 1.0 ± 0.4 ng g^{-1} was also detected in the cryoconite samples, revealing the surface of cryoconite could act as a potential site for Hg methylation in alpine environments. Our preliminary estimate suggests a storage of

~34.3 ± 17.4 and 0.65 ± 0.28 kg of HgT and MeHg from a single year of formation process in the glacier cryoconite. [Submove 3.2] Therefore, glacier cryoconite could play an important role in Hg storage and transformation, which may result in downstream effects on glacier-fed ecosystems in a climate warming scenario. (*220 words*)
SUBJECTS: *Ablation, Melting, Organic reactions, Deposition, Mercury*

Error Terms

Error terms, such as standard deviations and standard errors, should be included in the abstract, when appropriate. Note there is a space before and after the "±" symbol (e.g., 17 ± 9 mL). The units appear only once after the error term.

Exercise 7.3.

Each of these sentences, from abstracts, reports numerical values quantitatively; however, each sentence contains at least one error. Find and correct these errors.

a. Indoor concentrations of both virus-like and bacterium-like particles were ~10^5 particles m^{-3}, and the virus:bacteria ratio was 0.9±0.1 (mean±standard deviation across different locations). (Adapted from Prussin et al., 2015)
b. It was also revealed that H_2O is essential in the electrolyte to achieve an improved cycling efficiency (65% ± 2%). (Adapted from Periyapperuma et al., 2018)
c. A mean unoccupied background concentration of 32 ppb was observed, with further enhancements of ammonia occurring during cooking, cleaning, and occupancy activities, reaching maximum concentrations during these activities of 130 ppb, 1592 ppb, and 99 ppb, respectively. (Adapted from Ampollini et al., 2019)
d. Mean sediment and aqueous neonicotinoid concentrations were 1.19 μg kg^{-1} ± 0.2 μg kg^{-1} and 0.03 ± 0.01 μg L^{-1}, respectively. (Adapted from Kuechle et al., 2019)

The abstract for the article on theoretical alternatives for SF_6 includes all moves and submoves in Figure 7.1. Both the abstract and abstract graphic are shown in excerpt 7D. The first sentence introduces the research area and stresses the need for SF_6 alternatives, given that SF_6 is a potent greenhouse gas (submoves 1.1 and 1.2). The gap statement (submove 1.2) starts near the end of the first sentence ("but without success") and continues into the second sentence, suggesting that finding an alternative superior to SF_6 has been nearly impossible to achieve. Move 2 is a single sentence, integrating what the researchers did (proposed a hybrid model) with how they did it (by full or partial integration of the parent molecules). The graphic serves to elucidate the proposed hybrid model. Qualitative findings are reported (submove 3.1), highlighting the desirable (and paradoxical) properties that the hybrid molecules achieve. The authors conclude by stating that their theoretical predictions can lead to a new design for perfluorinated compounds and thereby promote their synthesis and field testing (submove 3.2).

Excerpt 7D (adapted from Yu et al., 2018)

ABSTRACT: [Submoves 1.1, 1.2] Because SF_6 is the most potent greenhouse gas, the search for a viable alternative has taken on great urgency for several decades [submove 1.3] but without success. The demanding combination of performance, safety, and environmental properties for a compound superior to SF_6 was thought to be nearly impossible to achieve. [Move 2] In contrast to commonly used mixtures of two or three individual gases, we propose a hybrid model to create new perfluorinated compounds with multiple unsaturated chemical bonds by means of full or partial integration of the parent molecules. [Submove 3.1] A unique combination of paradoxical properties—high dielectric strength and stability combined with a low boiling point and a significantly lower global warming potential—is achieved for the first time. [Submove 3.2] Our a priori theoretical predictions shed new light on a rational molecular design of perfluorinated compounds, predictions that will greatly inspire the experimental synthesis and field testing of these compounds and their dielectric properties. (*150 words*)

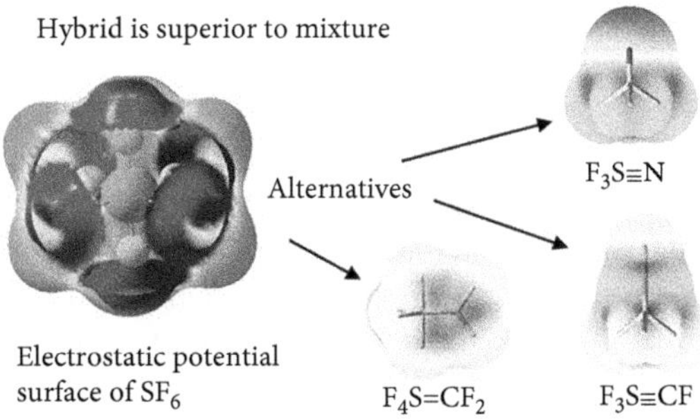

SUBJECTS: *Light absorption, Molecules, Insulators, Transition temperature, Gases*

Exercise 7.4.

Using Figure 7.1 as a guide, assign the sentences in the following abstract (adapted from Cao et al., 2019) to the correct move or submove (1.1, 1.2, 1.3, 2, 3.1, or 3.2). For convenience, we have labeled the sentences a-i in brackets.

ABSTRACT: [a] Heavy metal pollution elicits environmental concern and causes severe health problems worldwide. [b] Mercury is considered to be a ubiquitous pollutant due to its versatile applications in medicine, industry, and cosmetics. [c] Long noncoding RNAs (lncRNAs) are transcripts greater than 200 nt without a protein-encoding function; however, little is known about the mechanism of heavy metal-induced noncoding RNA changes in aquatic organisms. [d] To reveal the epigenetic mechanism of mercury toxicity in zebrafish embryos and explore novel mercury-specific toxicological biomarkers, several well-studied lncRNAs were screened by real-time PCR, and the spatial–temporal expression of the lncRNA biomarker was evaluated by in situ hybridization. [e] The nerve systems of zebrafish embryos were evaluated by detecting locomotor behavior and the expression of neuro-genes. [f] We identified a mercury responsive lncRNA among five candidate lncRNAs: a metastasis-associated lung adenocarcinoma transcript 1 (malat1). $HgCl_2$, MeHg, $PbCl_2$, $CdCl_2$, and K_2CrO_4 exposure assays indicated that malat1 was a mercury-specific induced lncRNA. [g] Malat1 was highly expressed in the brain region, eyes, and notochord of developing zebrafish embryos after exposure to mercury compounds. [h] $HgCl_2$ showed neurobehavior disturbance and changed the neuro-genes expression pattern in zebrafish larvae. [i] This study provides a biological method to detect inorganic or organic mercury using malat1 as a novel biomarker of mercury contamination and offers clues for the exploration of a neurotoxicity mechanism of mercury compounds.

We next consider the moves in three abstracts from *The Journal of Organic Chemistry*. Abstracts in this journal are short (the limit is 200 words); hence, authors often omit optional moves. This journal also requires a graphic. The graphic is often a reaction mechanism or scheme but may also be an illustrative drawing, graph, or spectrum. The graphic must be entirely original and should not duplicate a graphic already in the text. The graphic has no title or caption but may include text for labeling purposes. Long phrases or sentences should be avoided in the graphic.

We first consider the abstract for the article on allylic Grignard reagents (excerpt 7E). In the first sentence, the authors introduce the research area (addition reactions of allylic Grignard reagents) by describing their reactivity in addition reactions to carbonyl compounds (submove 1.1). They do not address why this reaction is important (submove 1.2) or include a gap statement (submove 1.3). Move 2 is accomplished in a single sentence, stating what they did (studied addition reactions of allylic Grignard reagents) and how they did it (with prenylmagnesium chloride). Submove 3.1 is the longest (three sentences), where two possible pathways for the addition reaction are proposed. These pathways are further illustrated in the abstract graphic. The abstract ends without a concluding remark (submove 3.2).

Excerpt 7E (adapted from Bartolo and Woerpel, 2018)

ABSTRACT: [Submove 1.1] Allylic Grignard reagents exhibit high reactivity and low selectivity in addition reactions to carbonyl compounds. [Move 2] Additions of allylic Grignard reagents to carbonyl compounds were investigated using prenylmagnesium chloride as a mechanistic probe. [Submove 3.1] When the carbonyl group is relatively unhindered, the addition proceeds through a six-membered transition state with allylic transposition. This process generally occurs with no diastereoselectivity because the reaction rates approach the diffusion limit. With hindered ketones, however, this pathway is disfavored, and the addition proceeds through a transition state resembling that of other Grignard reagents. (*87 words*)

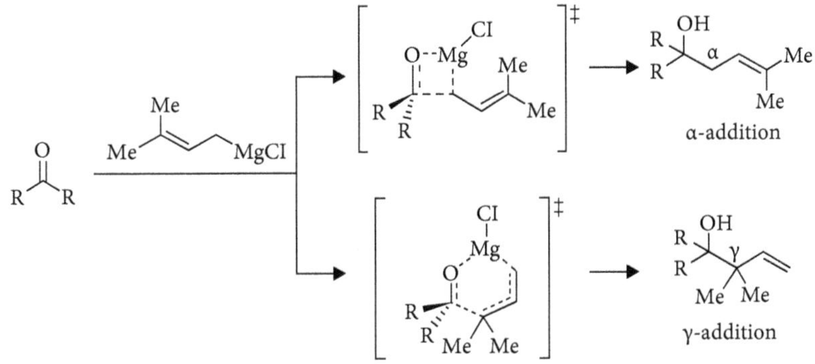

SUBJECTS: *Reagents, Anions, Transition states, Ketones, Carbonyls*

The last two excerpts in this section (7F and 7G) illustrate just how concisely the required moves can be accomplished in an abstract. Both abstracts are from articles that describe a new synthesis. Each abstract has only two sentences: the first sentence highlights the reaction that was studied, introducing both the research area (submove 1.1) and what was done and how (move 2). The last sentence highlights major results (submove 3.1). The accompanying graphics summarize key steps in the proposed syntheses. Note that quantitative information about product yields and/or diastereomeric ratios (dr) are included in both abstracts, either in the graphic (excerpt 7F) or in both the abstract and graphic (excerpt 7G).

Reporting Product Yields

Product yields should be reported quantitatively in abstracts (e.g., 85–90%), not qualitatively (e.g., "The yields were good.").

Excerpt 7F (from Demko and Sharpless, 2001)

ABSTRACT: [Submove 1.1, Move 2] The addition of sodium azide to nitriles to give 1H-tetrazoles is shown to proceed readily in water with zinc salts as catalysts. [Submove 3.1] The scope of the reaction is quite broad; a variety of aromatic nitriles, activated and unactivated alkyl nitriles, substituted vinyl nitriles, thiocyanates, and cyanamides have all been shown to be viable substrates for this reaction. (*57 words*)

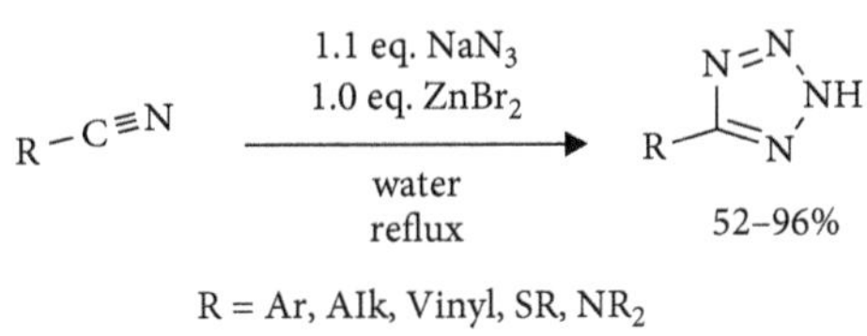

SUBJECTS: *Azides, Zinc, Nitrogen compounds, Solvents, Sodium*

Excerpt 7G (from Xiao et al., 2019)

ABSTRACT: [Submove 1.1, Move 2] A palladium-catalyzed ring-opening oxo-formal [3 + 2]-cycloaddition reaction of novel donor–acceptor spirovinylcyclopropyl oxindole with 3-oxindole is described. [Submove 3.1] The developed protocol provides facile access to oxo-bispirooxindole derivatives in good yields (up to 82% yield) with excellent diastereoselectivities (up to 20:1 dr). (*41 words*)

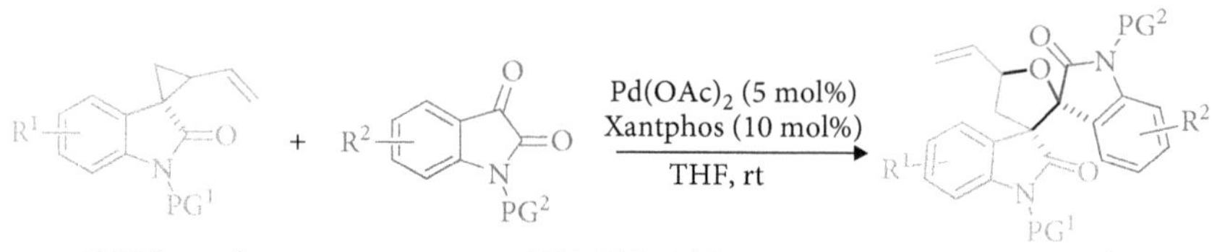

SUBJECTS: *Reaction products, Ligands, Oxindoles, Stereoselectivity, Molecular structure*

Exercise 7.5.

Browse through several abstracts in your favorite chemistry journal. How well do the authors follow the move structure in Figure 7.1? What variations do you notice?

Exercise 7.6.

The following abstract (adapted from Hageman et al., 2006) has been determined to be too long by its authors (225 words). The authors would like to reduce it to fewer than 200 words and, if possible, to 180 words. Rewrite the abstract to make it more concise. (Hint: in addition to some unnecessary words within sentences, there are two entire sentences that can be deleted.)

ABSTRACT: The U.S. National Park Service conducted a very impressive study of the atmospheric deposition and fate of semi-volatile organic compounds in the alpine, sub-Arctic, and Arctic ecosystems of the Western U.S. Results for the analyses of pesticides in seasonal snowpack samples collected in spring 2003 from seven national parks are presented herein, in the Results section of this paper. From a target analyte list of 47 pesticides and degradation products, the most frequently detected current-use pesticides were dacthal, chlorpyrifos, endosulfan, and γ-hexachlorocyclohexane, whereas the most frequently detected historic-use pesticides were dieldrin, α-hexachloro-cyclohexane, chlordane, and hexachlorobenzene. The results of several tests that we performed are described in this paper to help to explain what sources are responsible for the pesticide results that we observed. Correlation analysis with latitude, temperature, elevation, particulate matter, and two indicators of regional pesticide use reveals that regional current and historic agricultural practices are largely responsible for the distribution of pesticides in the national parks studied. Pesticide deposition in the Alaskan parks is attributed to long-range transport because there are no significant regional pesticide sources. The percentage of total pesticide concentration due to regional transport (%RT) was calculated for the other parks; %RT was highest at parks with higher regional cropland intensity and for pesticides with lower vapor pressures and shorter half lives in air. These results will have many important implications. (*225 words*)

Before moving onto titles, we briefly consider the use of citations and abbreviations in abstracts. If you glance back at excerpts 7A–7G, you will see that none included citations to the literature. Indeed, authors are typically cautioned against referring to others' works in the abstract. In most cases, abstracts avoid mentioning others' works entirely; hence, citations are not necessary. A few authors hint at others' works, but in a general way that does not necessarily require a citation. For example, in excerpt 7B, the authors allude to others' field-based ecotoxicology studies, and in excerpt 7D, the authors contrast their hybrid approach with others' use of gas mixtures. These other works will be cited in the actual article, but they are not cited in the abstract.

Cite in an Abstract?

In general, most journals prefer that you not cite others' works in the abstract.

Alternatively, abbreviations and acronyms are commonly used in abstracts for terms that are repeated in the abstract. As in the journal article, the abbreviated term is placed in parentheses following its first spelled-out occurrence in the abstract. Consider P1, a passage from the abstract for the article on the rapid detection of Ebola virus. The authors introduce five abbreviations (EBOV, NPs, RNs, Ab, and SA); each abbreviation is then used on its own in the rest of the abstract (not shown). Note that all abbreviations defined in the abstract must be defined again in the main body of the journal article.

P1 We developed a sensitive and quantitative lateral flow assay for detection of Ebola virus (EBOV) glycoprotein by adhering Au nanoparticles (NPs) onto red fluorescent nanospheres (RNs) forming RNs@Au. . . . Antibody (Ab) and streptavidin (SA) were simultaneously modified onto the RNs@Au to label the target and act as a signal enhancer. (From Hu et al., 2017)

Abbreviations for most instrumental techniques are also defined in abstracts. For example, abbreviations for Fourier transform infrared spectroscopy and thermogravimteric analysis are defined in P2; the abbreviated terms are then used in the rest of the abstract (not shown). Common abbreviations (e.g., NMR and GC) are sometimes defined and sometimes used without definition in an abstract, depending on the journal.

P2 This work correlates traditional wet-chemistry data with those from three rapid instrumental techniques, Fourier transform infrared (FTIR) spectroscopy, conventional thermogravimteric analysis (TGA), and high-resolution TGA (MaxRes TGA), to predict the lignin content of diets and feces from digestibility trials. (Adapted from Lyons et al., 2018)

Abbreviated Terms in Abstracts

Abbreviations and acronyms should be defined in an abstract at first use only if the abbreviated term is used again in the abstract. These abbreviated terms should be introduced and defined inside parentheses again at first use in the text. Common abbreviations (e.g., NMR, IR, and GC) are sometimes defined and sometimes used without definition in an abstract, depending on the journal.

Analyzing Titles

Titles are often the last part of a journal article to be written. The title of a journal article must be as concise, specific, and informative as possible. Also, because the title is written for an expert audience, the title should be formal. Catchy titles, although informal, are appropriate in many genres (e.g., in newspaper headlines and popular science articles), but they are inappropriate in journal articles.

A common organizational pattern for titles (X of Y by Z) was presented in Table 7.1. Following this pattern, titles often begin with a phrase or nominalization (X) that modifies what was studied (Y). Effective titles do not begin with empty words (e.g., "The," "A," "An") or redundant phrases (e.g., "A Study of"; "Research on"); such redundancy will make your title wordy, without adding relevant information.

Wordy The Preparation of 5-Substituted 1*H*-Tetrazoles . . .
A Study of the Cancer-Protective Properties of . . .
An Analysis of Aldehydes in Beer Using . . .

Better Preparation of 5-Substituted 1*H*-Tetrazoles . . .
Cancer-Protective Properties of . . .
Analysis of Aldehydes in Beer Using . . .

Most titles are written in title case, where all main words (nouns, pronouns, verbs, adjectives, and adverbs) are capitalized but not prepositions, articles (*a*, *an*, *the*), or any of these words: *and*, *but*, *nor*, *or*, *so*. In some journals, however, titles are written in sentence case, where only the first word of the title and proper nouns are capitalized. Because of its prevalence, title case is used throughout this chapter.

Two- and three-word modifiers (often requiring the use of hyphens) are common in titles because they can make the title more concise. The hyphenated words may be used in the X, Y, and/or Z parts of the title. Three titles with hyphenated modifiers are shown here. (Note that none of these titles begins with the word *The*.)

- *Single-Particle Tracking* with *Scattering-Based Optical Microscopy* (from Ye et al., 2019)
- Atmospheric Deposition of *Current-Use* and *Historic-Use Pesticides* in Snow at National Parks in the Western United States (from Hageman et al., 2006)
- Synthesis of Arabinoxylan Oligosaccharides by *Preactivation-Based Iterative Glycosylations* (from Underlin et al., 2019)

Two- and Three-Word Modifiers

See Chapter 16, "Language Tips," Part 5: Punctuation, for tips on hyphenated two- and three-word modifiers.

Colons are also common in titles. If you use a colon, be sure that the segment of the title that precedes the colon can stand alone as the full title. The segment after the colon may or may not stand alone. It is customary to capitalize the first word after the colon. Here are three examples:

- Triplet Decay Dynamics in Sulfur-Substituted Thymine: How Position of Substitution Matters (from Manae and Hazra, 2019)
- Identifying Fraudulent Natural Products: A Perspective on the Application of Carbon-14 Analysis (from Gershon et al., 2019)
- Developing a Drug Screening Platform: MALDI–Mass Spectrometry Imaging of Paper-Based Cultures (from Tobias et al., 2019)

Colons

See Chapter 16, "Language Tips," Part 5: Punctuation, for tips on colons (and semicolons).

Most journals discourage authors from using abbreviations or acronyms in their titles, especially if the abbreviated terms are not well known. Exceptions are sometimes observed for common abbreviations; however, even for well-known terms, abbreviations are typically avoided. For example, in the following titles, both polycyclic aromatic hydrocarbons (PAHs) and gas chromatography/mass spectrometry (GC/MS) are spelled out:

- What Will Photo-Processing of Large, Ionized Amino-Substituted Polycyclic Aromatic Hydrocarbons Produce in the Interstellar Medium? (from Burner et al., *J. Phys. Chem. A*, 2019)
- Comprehensive Monosaccharide Composition Analysis of Insoluble Polysaccharides by Permethylation to Produce Methyl Alditol Derivatives for Gas Chromatography/Mass Spectrometry (from Black et al., *Anal. Chem.*, 2019)

Abbreviated terms can also be introduced in titles, as shown in these examples:

- Organic Hydroxy Acids as Highly Oxygenated Molecular (HOM) Tracers for Aged Isoprene Aerosol (from Jaoui et al., 2019)
- Derivatives for Application in a Dye-Sensitized Solar Cell (DSSC): A Theoretical Study (from Portillo-Cortez et al., 2019)

Abbreviated Terms in Titles

With a few exceptions (most notably NMR and DNA), abbreviated terms should be avoided in titles. See Chapter 16, "Language Tips," Part 3: Writing Conventions, for additional tips on abbreviations and acronyms.

Exercise 7.7.

Using only the information provided, propose titles for these journal articles. Use the X of Y by Z pattern (Table 7.1), whenever possible.

a. A paper that reports the antioxidant activity of the phenolic fraction in extra virgin olive oil (EVOO) using electrochemical methods
b. A paper that describes a novel method used to synthesize quinolines by incorporating allenyl cations in a catalytic intermolecular Friedel–Crafts reaction
c. A paper that describes the results of an inhalation study using Sprague–Dawley rats to investigate the toxicity of low levels of 1,3-butadiene
d. A study of how fermentation and distillation affect the oxygen-18/oxygen-16 isotope ratio in ethanol.

Perfecting Your Writing

Verb tense and voice in abstracts are generally consistent with conventions used in other sections of the journal article. Common verb tense–voice combinations and their functions are presented in Table 7.2.

Exercise 7.8.

Consult Table 7.2 as you complete these tasks.

a. Review passages P1 and P2. Which of these passages includes the word *we*? When *we* is included, what verb tense–voice combination is used immediately after *we*?

b. Authors use various verb tense–voice combinations to highlight results (submove 3.1). Review excerpts 7B–7E to find at least one example of each of these verb tense–voice combinations: past–active, past–passive, present–active, and present–passive.

Table 7.2. Common Verb Tense–Voice Combinations for Key Functions in the Abstract of a Journal Article (other combinations are possible)

function	tense-voice combination	example
a. to identify research area	present–active	Cryoconite *is* a granular aggregate . . .
		Allylic Grignard reagents *exhibit* high . . .
b. to stress importance	present–active	Field-based studies *are* invaluable for . . .
		MicroRNAs *play* a significant role in . . .
c. to mention a gap	present–active	However, large-scale exposures *are* quite rare.
d. to state what was done and how	past–passive	Laser ablation and ESI-MS *were used* . . .
		Reactions *were investigated* using . . .
	past–active	*We developed* a sensitive and quantitative . . .
	(with *we*)	Here, *we evaluated* the viability of a . . .
e. to highlight results	past–active	The lake sample *contained* the most . . .
		Concentrations *decreased* in the . . .
	past–passive	The highest content (112 μg/g) *was found* in . . .
	present–active	Preliminary estimates *suggest* that . . .
	present–passive	This mechanism *is favored* over . . .
f. to conclude	present–active	This method *offers* a promising technique to . . .
		These predictions *shed* new light on . . .

Finalizing on Your Own: Journal Article

Visit the *Write Like a Chemist* (2nd ed.) website for tips on finalizing your journal article, now that you have written your title, abstract, Introduction, Methods, Results, and Discussion.

Chapter Review

As a self-test, check what you've learned in this chapter by explaining the purpose and characteristics of each of these terms, in the context of this chapter, to a friend who is new to the field:

abstract	Keywords	word count
abstract graphic	quantitative results	X of Y by Z pattern
error terms	Subjects	

Also explain the following to a friend in chemistry who has not yet given much thought to the final steps of drafting a journal article, specifically writing an abstract and composing a title:

- Moves of an abstract
- Verb tense–voice combinations in an abstract
- Use of capitalization in titles
- Use of abbreviated terms in titles

Module 2

Conference Abstract and Scientific Poster

8 *Writing the Conference Abstract and Title*

All too familiar words from research mentor to student: "I'd like you to submit an abstract to the upcoming ACS conference on your research. The submission deadline is in two weeks. Unfortunately, I'll be out of town, but I'm sure you can write it on your own."

This chapter focuses on writing a conference abstract, not to be confused with a journal article abstract (addressed in Chapter 7). A conference abstract is written in response to a Call for Abstracts issued by conference organizers, typically 4–6 months before the conference takes place. Abstracts are usually submitted to a specific division of chemistry for review (e.g., Analytical Chemistry, Environmental Chemistry, Organic Chemistry). The abstract is reviewed and, if accepted, the authors are able to present their work (as an oral or poster presentation) at the conference. Undergraduates typically present posters, and at large conferences (such as national American Chemical Society [ACS] conferences) there is an Undergraduate Research Poster Session in each major division of chemistry. The goal of this chapter is to guide you in preparing an abstract for this type of poster session. By the end of this chapter, you will be able to do the following:

- Describe the major purposes of the conference abstract
- Write your conference abstract and title following typical organizational patterns
- Use common writing conventions in your conference abstract and title

Writing on Your Own: Conference Abstract and Title

Visit the *Write Like a Chemist* (2nd ed.) website for step-by-step guidance in writing your own conference abstract and title.

Write Like a Chemist. Marin S. Robinson and Fredricka L. Stoller, Oxford University Press. © Oxford University Press 2022.
DOI: 10.1093/oso/9780190098940.003.0008

The conference abstract is often the first professional genre that novice writers tackle on their own with little to no guidance from their research mentors. This chapter is intended to help you with this challenge by focusing on the text of the abstract. (If your abstract is accepted and results in a poster presentation, see Chapter 9 for guidance in preparing the poster.)

Like the journal article abstract, most conference abstracts have strict word limitations (typically 200 to 300 words). The word limit for national ACS meetings is 300 words (including figures and tables). Adherence to this limit is expected; your abstract will likely be rejected if the word limit is exceeded. Many other formatting specifications must be followed. For example, most instructions include rules for formatting your title, author list, graphics, and special characters (e.g., µm, α, $\leq$, m^3). The reason for such specificity is that your abstract, if accepted, will be published (as is) in the conference's program or **conference proceedings**. Adherence to common guidelines by all authors ensures that the proceedings appear uniform and professional. To achieve these standards, your abstract will likely require multiple revisions and (just like Snoopy's novel) "a little editing" (Figure 8.1).

Figure 8.1. Snoopy realizes that his first draft is going to need a little editing!
@Copyright 2021 Andrews McMeel Universal.

Conference Proceedings

The published program for a conference or symposium that includes written abstracts for all oral and poster presentations.

Reading and Analyzing Writing

Rather than including an actual abstract for you to read and analyze, we include instead *Anatomy of an Awesome Abstract* (excerpt 8A). This infographic was developed by writers at *inChemistry*, an ACS student member magazine, to help undergraduates write an abstract for an ACS conference presentation. Complete exercise 8.1 as you read the infographic.

Excerpt 8A (Image credit: *inChemistry* magazine, ©2019 American Chemical Society)

Anatomy of an Awesome Abstract

A professional abstract summarizes the scope of your research and shows why your study is worth learning more about.

HPLC analysis of contaminants on baby pacifiers

Dawn Young[1], Joey Fresh[1], Grey Oldman[1]

1. The Pennsylvania State University, Department of Chemistry

Studies have shown that a large fraction of baby pacifiers have caffeine on them, especially those of first-born children. Because the caffeine is believed to originate with the parents, it is hypothesized that parents of multiple children might seek out pain relief in addition to stimulants. However, the most common analgesic—acetylsalicylic acid (aspirin)—can trigger Reye's syndrome in infants, leading to brain and liver damage. To determine the amount of contamination, pacifiers were collected from a local day care center. Organic contaminants were removed from the pacifiers by ethanol extraction. The resulting solutions were analyzed by HPLC. More than 90% of the pacifiers were found to be contaminated with acetylsalicylic acid. The complete results for all pacifiers will be presented.

Abstract title

* Write a short, descriptive title in sentence case.
* Capitalize only the first word, acronyms, and proper nouns.

Authors and affiliations
(you, faculty advisor(s), co-workers(s))

* List the names and affiliations of everyone who contributed to the work.
* Be sure the lead author is the person who conducted the study or will be presenting the work (usually you).
* The affiliation should be the complete name of your school, without any abbreviations.

Abstract body
Write a concise, high-level summary of 2000 characters or shorter. Use the third person and passive voice, and be sure to correct any grammar or spelling errors.

Key elements to include are as follows:

* Importance of your research
* Problem your work attempts to solve
* Approaches or methods used
* Overview, not details, of results

This abstract is for educational purposes and is not based on a real study

ACS Chemistry for Life

in chemistry
The ACS Student Member Magazine

Exercise 8.1.

Complete these tasks based on the abstract in excerpt 8A.

a. What are the two main purposes of the abstract? Describe how those goals are achieved in the abstract.

b. Using the title in excerpt 8A as a guide, rewrite the following title so that it is in sentence case, does not begin with "The," does not end with a period, and has no more than 8 words.

   The Use of ICP-MS to Analyze Chromium Concentrations in Drinking Water.

c. The infographic comments on the use of passive voice, as does the ACS website, which states, "It is best to stick to the passive voice, even if it makes your English teacher cringe." Give three examples of passive voice in the abstract.

d. Based on both the abstract body and the four key elements listed in the sidebar, propose a move structure (organizational structure) for the conference abstract.

Analyzing Audience and Purpose

It is important to make a good first impression when submitting your conference abstract. Follow guidelines exactly, or you may risk being eliminated from the start.

—Hofmann (2020)

A conference abstract is written for two audiences. Conference organizers represent the first audience; they read the abstract and decide whether to accept it (or reject it) for a conference presentation. Several factors influence this decision, not the least of which is the quality of writing in the abstract. A poorly written abstract often leads to a poorly prepared conference presentation; hence, such abstracts are easy for reviewers to dismiss. Other factors include the appropriateness of the topic (does it fit the theme of the conference session?), the quality of the science, and the stage of the work (is it far enough along to be ready to present?). Conferences are forums for presenting new, even late-breaking, results. However, an abstract may be rejected if the work is too new or appears to lack sufficient data to ensure a quality presentation by the time of the conference.

Alternatively, an abstract should not be submitted if the content is too old. For example, an abstract describing work that is already published should not be submitted. Similarly, abstracts describing results that have already been presented at other national conferences should not be submitted, unless significant new progress has been made. Some overlap with previous presentations is allowed (after all, new science builds on past accomplishments), but a good rule of thumb is that at least 75% of the content should be new. (Note: This rule is often relaxed if the conference is local, university-based, or student-based.)

The second audience comprises conference attendees, who read abstracts to determine whose talk to attend or poster to view. Because national conferences typically have multiple concurrent sessions, attendees rely on abstracts to make their decisions about what sessions to attend and when. Individuals who read your abstract are likely to be in a related field of chemistry, but may not be in your specific area of chemistry. Thus, it is important to keep your abstract general enough to be readily understood across different areas of science, thereby targeting primarily a scientific audience. In this regard, the conference abstract is more similar to the Introduction section of a journal article (Chapter 6) than to the abstract of a journal article, the latter written for an expert audience.

Analyzing Organization

Conference Abstract

Here we consider how a typical conference abstract is organized. If you completed exercise 8.1 (item d), you likely came up with a four-step move structure, based on the four key elements outlined in excerpt 8A. These elements are specified for ACS conference abstracts, but they are also true for conference abstracts in general. Thus, we adapted these elements to create the move structure shown in Figure 8.2 for a typical conference abstract. The first move, in two submoves, identifies the research topic and addresses the importance of the topic, paralleling the first two submoves in the Introduction of a journal article (Figure 6.1 in Chapter 6). Unlike the journal article Introduction, however, the conference abstract omits citations to others' work. Recall, the conference abstract is designed to tell readers about *your* work and to help them decide if they should visit *your* poster; summarizing others' works is of little value in this regard. Moreover, if you include a citation, you also need to include the reference; hence, citations and references are exceedingly rare in conference abstracts.

Move 2 also parallels move 2 of the journal article Introduction: identifying a gap in the field. In a conference abstract, a gap statement points out limitations or obstacles that have been encountered in the past and thereby identifies the problem

1. Introduce the Research Area
1.1 Identify the topic
1.2 Highlight the importance of the research

2. Identify a Gap in the Field or Problem to Be Solved

3. Preview the Work to Be Presented
(i.e., the methods used and the results obtained or to be obtained)

Figure 8.2. A suggested move structure for a typical conference abstract.

(or problems) that your work attempts to solve (a key element in the "awesome abstract" in excerpt 8A). Move 3 previews the work to be presented and typically includes both methods and results (paralleling the last two elements in excerpt 8A). Numerical results can be reported (e.g., in excerpt 8A, 90% of the pacifiers were found to be contaminated), but be careful not to provide too much detail. More than likely, you are still completing parts of the work when writing your abstract; if this is the case, state what results will be presented at the conference (e.g., "Arsenic levels from two different lakes will be reported.").

Conference Abstract Title

The title for a conference abstract follows the same organizational structure (X of Y by Z pattern) as the title of a journal article (see Table 7.1 in Chapter 7). Note, too, the formatting requirements for the ACS abstract title: the title should be in sentence case, with no period at the end. The title should not begin with "The" or "A" or "An". A complete list of authors (not just the presenting author) and their affiliations are also submitted with the conference abstract. Typically, first and last names of all authors are included, and the author giving the presentation (either oral or poster) is often listed first. Of course, these guidelines vary for different conferences, so be sure to check the instructions for your specific conference.

Analyzing Conference Abstracts and Titles

We now analyze three conference abstracts (with titles) published in the programs of the Spring 2019 and Fall 2019 National Meetings & Expos of the American Chemical Society. The abstracts are from three ACS divisions: Division of Organic Chemistry (ORGN), Division of Analytical Chemistry (ANYL), and Division of Inorganic Chemistry (INOR). Although abstracts in other areas of chemistry or for other conferences will vary slightly, these three examples should serve as useful templates for your own writing.

We first examine how the three moves shown in Figure 8.2 are accomplished in these conference abstracts. To facilitate analysis, we consider the entire abstract at once. Using brackets, we identify moves 1, 2, and 3 in each abstract and include the total word count at the end. In this way, you can see for yourself how much emphasis the authors place on each move. The word limit for each of these abstracts was 300 words (~2000 characters), and they vary in length between 73 and 291 words.

In the first abstract (excerpt 8B), move 1 is accomplished in a single sentence. In just a few words, the topic is identified (solubility of electrolytes in nonaqueous redox flow batteries) and its importance is stressed (solubility is critical to energy density). The second and third sentences (connected with a semicolon) suggest a gap in previous efforts (move 2). The last two sentences are devoted to move 3. The authors describe their methods (predictive quantitative structure-property relationship [QSPR] models combined with molecular engineering) and highlight their achievement (stable cycling in an unprecedented high-concentration flow cell). With only 73 words, excerpt 8B serves as a useful reminder that abstracts need not be lengthy to be accepted. Moreover, the title of the abstract is informative and adheres to the X of Y by Z pattern:

X	(of)	Y	(by)	Z
Development	of	highly soluble cyclopropenium salts	leading to	a highly concentrated non-aqueous flow battery

Excerpt 8B (adapted from Yan et al., 2019)

ORGN 157: Development of highly soluble cyclopropenium salts leading to a highly concentrated non-aqueous flow battery

Abstract: [Move 1] Solubility of electrolytes in all redox states is one of the critical properties determining energy density in nonaqueous redox flow batteries. [Move 2] Cyclopropenium salts were previously reported as a promising catholyte for redox flow batteries; however, these salts suffer from low solubility at oxidized states. [Move 3] Herein, predictive QSPR models and molecular engineering are combined to identify highly soluble cyclopropenium salts. The most soluble cyclopropenium salt evinces stable cycling in an unprecedented high-concentration flow cell. (*73 words*)

In contrast to excerpt 8B, the authors of except 8C use two paragraphs for move 1 and a third paragraph for moves 2 and 3. In the first paragraph, the research area is introduced (mycotoxins and their detection); in the second paragraph, a promising method for detection is introduced (SERS nanoaptasensors). The third and final paragraph begins with a gap statement (move 2), listing challenges others have faced when using SERS sensors. The paragraph ends with move 3, describing how nanoaptasensors can be used successfully to discriminate mycotoxin content at the ppb level. Once again, the title is informative and follows the X of Y by Z pattern:

X	(of)	Y	(by)	Z
Accessible mycotoxin identification	by	label-free SERS nanoaptasensors	in	solution

Excerpt 8C (adapted from Galarreta et al., 2019)

ANYL 353: Accessible mycotoxin identification by label-free SERS nanoaptasensors in solution

Abstract: [Move 1] Mycotoxins are secondary metabolites produced by different species of fungi. Mycotoxins colonize and contaminate several varieties of food, either in the field or after harvest; hence, they have a huge economic impact globally. Because mycotoxins are small (MW around 700) and have a variety of structures, it is impossible to use one standard technique for their analysis and/or detection. Additional challenges to routine analysis include the need for specialized laboratory facilities and highly sensitive detection methods.

[Move 1] Aptamer-based biosensors, when combined with gold nanoparticle optical transducers, can provide simple, rapid, and reliable detection methods that can be easily integrated into portable and accessible optical devices. One attractive optical device involves label-free surface-enhanced Raman scattering (SERS). SERS signals from aptamors adsorbed on gold nanoparticle surfaces can be enhanced by several orders of magnitude due to the surface-enhanced effects of the gold nanostructures. The enhanced signal can then be used in the selective determination of mycotoxins such as ochratoxin A (OTA) and aflatoxin B1 (AFB1) without requiring the presence of a Raman dye.

[Move 2] Label-free SERS sensors for mycotoxin detection have received little attention. The studies that do exist still need to overcome accessibility issues, as they rely on expensive equipment, benchtop spectrometers, and a clean-room facility to prepare the substrates. [Move 3] Herein, we introduced a liquid-phase gold nanoaptasensors system based on gold nanotriangles for label-free determination of OTA and AFB1 by SERS in a simple and accessible manner. Batch-to-batch reproducibility was confirmed by testing the performance of seven independently prepared sensors. Multivariate analysis methods were used to process the Raman spectra and to selectively discriminate mycotoxin content down to the ppb level. Finally, the solvent effect was evaluated to corroborate its influence on the sensitivity and reproducibility of the assay when dealing with real samples. (*291 words*)

In the third example (excerpt 8D), move 1 is accomplished in two sentences, introducing the research area (iron oxide nanocrystals and their applications) and the importance of controlling the oxidation state of iron in these nanocrystals. Move 2 is accomplished in a single sentence, suggesting that although oxidative pathways are known, ways to reduce γ-Fe_2O_3 to Fe_3O_4 are rare. In move 3, the authors fill this gap by using UV irradiation to photoreduce the nanocrystals. The effects of photoreduction on the crystal and the size and shape of the crystals will be presented at the conference. The title follows a reduced X of Y by Z pattern (without Z):

X	(of)	Y	(by)	Z
Photoreduction	of	ion oxide nanocrystals		

Excerpt 8D (from Jung et al., 2019)

INOR 364: Photoreduction of iron oxide nanocrystals

Abstract: [Move 1] Magnetite (Fe_3O_4) and maghemite (γ-Fe_2O_3) nanocrystals have been extensively studied for decades due to their superparamagnetic behavior and potential applications in magnetic storage devices, catalysis, and medical diagnosis. Because the ratio of Fe^{3+}/Fe^{2+} determines the phase and magnetic properties of iron oxide nanocrystals, it is important to control the oxidation state of Fe. [Move 2] Although Fe_3O_4 can be slowly oxidized to γ-Fe_2O_3 in air or more rapidly by applying heat or adding oxidants, the reduction of γ-Fe_2O_3 to Fe_3O_4 is relatively rare. [Move 3] We report the reduction of colloidal γ-Fe_2O_3 nanocrystals by UV irradiation in the presence of a sacrificial reductant. Photoreduction is followed using UV/vis/NIR electronic absorption spectroscopy in combination with superconducting quantum interference device magnetometry. X-ray diffraction and transmission electron microscopy are used to monitor the crystal structure and the size/shape of the nanocrystals, respectively. The effects of photoreduction on the crystal and electronic structures of the iron oxide nanocrystals will be discussed. (*153 words*)

Abbreviations in Conference Abstracts

Abbreviations that are widely known by the targeted audience (e.g., UV–vis and NIR) can be used alone, without being defined at first use; however, abbreviations that are not well known should be defined at first use. Note, too, the inclusion of the noun *spectroscopy* (in excerpt 8D) following UV–vis and NIR.

Exercise 8.2.

Read and analyze the following abstract and title from the Division of Medicinal Chemistry (MEDI) and complete these tasks.

a. Consider the key elements that should be included in an "awesome abstract," according to excerpt 8A. Briefly explain how the abstract achieves each of these elements.

b. Indicate where submoves 1.1 and 1.2, move 2, and move 3 begin in the abstract.

c. Identify X, Y, and Z in the abstract title.

MEDI 171: Strategies for targeting aberrant microRNA activity in cancer (adapted from Garner, 2019)

Abstract: MicroRNAs (miRNA) are an emerging class of small RNAs that play critical roles in human development and disease. These micromanagers function at the level of post-transcriptional gene regulation, and alterations of miRNA expression levels have been implicated in many human diseases. In cancer, miRNAs can have both oncogenic and tumor suppressive functions, indicating that multiple strategies will be necessary to target miRNAs in cancer for therapeutic development. Using our laboratory's innovative high-throughput screening approach, catalytic enzyme-linked click chemistry assay or cat-ELCCA, we have developed platforms for assaying RNA-small molecule and RNA-protein interactions toward our goal of developing small molecule-based modulators of miRNA biology. Our progress toward the goals of selectively targeting Dicer-mediated pre-miRNA processing and miRNA-miRNA-binding protein interactions with small molecules and natural products will be discussed. (*127 words*)

Although excerpts 8B–8D all follow the move structure in Figure 8.2, deviations from this move structure are sometimes observed. One variation is to omit the gap statement (move 2); another variation is to end the abstract with implications or applications of the work, similar to the conclusion of a journal article. A graphic (e.g., a figure or illustration) is sometimes included in a conferences abstract, although it is not required. If you do include a graphic, be sure to follow the instructions for how to prepare and submit it. Note, too, the word limit will likely be reduced to make room for the graphic.

Exercise 8.3.

Passages (a)–(f) are from conference abstracts. All but one of the passages includes moves depicted in Figure 8.2. Identify the move (or moves) that are present in each passage. Also, identify the alternative move and its purpose.

a. Relatively high levels of DEHP have been found frequently in samples of vegetables and fruits, for example, 899 ng/g in rutabagas, 713 ng/g in peas, 676 ng/g in mushrooms, 675 ng/g in cucumbers, 657 ng/g in lettuce, 600 ng/g in apples, 383 ng/g in bananas, 324 ng/g in cherries, etc. However, the sources of DEHP in the vegetable and fruit samples are not clear. (Adapted from Cao, 2019)
b. Phthalates have become ubiquitous environmental contaminants due to their widespread applications in polyvinyl chloride (PVC) products (as plasticisers) and in consumer and personal care products. (From Cao, 2019)
c. Here, we find that perovskite nanocrystals (NCs) of $APbBr_3$ are exceptional candidates as photocatalysts. Under visible light illumination, readily prepared $APbBr_3$ colloids (*ca.* 2–100 nm) effectively photocatalyze fundamental C–C, C–O, and C–N bond-formation reactions. (Adapted from Zhu, 2019)
d. Anguidine, and other related trichothecenes, have been actively pursued due to their varied biological activities that depend on the respective functional groups on the tricyclic backbone. Anguidine itself has been shown to inhibit the initiation of protein synthesis resulting in the death of rapidly proliferating cells such as tumors. (From Zhao and Manetsch, 2019)
e. From lubricants to agrochemicals, ether linkages are important motifs in chemical entities throughout all chemical sectors. The most common method to form ethers is via the Williamson method, a simple and reliable method utilizing a metal hydride and an alkylating agent. However, safety issues associated with the reagents and the large amount of inorganic waste products render this methodology undesirable for large scale applications. (Adapted from Whitaker et al., 2019)

f. These first steps in unraveling the nature of organic compounds on Mars and the multiple mechanisms responsible for their degradation will pave the way for a more direct search for extinct or extant life on this planet. (Adapted from Mahaffy, 2019)

Perfecting Your Writing

Many of the writing conventions used in conference abstracts are those used in other chemistry genres. Important conventions are summarized here.

- **Abbreviations and acronyms:** Abbreviated terms can be introduced, in parentheses, following the first use of multiword terms that occur more than once in the abstract (e.g., SERS in excerpt 8C). Abbreviated terms may be used without definition if they are not considered essential for understanding the work (e.g., the predictive QSPR model in excerpt 8B).
- **Personal pronouns:** The word *we* often appears in conference abstracts, particularly in move 3 (e.g., "*We* have used" and "*We* will present"). In abstracts written by a single author, *I* is used.
- **Verb tense–voice combinations:** Verb tense–voice combinations commonly used in conference abstracts are summarized in Table 8.1.

Exercise 8.4.

Re-read excerpts 8B–8D and answer these questions. Consult Table 8.1, if needed.

a. What verb tense and voice are used in the first sentence of each abstract? Explain why excerpt 8D is different from excerpts 8B and 8C.
b. What verb tense and voice are used in the last sentence of each abstract? Explain each choice.
c. How often is *we* used in these excerpts? What move(s) does it appear in?

Finalizing on Your Own: Conference Abstract

Visit the *Write Like a Chemist* (2nd ed.) website for tips on finalizing your written work, now that you have completed your conference abstract and its title.

Table 8.1. Common Verb Tense–Voice Combinations for Key Functions in Conference Abstracts (other combinations are possible)

function	tense–voice combination	example
a. to introduce research area and establish importance	present–active	Mycotoxins *colonize* and *contaminate* several varieties of food globally.
	prt pf[a]–passive	Fine particulate matter *has been studied* extensively due to its ability to . . .
b. to identify a gap or problem to be solved	present–active	However, these reagents *suffer* from . . .
		Studies still *need* to overcome . . .
	prt pf[a]–active	Such sensors *have received* little attention.
c. to preview work to be presented	present–active	The most effective assay *demonstrates* . . .
		We *report* . . .
	present–passive	Predictive QSPR models and molecular engineering *are combined* to . . .
	past–active	We *introduced* a liquid-phase . . .
	past–passive	Multivariate analysis methods *were used* . . .
	future–active	We *will begin* with a study of . . .
	future–passive	The consequences of . . . *will be presented*.

[a] prt pf = present perfect

Chapter Review

Check your understanding of what you've learned in this chapter by defining each of these terms, in the context of this chapter, for a friend or colleague who is new to the field:

abstract title
author list
conference abstract
conference proceedings

Also, answer these questions for a friend who has just been encouraged to submit a conference abstract:

- What is the purpose of a conference abstract?
- What audience(s) should be addressed in a conference abstract?
- How are conference abstracts and journal article Introductions similar? How are they different? What are the similarities and differences between conference and journal article abstracts?
- What are the typical moves of a conference abstract? What are some common variations of these moves?
- What are common writing conventions regarding the use of verb tense, voice, and personal pronouns in a conference abstract?

9 *Preparing a Poster*

At chemistry conferences, poster sessions are nearly always more engaging than talks. As proof, consider how few people fall asleep while viewing a poster! This is because the poster, by its very nature, is interactive, inviting people in to talk with you about your science.

Congratulations! If you are reading this chapter, you are likely preparing a poster for a scientific conference. This means that your conference abstract (Chapter 8) was accepted and you now need to prepare a poster. In this chapter, we focus on organizing and formatting your poster and writing each of the poster sections. By the end of this chapter, you will be able to do the following:

- Format your poster
- Write the major sections of your poster
- Compose a title and acknowledgments for your poster
- Compile and format references appropriately for your poster

Preparing Your Own Poster

Visit the *Write Like a Chemist* (2nd ed.) website for step-by-step guidance in preparing your own poster.

Reading and Analyzing Writing

We begin by asking you to read and analyze excerpt 9A: *Anatomy of an Ace Research Poster*. This infographic was created by writers at *inChemistry*, the ACS student member magazine, to help undergraduates prepare their first ACS conference posters. Although the *Ace* poster specifically targets ACS conferences, the guidance

Write Like a Chemist. Marin S. Robinson and Fredricka L. Stoller, Oxford University Press. © Oxford University Press 2022.
DOI: 10.1093/oso/9780190098940.003.0009

it provides is applicable for many other scientific conferences. Exercise 9.1 guides you in the analysis of the poster.

Exercise 9.1.

Browse through the *Ace* poster in excerpt 9A and complete these tasks:

a. Identify the major sections of the poster. What headings are used to signal these sections? What types of information are included (and excluded) in each section?
b. Comment on the balance between text and graphics in the poster.
c. What audience appears to be targeted in this poster?
d. What sections, typically in a journal article, are missing from the poster? Suggest a reason for their omission.

Excerpt 9A (Image credit: *inChemistry* magazine, ©2019 American Chemical Society)

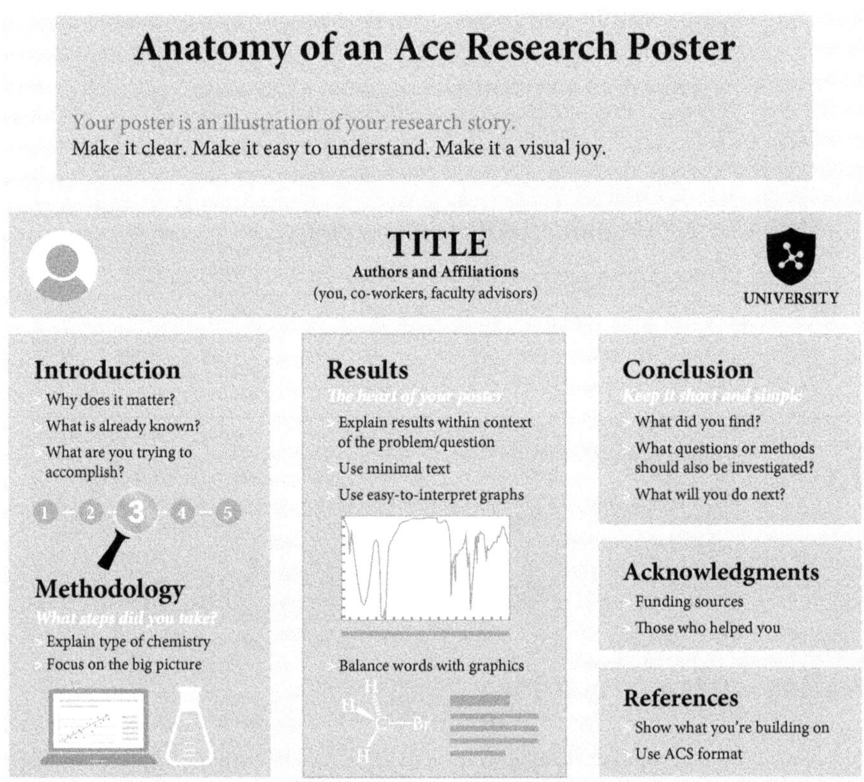

Analyzing Audience and Purpose

To benefit most from a poster presentation, you should consider it primarily an opportunity for an exchange of ideas and dialogue, rather than merely a forum for data presentation.

—Anholt (2006)

There is no one right audience to target in a poster; rather, the correct audience will vary with the conference. Some conferences are primarily for experts; others, such as the national meeting of the American Chemical Society, attract a wide range of individuals, from high school teachers and undergraduate students to research professors and Nobel Laureates. Sometimes the official name of the conference suggests the predominant audience. For example, a conference named *Colloidal, Macromolecular, and Polyelectrolyte Solutions* is likely for experts, whereas *The National Conference on Undergraduate Research* is clearly for university students. Most conferences send poster guidelines to presenters; information about audience is often addressed in those guidelines.

Poster Audience

The audience for a poster depends on the conference. Some conferences are more specialized than others. In general, however, it is best to assume that your audience knows little about your topic.

If, after reviewing the conference name and poster guidelines, you are still unsure about the correct audience for your poster, it is best to target a scientific audience (e.g., graduate students, faculty, and professionals who are in different disciplines or subfields of chemistry). Don't assume that most viewers will be knowledgeable about your research area; more than likely, only a few will have such specialized knowledge. Thus, prepare your poster for your most probable audience: chemists and chemistry students with a research emphasis different from your own. The infographic presented in excerpt 9A is designed for such an audience. If viewers want more detailed information, they can ask you questions. Your thoughtful answers to their questions will satisfy their need for more information. Indeed, the best part of poster sessions is the exchange of ideas that takes place when passersby stop and ask you questions about your work.

Analyzing Poster Organization

Many people believe scientists do not need to learn writing skills. I would argue that scientists need these skills more than most professionals. Concise, informative, flowing, and intelligent verse can make the difference between a funded and rejected grant, a published and rejected scientific publication, or a well-received and botched poster. Writing skills are not only encouraged, they are mandatory for success in all scientific fields.

—Bradley F. Schwartz, Southern Illinois University School of Medicine

As shown in excerpt 9A, a poster is organized only slightly differently than a journal article. Both posters and journal articles have an Introduction, Methods, and Results section; however, unlike the journal article, most posters do not have a Discussion section. Instead of a Discussion section, the poster has a short section entitled Conclusions (or Conclusion). The last two sections, Acknowledgments and References, are not required but are included in most posters. Note, too, that the poster in excerpt 9A has no abstract. Abstracts are omitted because of space limitations and the need to minimize unnecessary duplication. (Recall, the abstract is commonly published in conference proceedings, as explained in Chapter 8.) However, some conferences do recommend that posters include abstracts, so check conference guidelines before preparing your poster.

Analyzing Poster Format

Posters must be formatted according to conference guidelines (which state poster dimensions, font size, font style, etc.). Although guidelines vary, typical formatting conventions for a poster are highlighted in Figure 9.1. Note that poster sections are organized in columns (rather than rows). Columns make it easier for multiple viewers, standing side by side, to read your poster at once. The viewer reads one column, reaches the bottom, and moves over to the top of the next column. Poster sections can be continued into the next column as needed (e.g., in Figure 9.1, the Methods section begins in column 1 and ends in column 2). The title, author list, and authors' affiliations go across the top of the poster and can be left justified or centered. They are centered in Figure 9.1.

Formatting Your Poster

Visit the *Write Like a Chemist* (2nd ed.) website for a full-color and downloadable version of Figure 9.1 in PowerPoint. It can be used as a template for formatting your poster.

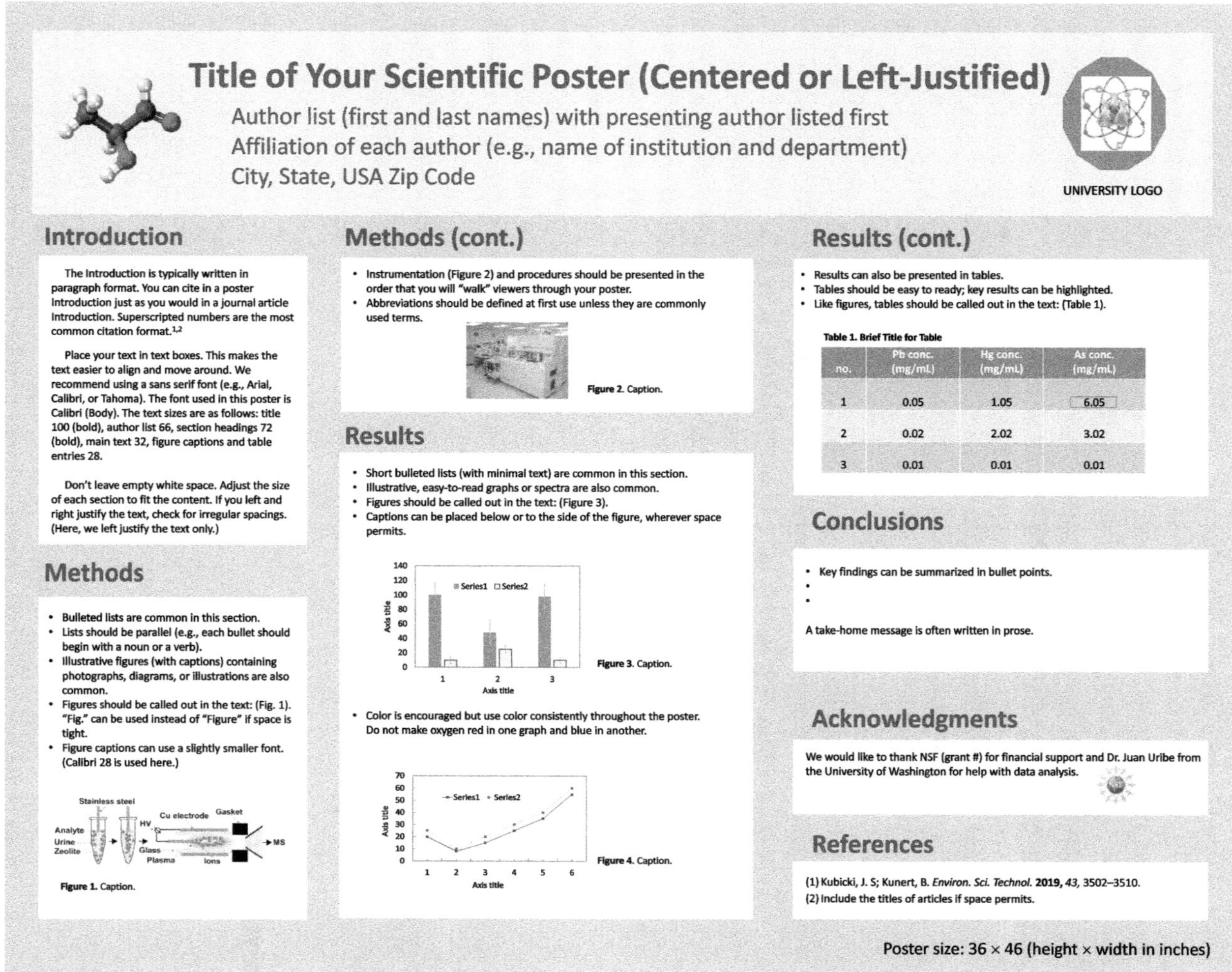

Figure 9.1. A generic poster template.

The dimensions of your poster will be limited by the physical size of the poster board that is available at your conference; hence, conference specifications must be followed. Unless stated otherwise, it is best to assume that the poster board includes a frame, so your poster must be smaller than the specifications given (e.g., for an ACS conference, the poster must be smaller than 4 ft × 6 ft). Remember, too, that the printed poster will have margins; the size of those margins is determined by the size of your plotter and paper. Hence, unless you want to trim the margins, you will need to adjust the size of your poster for this space. Common poster dimensions (without margins) are 3 × 4, 3 × 5, or 4 × 6 (height × width in feet). The template poster in Figure 9.1 is scaled on the small side (36 in × 46 in) so that it can be printed on most university printers.

After you have determined the size of your poster, you can begin laying out the various sections. In Figure 9.1, we have used solid colors, without patterns or designs, for the poster background and section boxes. Although once popular, background patterns or designs make the text difficult to read. We have also placed the text within each section in text boxes, making the text easy to align and move around. The text size should be large enough to be read from a distance of 2 to 3 feet. Sans-serif fonts are preferred (e.g., Arial, Calibri, Tahoma). The font used in Figure 9.1 is Calibri (Body). The font sizes are as follows: 100 for the title (bolded), 66 for the author list and affiliations, 72 for the headings (bolded), 32 for the major text and table entries, and 28 for figure captions. The text can be left-justified or left- and right-justified, but if you use the latter, be sure to manually correct for any irregular spacings that may result. (We use only left-justified text in Figure 9.1.) Color schemes should be used cohesively throughout the poster (e.g., various shades of blues and yellows); avoid distracting colors or combinations of reds and greens, as these colors can be difficult for color-blind individuals to discriminate. In graphs, use colors consistently, always using the same color to represent the same sample or compound.

Analyzing Excerpts

We are now ready to read and analyze excerpts representing each major section of the poster (Introduction, Methods, Results, and Conclusions). This order reflects how you will walk viewers through your poster, not necessarily the order in which you will write these sections. (Most authors begin posters with sections that contain graphics, usually the Methods and Results sections.) In contrast to journal articles, posters are not published; hence, it is not possible to consider authentic poster excerpts. Instead, for the purposes of this chapter, we have created our own "model" excerpts by adapting text from the article on imazalil distribution in apples (Pereira et al., 2019) introduced in Module 1. For each major poster section, we suggest a move structure, provide excerpt(s) to illustrate that move structure, and point out general writing practices for that section. We also briefly consider examples of poster Acknowledgments, References, and Titles. We conclude the chapter with a complete poster, created from excerpts in the chapter and formatted following the template provided in Figure 9.1.

Walking Viewers through a Poster

At most conferences, there is a block of time (usually 2 h) when you are expected to present your poster. During this time, you have the opportunity to "walk" viewers through your poster and engage in dialogue with them.

Poster Introduction

The major difference between sections in a poster and those in a research article is that poster sections concentrate only on the main points and present these sections briefly and visually.

—Hofmann (2020)

The *Ace* poster in excerpt 9A suggests that a poster Introduction should answer three questions about your research: (1) Why does it matter? (2) What is already known? and (3) What are you trying to accomplish? The move structure in Figure 9.2 addresses these questions in three corresponding moves. Move 1 introduces the research area and establishes its importance (i.e., why it matters). Move 2 provides background information (i.e., what is already known); this move often includes citations to the literature, but if citations are not included, you should still have a few citations in mind (in case someone asks). Move 3 previews accomplishments, thereby giving viewers their first glimpse at what the authors are trying to accomplish. Move 3 is often signaled with a subheading (e.g., **Research**

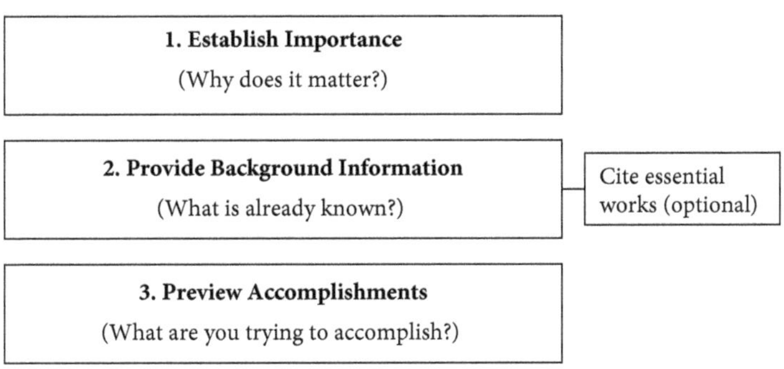

Figure 9.2. A suggested move structure for a poster Introduction section.

Goals or **Objectives**) to help viewers locate a concise description of the work's accomplishments.

Examples of these three moves (noted in brackets) are shown in excerpt 9B. We composed these two paragraphs (115 words) from eight paragraphs (865 words) in the Introduction of the original article. In general, poster Introductions should have fewer than 150 words. Keeping your Introduction short is important because most viewers will spend only a few minutes at your entire poster; hence, you must learn to present only the "short story" version of your work. To help you with this task, we include examples of what we deleted from the original article as we analyze excerpt 9B.

Conciseness

For the poster Introduction, a good rule of thumb is to stay under 150 words.

Let's begin with move 1 in excerpt 9B. The importance of the research is established in a single sentence: the use of pesticides on fruits is necessary to increase yield, but pesticide levels must be monitored to minimize human exposure. The journal article goes on to describe how pesticides are used on vegetables, how pesticides increase storage lifetime, and what the environmental impacts of pesticides are, all topics omitted from the poster. The poster includes only one citation, whereas eight articles were cited in the journal article.

In move 2, both the poster and journal article provide background information on imazalil, the fungicide investigated in the study. The poster addresses the fungicide's possible genotoxic and teratogenic effects; the journal article also mentions reproductive and neurobehavioral concerns. The poster and journal article both describe how farmers apply imazalil to fruits and in what concentrations, as corresponding procedures are followed in the study. Included in the journal article but omitted from the poster are details on how imazalil controls and targets specific fungi; information on imazalil's high polarity, which may explain its fast diffusion rate, is also omitted. In five additional paragraphs, the journal article also provides information on previous methods used to study pesticides in fruits and ways to dissect plant tissue. This information was omitted entirely from the poster. In all, the journal article cited 46 articles in move 2; the poster cites only two.

The authors begin move 3 (preview accomplishments) with a bolded subheading (**Goal**), which is followed by a description of how imazalil was monitored. Here MALDI, IR, and ESI–MS are mentioned for the first time; hence, these terms are defined. (These key terms were defined earlier in the Introduction of the journal article.) Neither the journal article nor the poster cites the literature in this move.

Excerpt 9B (a poster Introduction based on Pereira et al., 2019; see full poster in Figure 9.6)

Introduction

[Move 1] Although pesticides can benefit fruits by increasing yield and preventing damage from pests or fungi, pesticide levels must be monitored to minimize human exposure.[1] [Move 2] A commonly used pesticide is imazalil, a fungicide with possible genotoxic and teratogenic effects.[2] Imazalil is commonly applied to fruits postharvest by dipping, drenching, or spraying in concentrations of 1000 and 2000 ppm.[3]

Goal. [Move 3] The distribution of imazalil in apples was monitored post application. Using matrix-assisted laser desorption ionization (MALDI) imaging, infrared (IR) laser ablation, and offline electrospray ionization–mass spectrometry (ESI–MS), we monitored both the distribution and concentration of imazalil in apples for 7 days after application. Imazalil was detected at 6 mm inside the apple after 7 days. (*115 words*)

Excerpts 9C and 9D offer two additional examples of poster Introductions, with details drawn from journal articles. As you read these excerpts, complete exercise 9.2.

Exercise 9.2.

Read excerpts 9C and 9D. Explain how each excerpt addresses the three questions an *Ace* research poster should answer in its Introduction:

- Why does this research matter?
- What is already known?
- What are the authors trying to accomplish?

Excerpt 9C (a poster Introduction based on Llompart et al., 2001)

Introduction

The pollutants known as polychlorinated biphenyls (PCBs) are widely distributed in the environment due to their extensive use in the past, lipophilic (fat-loving) character, and general chemical stability. Moreover, PCBs tend to accumulate in the food chain. Thus, several countries have established recommended maximum limits for PCBs in food products. Among these products, milk is especially important because of its extensive and widespread consumption by humans.

Objective. To date, there is no simple or rapid procedure for testing PCBs in milk. Headspace solid-phase microextraction (HSSPME) offers a promising approach. Here, we present a simple and rapid saponification-HSSPME procedure for extracting PCBs from full-fat and skim milk. (*106 words*)

Excerpt 9D (a poster Introduction based on Vesely et al., 2003)

Introduction

American packaged beers lose their flavor and become stale during storage. Carbonyl compounds, particularly aldehydes, may be involved in this process. Aldehydes are formed in beer through one of several possible mechanisms including Strecker degradation of amino acids or linoleic acid.[1] In most cases, their concentrations are very low. As a result, it has been difficult for brewing chemists to monitor their levels in beer and thereby deduce their role in affecting beer flavor.

Research Objective. Using a newly optimized derivatization and solid-phase microextraction (SPME) process, we monitored low-level aldehyde concentrations in beer stored for 4, 8, and 12 weeks at 30 °C. Although all aldehydes increased in concentration, no aldehyde exceeded its flavor threshold. (*115 words*)

Exercise 9.3.

Glance again at excerpts 9C and 9D. How well do the Introductions adhere to the move structure in Figure 9.2? What generalizations can you make about sentence types, citation format, and verb tense–voice combinations?

Perfecting Your Writing in Poster Introduction Sections

Here we highlight some writing conventions that are characteristic of the poster Introduction section:

- **Bulleted or numbered lists:** Lists are uncommon in poster Introductions. More often, the Introduction is written in paragraph form with the text left-only or right-and-left justified.
- **Citations:** In-text citations (typically fewer than four) are often included in posters, most commonly in the Introduction. To conserve space, the superscripted number format is used most often. When citations are used, a References section must also be included.
- **Verb tense–voice combinations:** A variety of tense-voice combinations can be used in poster Introductions. Several possible combinations are shown in Table 9.1.
- **Personal pronouns:** *We* can be used in the Introduction to signal the current work (e.g., Here, *we* present a . . .).

Poster Methods Section

As shown in the *Ace* poster in excerpt 9A, the poster Methods section should emphasize *the big picture* and answer the question: *what steps did you take?* We translate this question into a single move: highlight essential steps (Figure 9.3). Most often, essential steps include an overview of materials, instrumentation, and/or procedures, but these items will vary from poster to poster. Avoid complicated descriptions, which the average viewer cannot possibly comprehend in the few minutes that they will spend

Table 9.1. Common Verb Tense–Voice Combinations for Key Functions in Posters (other combinations are possible)

function (by section)	tense–voice combination	example
Introduction		
a. to establish importance	present–active	Aldehydes *are* important for their . . .
		Cr^{60} *induces* DNA single-strand breaks.
b. to provide background information	present–active	Fungicides *have* possible genotoxic effects.
	present–passive	Imazalil *is* commonly *applied* by . . .
	prt pf[a]–active	Many countries *have established* . . .
c. to preview accomplishments	present–active	Here, we *present* an efficient . . .
	past–active	We *monitored* low-level ketone . . .
	past–passive	Concentrations *were monitored* using . . .
Methods		
d. to highlight essential steps	past–passive	The fiber *was placed* in the headspace.
		Samples *were ablated* . . .
Results		
e. to share results	past–active	Initial experiments *confirmed* . . .
	past–passive	High sensitivity *was achieved*.
Conclusions		
f. to summarize key results	past–passive	Imazalil *was detected* at 3 mm . . .
	past–active	We *found* that zebra mussels accumulate . . .
g. to state take-home message	present–active	This work *demonstrates* the utility of . . .
		The amide **1** *is* an excellent . . .
h. to describe ongoing or future work	present–active	Work *is* underway to . . .
		In the future, we *plan* to . . .

[a] prt pf = present perfect

at your poster. If viewers want additional information, they can ask for it. Whatever steps you decide to highlight, be sure to arrange them in a logical order (from the viewer's perspective), adhering to the sequence that you will follow when you "walk"

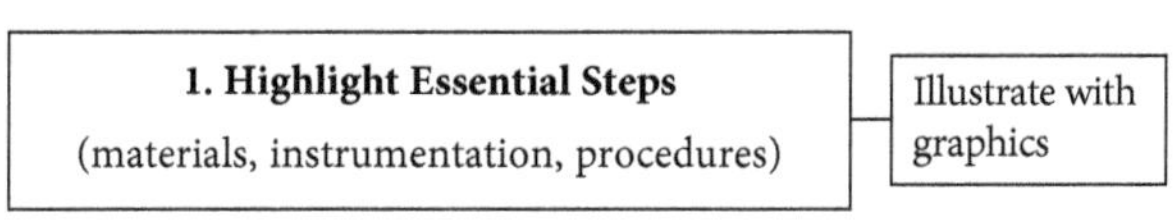

Figure 9.3. A suggested move structure for a poster Methods section.

individuals through your poster. Moreover, we strongly encourage you to use figures and illustrations in this section, to help viewers better visualize *your* methodology.

With this move in mind, let's examine the Methods section in our model poster regarding imazalil treatment of apples (excerpt 9E). In the journal article, the Methods section comprises five subsections: Chemicals and Materials, Sample Preparation, MALDI Imaging, Laser Ablation Sampling, and ESI–MS. In the poster, we condense this information into two subsections: Preparation and Analysis. The Preparation subsection uses a bulleted list to describe how the apples were prepared for analysis. We chose to use full sentences in the bulleted list (each sentence ending with a period), but fragments (without periods) can also be used. As with all lists, parallel language is strongly encouraged because it is easier for viewers to read and comprehend quickly.

We also created an illustration (Figure 1) to accompany the bulleted list in the Preparation subsection of excerpt 9E. The presenter can point to this figure, rather than the text, as procedures are described to viewers. The use of pictures, photographs, and other computer-generated graphics is strongly encouraged in posters. Graphics are useful both for helping viewers visualize concepts quickly and for adding color and visual appeal. Two additional images are included in the Analysis subsection of excerpt 9E (Figures 2 and 3). Both images are from the original journal article.

Excerpt 9E (a poster Methods section based on Pereira et al., 2019; see full poster in Figure 9.6)

Methods

Preparation. Test apples (12) and control apples (4) were prepared as follows (Figure 1):

- On Day 0, test apples were submerged in imazalil (20 min, 1000 ppm) and stored.
- On Days 0, 1, 4, and 7, three test apples and one control apple were sliced (skin to core) and mounted on glass slides.

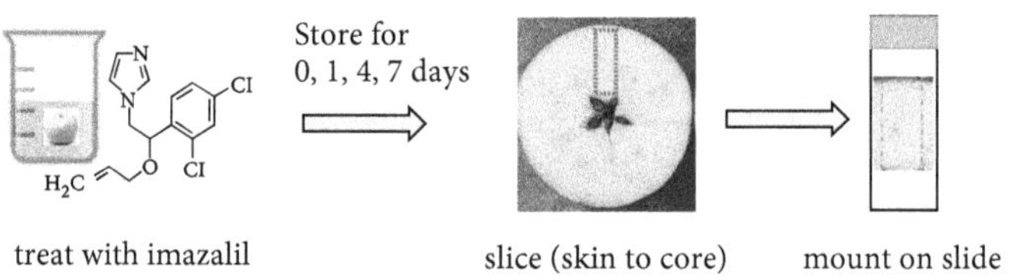

Figure 1. Preparation of apples for analysis.

Analysis. Apples were analyzed by MALDI imaging and ESI–MS (Figure 2):

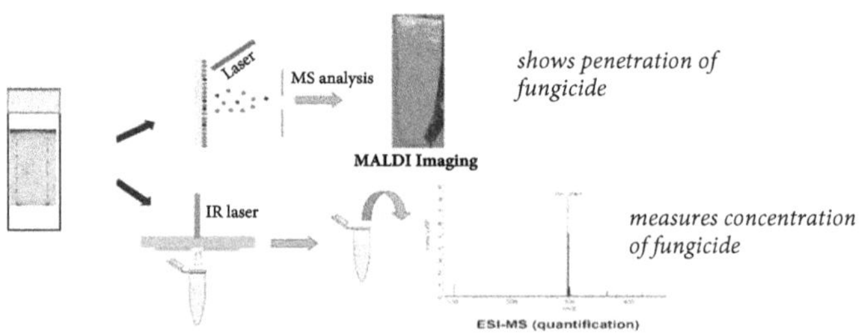

Figure 2. (Top) MALDI analysis and (bottom) ESI–MS quantification.

- **MALDI imaging.** Spectra were recorded using a tandem time-of-flight mass spectrometer (UltrafleXtreme, Bruker). Images were reconstructed using FlexImaging software (Bruker). Protonated imazalil was used to create images.
- **IR Laser Ablation/ESI–MS.** Apples were ablated from skin (0 mm) to core (7 mm) (Figure 3). Ablated material was collected, dried, and resuspended in acetonitrile for ESI–MS analysis (Agilent 6230 TOF–MS). Protonated imazalil (*m/z* 297.0561) was used for quantification. Concentrations were determined by comparison to a seven-point calibration curve.

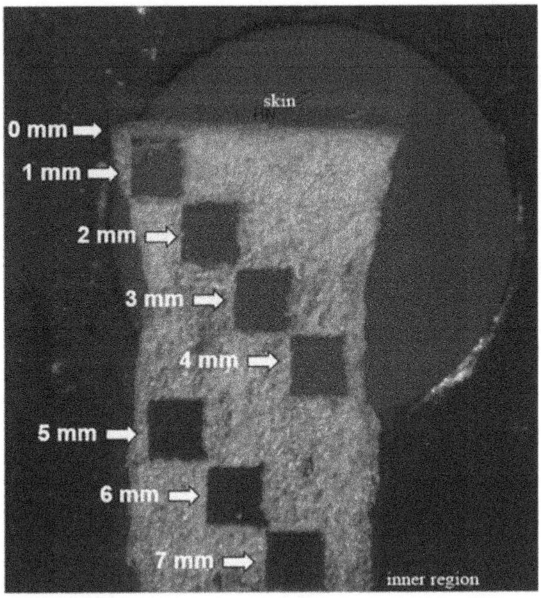

Figure 3. Ablated regions of apple from skin (0 mm) to inner region (7 mm).

Exercise 9.4.

The following bulleted list describes a synthesis for a poster Methods section. Each individual item in the list is appropriate for a poster Methods section; however, the language is not parallel across all four items.

a. Rewrite the list to make the language of all items parallel to the first item.
b. Rewrite the list to make the language of all items parallel to the second item.
c. What are advantages of each approach? Which list style do you prefer?

- A stirred suspension of **1** (400 mmol) in H_2O (400 mL) was added to pivaldehyde **2** (419 mmol) and stirred for 30 min.
- added NaCN/HOAc at 23–28 °C, heated the mixture to 70 °C, stirred for 24 h

- cooled to 30 °C, filtered, washed (500 mL H_2O), dried, analyzed by ^{1}H NMR
- (*R*,*S*)-**3** was formed as a colorless solid (92.4%, dr > 99/1).

Bulleted Lists

Start full sentences with a capital letter and end them with a period. Start sentence fragments with either a capital or lowercase letter; a period is not needed. If possible, use one list style consistently within a poster.

Perfecting Your Writing in Poster Methods Sections

Here we highlight some writing conventions that are characteristic of the poster Methods section:

- **Abbreviations:** Abbreviations should be defined in parentheses at first use if space permits, for example, "pentafluorobenzene (PFB)". However, abbreviations may be used without definition if an exact identity is not essential for understanding the poster. Common abbreviations may be used without definition (e.g., GC–MS, IR, aq); **unconventional abbreviations** that are easily understood may also be used if space is tight (e.g., Fig., temp, rt, &, wk). However, we caution against using texting abbreviations in your poster (e.g., B4 or B/C) as these abbreviations are not appropriate for a scientific audience. Also, use abbreviations consistently throughout the poster. For example, "eq" should not be used for "equation" and "equivalent" in the same poster.
- **Bulleted or numerical lists:** Lists are common in poster Methods sections. When possible, use the same types of lists (bulleted or numbered) throughout your poster.
- **Parallelism:** Parallel language should be used in bulleted and numerical lists in poster Methods sections. Not only is parallelism grammatically correct, but its repetitive pattern also helps viewers comprehend poster information more quickly. Pay particular attention to parallel subheadings and lists:

Nonparallel list	1. American lager samples were stored . . . 2. Load PFBOA on fiber . . . 3. Concentration of aldehydes on . . .
Parallel list	1. American lager samples were stored . . . 2. PFBOA was loaded on fiber . . . 3. Aldehydes were concentrated on . . .

Nonparallel subheadings	Chemicals Extracting the Filters Analyze using GC
Parallel subheadings	Chemicals Extraction Procedure GC Analysis

- **Numbers and units:** Conventional scientific units should be used in poster Methods sections (mL, µm, mol, M, etc.); however, some formatting conventions may be relaxed. For example, the conventional space between a number and its unit may be omitted in a poster, and the numerical form of a number may be used instead of its word form (even at the start of a sentence). If space allows, however, follow conventional practices. One convention that should never be relaxed is the use of leading zeros for numbers <1 (e.g., use 0.35 not .35).

Conventional formatting	2 µL, nine aldehydes, 30 m × 0.25 mm
Relaxed formatting	2µL, 9 aldehydes, 30m × 0.25mm

- **Verb tense–voice combinations:** Poster Methods sections are written predominantly in past tense–passive voice because they describe work done in the past and focus on the science rather than the scientist (see Table 9.1).
- **Personal pronouns:** In Poster Methods sections, *we* is typically avoided.

Unconventional Abbreviations

Some unconventional abbreviations are allowed in posters (e.g., Fig., temp); however, avoid abbreviations used in text messages (e.g., B4, B/C, THX).

Exercise 9.5.

Consider excerpts (a) and (b) taken from Methods sections of journal articles. Convert each procedure into a bulleted list of items that would be appropriate for a poster Methods section. Omit information that would not be included in a poster.

a. **2-(*p*-Toluenesulfonyl)-4′-methoxyacetophenone (2a)**. A mixture of 2-bromo-4′-methoxyacetophenone (45.8 g, 200 mmol) and *p*-toluenesulfinic acid sodium hydrate (35.6 g, 200 mmol) in ethanol (1 L) was heated at reflux for 1.5 h. The mixture was stirred and cooled to room temperature, and the resulting solid was collected, washed with ethanol (2 × 50 mL), and dried to give 54.6 g (90%) of pure **2a**. (Adapted from Swenson et al., 2002)

b. *Delphinium × cultorum* cv. Magic Fountains dark blue/white bee seeds were donated by Bodger Seed, Ltd., South El Monte, CA. The seeds were germinated in high porosity, peat-based growing mix in Styrofoam trays (4 in. × 6 in. × 2 in.) and grown until the first leaf appeared. The seedlings were then transplanted into pots (4 in. × 4 in. × 6 in.) and then into large clay pots (10 in. × 6 in. × 10 in.) and were grown until maturity. The plants were raised in the Bioactive Natural Products Laboratory Greenhouses, Michigan State University. The plants were subjected to a 12 h photoperiod, watered once daily, and maintained at 75 °F. Plant parts were harvested when 75–80% of the florets were fully expanded on each raceme and stored at –20 °C. (From Miles et al., 2000)

Poster Results Section

Like any scientific presentation, the poster should tell a story. . . . *A poster should include only material relevant to the storyline.*
—Anholt (2006)

The Results section is undoubtedly the most important section of the poster. The chance to share your results with others is the predominant reason for presenting your poster. However, you must be careful not to overwhelm your readers by including too many results. Focus on two or three key findings, targeting a reasonably narrow audience. Avoid the temptation to crowd too much information into a single poster. Limit yourself to results that can be digested in just a few minutes. Also, be sure to include only results that have been prepared for public view; "raw" data (e.g., spreadsheet calculations, uncalibrated measurements, original spectra before they have been baseline corrected) are rarely appropriate for a poster.

Most results in posters are presented in graphical form (e.g., figures, graphs, photographs, schemes, illustrations, spectra). Graphics are preferred over text because they allow the reader to comprehend the data more quickly (provided the authors do not overcrowd their graphics). The text accompanying the graphics serves largely to guide viewers through the data. In general, figures are preferred over tables because they are easier to read and understand at a glance; however, tables may be included if they provide useful information and/or illustrate important trends.

As shown in Figure 9.4, the organization of a poster Results section is centered around a single move: tell the research story. Like the journal article Results section, the poster Results section should be organized in a way that effectively conveys the story of scientific discovery. To facilitate this task, move 1 is divided into three submoves. In the first submove, the stage is set by presenting initial results, that is, results that lay the groundwork for your principal findings. For example, you can share a result that motivated your work (such as a spectrum that confirms the presence of a compound of interest) or builds confidence in your approach (such as a calibration curve). As you prepare submove 1.1, imagine

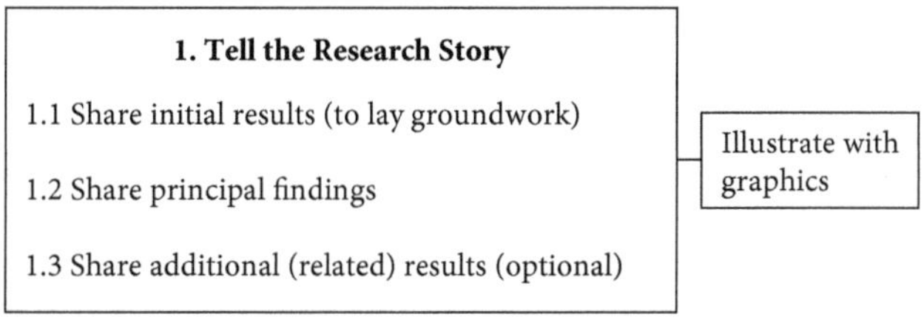

Figure 9.4. A suggested move structure for a poster Results section.

yourself pointing to your poster and saying to your viewers, "First, we wanted to be sure we could do ____." or "In previous work, we successfully prepared ____." or "This spectrum confirmed that ____ is present in our sample." By preparing your viewers for what lies ahead, you strengthen their confidence in themselves (to understand your results) and in you (to present data that they can trust).

After the foundation is laid, you are ready to share principal findings (submove 1.2). As recommended in the *Ace* poster (excerpt 9A), results should be shared in easy-to-interpret graphics with only a minimal amount of text. Labels and horizontal lines, although discouraged in journal articles, may be used in poster graphics to help viewers quickly grasp information. As in journal articles, each figure should have a caption, placed either below or to the side of the figure, whichever is more practical in the available space. Figures should also be called out in the poster text; however, Fig. can be used instead of Figure if space is tight (e.g., Fig. 2). Unlike journal articles, poster figures may also include titles placed above the figure.

Poster Figures

In journal articles, figures have captions but no titles. In posters, figures often use captions, titles, and other descriptive labels to highlight important points. Fig. (instead of Figure) can also be used in the text and figure captions.

Submove 1.3 (share additional results) is an optional submove that allows you to share one last result without overwhelming your viewers. The key is to share a result that is related to your principal findings. Viewers will grasp another set of data more easily if it supports, extends, or strengthens what they have just learned. To summarize, as you prepare your Results section, divide your results into three groups: (1) initial results, (2) principal results, and (3) additional (related) results. By categorizing your data in this way, a logical thread emerges, helping you tell your research story in the most effective way.

With this move structure in mind, let's consider a Results section for our model poster on imazalil treatment of apples (excerpt 9F). We begin with two mass spectra of protonated imazalil (*m/z* 297.06), confirming that imazalil can be detected successfully in both MALDI and ESI mass spectra (Figure 4). These spectra serve as initial results (submove 1.1) and validate the methods used in this work. You can imagine yourself saying to viewers, "First, we wanted to be sure that we could detect imazalil in the apple; fortunately, protonated imazalil was easily detected in both the MALDI and ESI mass spectra."

The next set of results reports principal findings (submove 1.2). In excerpt 9F, Figure 5 shows optical and MALDI images of the apples on Days 0, 1, 4, and 7. Key results are highlighted in the text, indicating that MALDI imaging could detect imazalil up to about 3 mm inside the apples on Day 7. Table 1 reports the concentrations of imazalil on these same days using ESI–MS data. This more sensitive technique was able to detect trace concentrations of imazalil (0.015 ± 0.008 ppm) at depths of 6 mm inside the apples on Day 7. The last set of results shares a related finding (submove 1.3). A bar graph (Figure 6) indicates that the overall extraction efficiency of ESI–MS for imazalil in apples is about 60%. This number suggests that the concentrations in Table 1 are about 60% of their actual values.

Excerpt 9F (a poster Results section based on Pereira et al., 2019; see full poster in Figure 9.6.)

Results

- Initial experiments confirmed that protonated imazalil (*m/z* 297.06) could be detected on apple slices in both MALDI and ESI mass spectra (Figure 4).

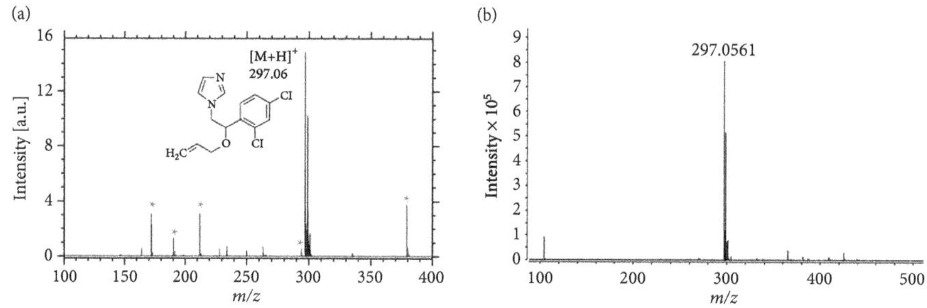

Figure 4. (a) MALDI mass spectrum with matrix peaks marked with asterisks. (b) ESI mass spectrum of protonated imazalil.

- MALDI (MSI) images showed penetration of imazalil into the apples (Figure 5).

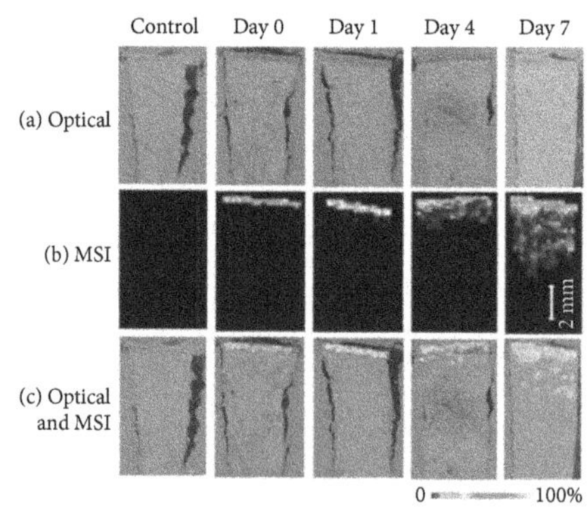

Imazalil distribution observed in apple:

Day 0 – skin only
Day 1 – skin only
Day 4 – 1 mm
Day 7 – 3 mm

Figure 5. Optical and MALDI images of apple on Days 0, 1, 4, and 7 after treatment.

- ESI mass spectra determined imazalil concentrations in the apples (Table 1).

Table 1. Imazalil Concentrations Inside Apple (0 to 7 mm)[a]

Distance from skin (mm)	Concentration (ppm) Day 0	Day 1	Day 4	Day 7
0	2.425 ± 0.105	2.161 ± 0.053	1.748 ± 0.093	1.252 ± 0.026
1	BLQ	0.024 ± 0.003	0.199 ± 0.016	0.415 ± 0.051
2	BLQ	BLQ	0.019 ± 0.002	0.181 ± 0.020
3	BLQ	BLQ	0.015 ± 0.002	0.059 ± 0.003
4	BLQ	BLQ	BLQ	0.022 ± 0.009
5	BLQ	BLQ	BLQ	0.019 ± 0.009
6	BLQ	BLQ	BLQ	0.015 ± 0.008
7	BLQ	BLQ	BLQ	BLQ

[a] BLQ = below the limit of quantification

- Extraction efficiency of imazalil from apples was about 60%; hence, the values in Table 1 underestimate the actual values.

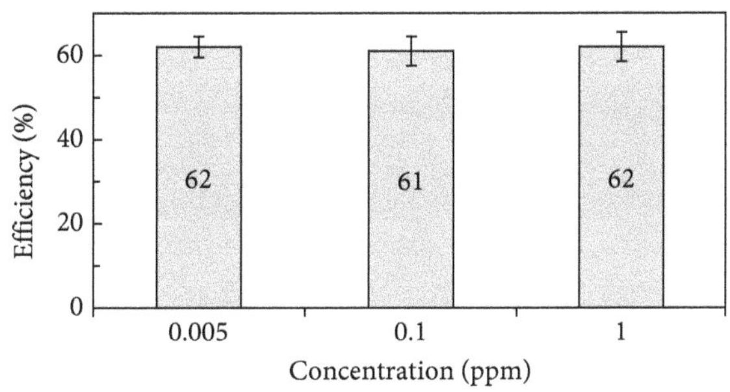

Figure 6. Imazalil extraction efficiencies.

Perfecting Your Writing in Poster Results Sections

Here we highlight some writing conventions that are characteristic of the poster Results section:

- **Abbreviations:** If space is limited, "Fig." may be used; otherwise, use "Figure." "Table" is not generally abbreviated. Other readily understood abbreviations may be used when space is tight (e.g., temp, conc, MeOH).
- **Bulleted or numerical lists:** Lists are frequently used when reporting results. Formatting guidelines are the same as in the poster Methods section. When possible, use consistent formatting throughout your poster.

- **Graphics and graphics labels:** In general, poster graphics follow the same formatting guidelines as journal articles (see Chapter 14). Three differences are that (1) title case may be used for column headings in poster tables; (2) titles may be used in poster figures (along with captions, which are required); and (3) color may be used freely in poster graphics (first to illustrate the science, but also to add visual appeal).

  As in journal articles, poster graphics should be labeled. Table titles are placed above the table. Figure captions are placed either below or beside the figure. "Dangling" graphics, that is, graphics that are never referenced in the text, should be avoided. Exceptions to this rule are graphics added primarily for visual appeal (e.g., a logo, photograph, or clipart). Photographs used to convey science (e.g., a photograph of an instrument or a map of a sampling site), however, should be labeled.
- **Verb tense–voice combinations:** In Poster Results sections, past tense (active and passive voice) are typically used to report results (see Table 9.1).
- **Personal pronouns:** In Poster Results sections, *we* is typically avoided.

Title Case

In posters, title case (e.g., Title Case) may be used for column headings in tables.

Poster Conclusion(s) Section

Many posters skip the Discussion section and go directly to the Conclusions. (Note: This section may be called Conclusion or Conclusions; we use Conclusions.) The Discussion section is omitted because, in posters, authors do not often describe their results in depth, compare their results to the literature, or propose new theories or mechanisms (interpretive functions of Discussion sections, as described in Chapter 5). The main emphasis in posters is to present, not interpret, results. This emphasis reflects both the newness of the data presented (late-breaking results may not yet be fully understood) and the interactive role of the poster (to promote an exchange of ideas and dialogue). Of course, some authors do include a short Discussion section, or use a combined Results & Discussion section, particularly in cases where the authors want to share interpretations or propose a reaction mechanism. If a Discussion section is included, there should still be a Conclusions, either at the end of the Discussion section (often denoted with a subheading) or as a separate section.

The Conclusions section is organized around one move: summarize the work (as shown in Figure 9.5). This move is accomplished in three submoves. In submove

1. Summarize the Work

(may be preceded by an optional Discussion section)

1.1 Summarize key results

1.2 Conclude with a take-home message

1.3 Describe ongoing or future work (optional)

Figure 9.5. A suggested move structure for a poster Conclusions section.

1.1, key results are summarized, often in a bulleted list. In submove 1.2, a **take-home message** is shared, reiterating what you want viewers to most remember about your work. In submove 1.3, you tell viewers about your work in progress or future plans. Because posters typically only tell part of the research story, submove 1.3 is quite common even though it is optional.

Take-Home Message

A succinct sentence or two that highlights what you most want people to remember about your work.

The Conclusions section for our model poster on imazalil distribution in apples is shown in excerpt 9G. There are no interpretive remarks; hence, a Discussion section is omitted. Key results are highlighted in two bulleted items and followed by a take-home message, which is written in prose (and sometimes italicized).

Excerpt 9G (a poster Conclusions based on Pereira et al., 2019; see full poster in Figure 9.6)

Conclusions

- Using MALDI imaging, imazalil was detected at 3 mm inside the apple after 7 days.
- Using laser ablation and ESI–MS, imazalil was detected at 6 mm inside the apple after 7 days (concentration = 0.015 ± 0.008 ppm).

This work demonstrates the utility of combining MALDI imaging with laser ablation/ESI–MS for monitoring the distribution of fungicides and pesticides in fruits and vegetables.

Unlike excerpt 9G, a mechanism is proposed for a reaction in excerpt 9H; hence, a Discussion section precedes the Conclusions. In excerpt 9H, the Discussion section is written in prose (not a bulleted list). Also note that hedging language is used in the Discussion section (e.g., *appears to be*), underscoring that the mechanism is interpretive. The Conclusions begins with a take-home message, summarizes the key result, and ends by describing work in progress (all in prose).

Hedging

See Chapter 16, "Language Tips," Part 2: Word Choice, for more details on hedging.

Excerpt 9H (a poster Discussion and Conclusion based on Boesten et al., 2001)

Discussion

A proposed mechanism is shown in Scheme 1. The *re*-facial attack of CN^- to the intermediate imine **4** appears to be preferred, forming (*R*,*S*)-**3**. (*R*,*S*)-**3** is less soluble and precipitates out of solution; (*R*,*R*)-**3** is more soluble and epimerizes in solution via the imine **4**.

Scheme 1

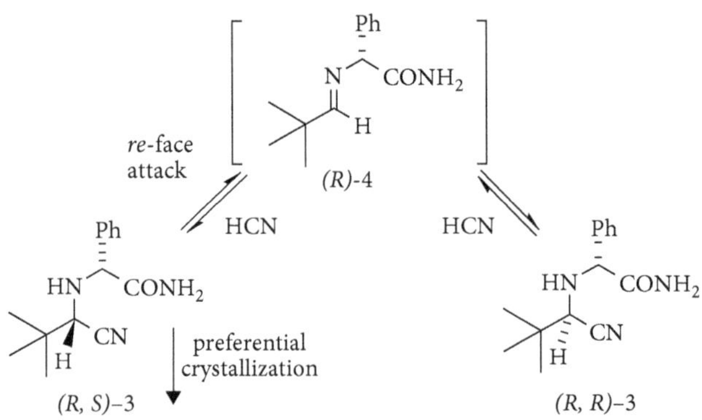

Conclusion

(*R*)-Phenylglycine amide **1** is an excellent chiral auxiliary in the asymmetric Strecker reaction of pivaldehyde **2**. In water at 70 °C, the (*R*,*S*)-**3** product was isolated in 93% yield and dr > 99/1. Work is underway to convert (*R*,*S*)-**3** to (*S*)-*tert*-leucine and thereby complete the asymmetric Strecker reaction.

Bullets or Prose?

Either bullets or prose can be used in Discussion and Conclusion sections, but interpretive remarks (in Discussion sections) and take-home messages (in Conclusion sections) are most often written in prose.

Exercise 9.6.

The following paragraph summarizes efforts to detect polychlorinated biphenyls (PCBs) in full-fat milk using solid-phase microextraction (SPME) with and without saponification (adapted from Llompart et al., 2001). (Saponification is a process that converts fats into water-soluble alcohols and salts.) Write a Conclusion section for a poster based on this information. Follow the move structure in Figure 9.5.

> Even at 100 °C, SPME without saponification was not efficient in extracting PCBs from full-fat milk. With saponification, extraction efficiency increased by 4 to 9 times depending on the PCB congener. Thus, saponification of milk fats to their corresponding glycerols and carboxylates enables a more efficient SPME process for the detection of PCBs in full-fat milk.

Perfecting Your Writing in Poster Conclusions Sections

Here we highlight some writing conventions that are characteristic of the poster Conclusions section:

- **Bulleted lists or prose:** Key results are often highlighted in list form; complete sentences are more common than fragments. The take-home message is usually written in prose. If a Discussion section is included, discussion items (i.e., mechanisms, interpretive remarks) are often written in prose, but bulleted lists may be used.
- **Hedging:** Hedging words should be used in posters (as in journal articles) to soften interpretive remarks, for example:

  > Synergistic effects among aldehydes *may* contribute to the stale flavor of aged beer.
  > The latter process is *possibly* caused by the less soluble amine.

- **Verb tense–voice combinations:** In poster Conclusions sections, past tense (active and passive voice) is often used to summarize key findings. Present tense–active voice is generally used to state the take-home message or describe ongoing work (see Table 9.1).
- **Personal pronouns:** In Conclusions, *we* should be used sparingly, but it can be used to signal a decision, interpretation, or conclusion drawn by the authors.

Poster Writing Conventions

For writing conventions used in multiple poster sections, see Chapter 16 for additional details on abbreviations (Part 3), numbers and units (Part 3), and parallelism (Part 4).

Additional Poster Sections

We briefly consider three additional poster sections: Acknowledgments, References, and Title and Author List.

Poster Acknowledgments Section

Funding agencies and any individuals who helped you with the work (who are not listed as authors) should be acknowledged. At a minimum, you need to include the name of the funding agency and the grant number. If space permits, consider including a funding agency logo. Acknowledgments can be written as full sentences or fragments, but keep the language formal:

Appropriate	We thank NSF (CHE-1709523) for financial support and Dr. Marcia Harris for useful conversations.
Appropriate	This work was supported by NSF (CHE-1709523).
Appropriate	NSF (CHE-1709523)
Inappropriate	THX to NSF for $$ and to Dr. Harris for making us think!

Poster References Section

A References section is needed if you cite others' works in your poster. Most often the References section is the last section of the poster and is placed in the bottom right-hand corner. If you used numerical citations in your poster, the references should be numbered in citation order. Follow the citation format of your discipline (e.g., ACS format), but if space is limited, you can use an abbreviated author list, omit the article title, and list only the first page (rather than inclusive page numbers). Do not omit the lead author, journal name, year, volume, and first page of the article. Consider the following examples of abbreviated references:

1. Davidson et al. *J. Agric. Food Chem.* **2019**, *51*, 6941.
2. Erickson. In *Anal. Chem. of PCBs*, Lewis Publ., 2018.

References and Citations

See Chapter 15 for more details on references and citations.

Poster Title and Author List

The poster title is submitted with the conference abstract (see Chapter 8) and is not typically altered in the final poster. That said, projects can change between the date of abstract submission and the actual conference, and you may decide to alter your title to reflect this difference. If this is the case, keep your changes minimal. The original title will be the one that appears in the conference proceedings, and a different title may make it difficult for interested individuals to find your poster. The same is true for the author list and affiliations; hence, if you make changes to the author list, they will not be reflected in the conference proceedings. The presenting author should be listed first on the poster; it is also customary to underline the name of the presenting author.

Final Poster

The visual appeal of a poster is important, but clever graphics and design are no substitute for good science.

—Frances Blanco-Yu, Penn State New Kensington

We conclude this chapter with the full poster (Figure 9.6) that we prepared using the poster template (Figure 9.1) and the excerpts adapted from the article on imazalil distribution in apples. Note the following features in the poster:

- Poster dimensions (with borders) are 36 in. × 46 in. (height × width).
- Calibri (Body) font is used throughout.
- Font sizes are as follows: title (100 bold), author list (46), sections headings (72 bold), main text (32), figure captions (28).

Chapter Review

Check your understanding of what you've learned from this chapter by defining each of the following terms, in the context of the chapter, for a friend or colleague new to the field:

bulleted lists	poster guidelines	sans-serif font
left- and right-hand justification	poster layout	

As a review, explain the answers to these questions to a friend or colleague who has not yet written text for a scientific poster:

- Who is the audience for a poster?
- What is the broad organizational structure of a poster?
- What are typical fonts, font sizes, and color schemes used in posters?

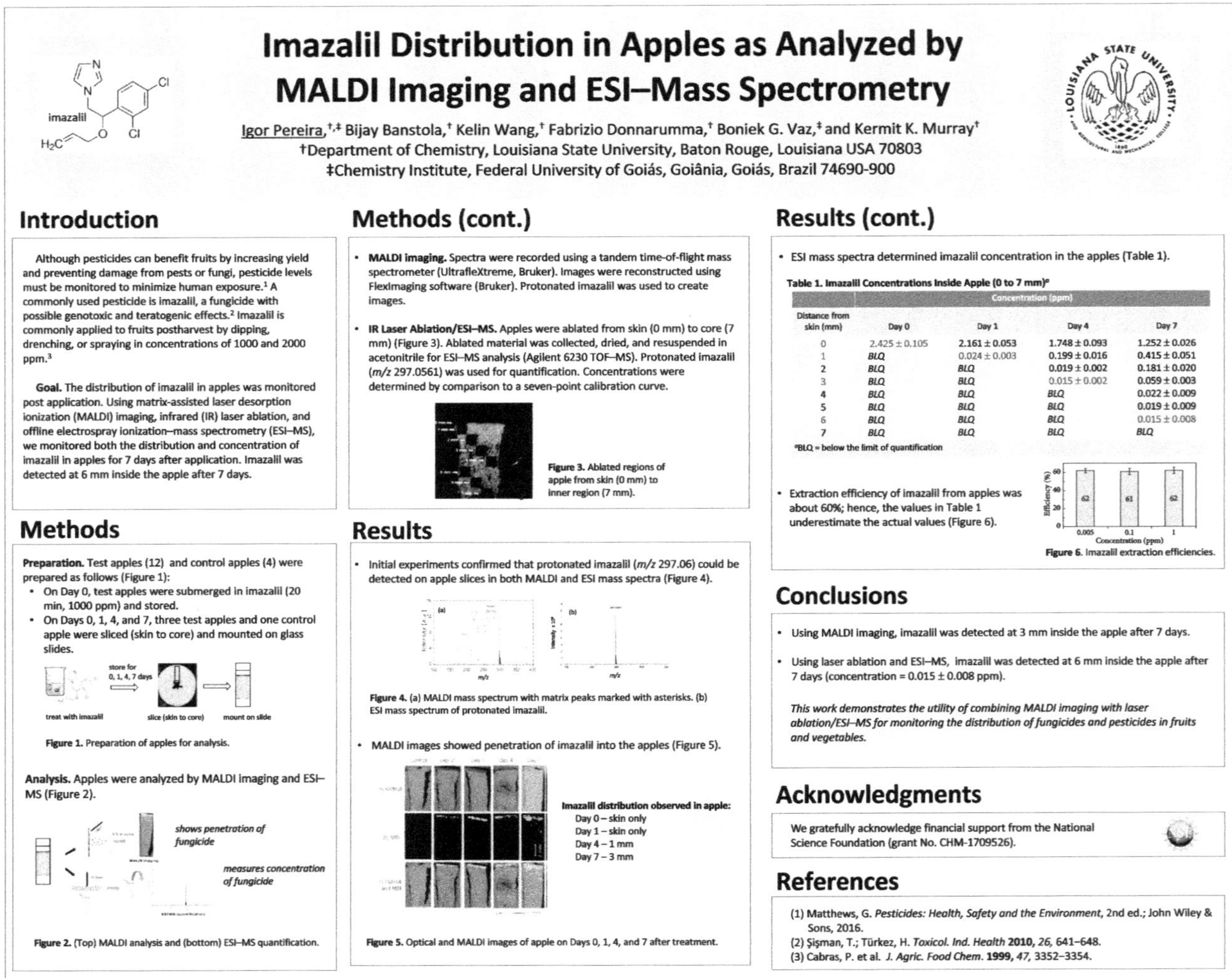

Figure 9.6. A poster based on Pereira et al., 2019. A full-color version of the poster can be viewed at the *Write Like a Chemist* (2nd ed.) website.

Module 3 *Research Proposal*

10 *Overview of the Research Proposal*

Proposal writing is a genre, just like whodonits or romance novels. Each proposal should be crafted so that it tells a coherent story.
—M J. Spires (in Friedland, Folt, & Mercer, 2018)

In this module, we focus on writing a **research proposal**, a document written to request financial support for an ongoing or newly conceived research project. Like the journal article (module 1), the proposal is one of the most important and most utilized writing genres in chemistry. Chemists employed in a wide range of disciplines including teaching (high school through university), research and technology, the health professions, and industry all face the challenge of writing proposals to support and sustain their scholarly activities.

Writing on Your Own: Getting Started

Visit the *Write Like a Chemist* (2nd ed.) website for step-by-step guidance in getting started on writing your own research proposal.

Before we begin, we remind you that there are many different ways to write a successful proposal—far too many to include in this book. Our goal is not to illustrate all the various approaches, but rather to focus on a few basic writing skills that are common to many effective proposals. To illustrate these skills, we include excerpts from successful proposals submitted to two separate funding programs: (1) the 2019 NSF Graduate Research Fellowship Program (GRFP) and (2) the NSF Faculty Early Career Development (CAREER) Program. We chose the NSF Graduate Research Fellowship (GRF) program because many of our readers are undergraduate or graduate students, and this short proposal (two pages) is an excellent place to begin developing your proposal writing skills. Recipients of this prestigious award are provided financial support to pursue a research-based master's or doctoral degree at an accredited U.S. institution. The grant supports graduate research in areas of science, technology, engineering, and mathematics, collectively known as the **STEM** disciplines, but our focus is specifically on chemistry.

Write Like a Chemist. Marin S. Robinson and Fredricka L. Stoller, Oxford University Press. © Oxford University Press 2022.
DOI: 10.1093/oso/9780190098940.003.0010

NSF Graduate Research Fellowship Program (GRFP)

Each year, the National Science Foundation (NSF) solicits proposals from individuals seeking full-time research-based master's and doctoral degrees in science, technology, engineering, and mathematics (STEM). For more information, visit the NSF Graduate Research Fellowship Program website.

STEM

An acronym for science, technology, engineering, and mathematics

We also decided to include excerpts from NSF CAREER proposals. This is a longer proposal (15 pages) written by "rising stars" in the STEM disciplines. This highly competitive award is given to early career (pre-tenure) faculty to support a five-year research initiative and serve as a foundation for a career in research and education. We include excerpts from successful NSF CAREER proposals (again, in chemistry) to provide aspirational examples of proposal writing at its best. After reading this chapter, you should be able to do the following:

- Explain the purpose of a Request for Proposals (RFP)
- Compare RFPs for NSF GRFP and NSF CAREER proposals
- Identify the major sections of a research proposal
- Describe what is meant by intellectual merit and broader impacts
- Write a title and references for a proposal

NSF Faculty Early Career Development (CAREER) Program

One of the U.S. National Science Foundation's most prestigious awards that supports early-career faculty who have the potential to (1) serve as academic role models in research and education and (2) lead advances in the mission of their department or organization. For more information, visit the NSF Faculty Early Career Development Program website.

Reading and Analyzing Writing

Consistent with the read–analyze–write approach used throughout this book, in this chapter we begin by asking you to read and analyze instructions used to solicit proposals for the NSF Graduate Research Fellowship Program (excerpt 10A) and the NSF Faculty Early Career Development award (excerpt 10B). Both solicitations include multiple components; however, we focus only on the section that asks applicants to describe their proposed research. In the NSF GRFP application (excerpt 10A), this section is called the **Graduate Research Plan**; in the NSF CAREER award (excerpt 10B), it is called the **Project Description**. Other components of proposals (such as the project overview, project summary, personal statement, and budget) are not addressed in this module.

The instructions shown in excerpts 10A and 10B are both examples of a Request for Proposals (**RFP**), a solicitation that agencies use to announce a funding opportunity and provide instructions to prospective authors on how to submit a proposal. Essentially all funding agencies use RFPs to solicit proposals. The number of funding sources is extensive and includes federal, state, and local government organizations, as well as public and private corporations, foundations, and individuals. In the United States, the most well-known funding institutions are referred to by their abbreviations, such as ACS, DOE, EPA, NIH, NSF, and USDA. Each funding agency, and every funding opportunity announced by that agency, will have its own RFP, but most RFPs will be similar to those shown in excerpts 10A and 10B.

RFPs and RFAs

RFP: Request for Proposals
RFA: Request for Applications (another term for RFP)

Abbreviations for Major U.S. Funding Institutions

ACS: American Chemical Society
DOE: Department of Energy
EPA: Environmental Protection Agency
NIH: National Institutes of Health
NSF: National Science Foundation
USDA: United States Department of Agriculture

As you read excerpts 10A and 10B, you will note key differences in how the instructions are formatted. The NSF GRFP application (2019) uses a fill-in template, designed to guide students in writing their research plan. Because we found this format particularly useful, we include it in excerpt 10A. (Note: future NSF GRFP applications may use different formats.) In contrast, the NSF CAREER plan application (2020) includes only a list of instructions (excerpt 10B). Most

RFPs only include instructions, not a template, but the template emphasizes for applicants (presumably novice proposal writers) the importance of adhering to proposal guidelines. A panel of reviewers reads and evaluates all proposals submitted. If applicants were to create their own rules regarding what to include in the proposal, or how to organize the content, it would be nearly impossible for reviewers to judge the proposals objectively and fairly. Indeed, most applications are returned without consideration if the rules are not followed. Thus, even if a formal template is not provided, you should adhere closely to proposal guidelines.

Similarly, applicants must follow specified page limitations and font requirements. For instance, the NSF GRFP application has a 2-page limit, in Times New Roman font (12 pt); the NSF CAREER award has a 15-page limit, in Times New Roman font (11 pt). Such rules are needed to guarantee that all applicants have the same amount of space in which to "make their case." As in writing a journal article, concise writing is essential. Indeed, most writers find it quite challenging to share their ideas within the allotted number of pages.

Exercise 10.1.

Read the RFPs in excerpts 10A and 10B and answer the following questions:

a. What are the page limitations for each proposal? What is the likely consequence if the page limitation is exceeded?

b. The main purpose of the NSF GRFP award is to help students pursue an original research project in graduate school. What is the main purpose of the NSF CAREER award?

c. Identify two items that applicants must address that are essentially the same in both RFPs (albeit the wording may be a bit different).

d. Paragraph two of the NSF GRFP RFP (excerpt 10A) instructs students to address the following topics in their Graduate Research Plan Statement:

   1. the research idea
   2. the proposed approach
   3. resources needed to accomplish the research goal
   4. the potential of the research to advance knowledge
   5. the potential of the research for broader impacts on society

   This module divides the proposal into three sections:

   I. Goals and Importance
   II. Approach and Methods
   III. Outcomes and Impacts

   Assign each topic (1–5) to its most likely section (I–III).

Excerpt 10A (from the NSF Graduate Research Fellowship Program Announcement, 2019)

Graduate Research Plan Statement

NOTE: This document has been formatted to the format requirements in the solicitation [Times New Roman, 12-pt, single-spaced, 1" margins, required separate sections for Intellectual Merit and Broader Impacts]. Sections can be any length within the 2 pages. Any deviation from this format will render your application non-compliant and it will be returned without review. (You can remove this text.)

Present an original research topic that you would like to pursue in graduate school. Describe the research idea, the proposed approach, and resources needed to accomplish the research goal, including access to national facilities or collections, collaborations, overseas work, etc. You may cite references. Address the potential of the research to advance knowledge and understanding within science as well as the potential for broader impacts on society. The research must be in a field listed in the Solicitation (Section X, Fields of Study). (You can remove this text.)

[space is provided for applicants to complete this section]

Intellectual Merit

State your Intellectual Merit (you can remove this text).

[space is provided for applicants to complete this section]

Broader Impacts

State your Broader Impacts (you can remove this text).

[space is provided for applicants to complete this section]

References (Times New Roman, 10-pt., single-spaced; you can remove this text)

Excerpt 10B (adapted from the Faculty Early Career Development [CAREER] Program solicitation NSF 20-525, 2020)

Project Description

The Project Description section should contain a well-argued and specific proposal for activities that will, over a 5-year period, build a firm foundation for a lifetime of contributions to research and education in the context of the Principal Investigator's organization. The proposed project should aim to advance the employee's career goals and job responsibilities as well as the mission of the department or organization. The Project Description may not exceed 15 pages.

The Project Description should include the following:

- a description of the proposed research project, including preliminary supporting data where appropriate, specific objectives, methods and procedures to be used, and expected significance of the results;

- a description of the proposed educational activities and their intended impact;
- a description of how the research and educational activities are integrated or synergistic;
- a description of other broader impacts, besides the education activities, that will accrue from the project; and
- results of prior NSF support, if applicable.

Exercise 10.2.

Log on to the NSF website. Search for active funding opportunities in chemistry. Jot down the titles of five funding opportunities that most interest you. (There were 63 titles listed at the time of this printing.)

Analyzing Audience and Purpose

The research proposal [is] an exercise in persuasion.

—Davis, Tyson, and Pechenik (2010)

A successful proposal must target the proper audience. For proposals, the audience typically comprises a panel of preselected reviewers. The task of the reviewers is to read and score the submitted proposals. Because the background and training of panel members will vary from proposal to proposal, you must consult the RFP for information about who your reviewers will be. For the NSF GRFP proposal (excerpt 10A), the RFP states that "applications are reviewed by disciplinary and interdisciplinary scientists and engineers and other professional graduate education experts." Similarly, the NSF CAREER proposal (excerpt 10B) is evaluated by reviewers who are experts in the field represented in the proposal. Thus, these proposals should be written largely for an expert audience.

If, however, you are writing a proposal to support an undergraduate research project, your audience will likely be different. Undergraduate research awards are typically judged by a broad cross section of faculty, including faculty from the physical sciences, life sciences, and social sciences. In this case, you should address both a disciplinary (chemistry) audience and a more general audience. If a biology or psychology professor learns some chemistry while reading your proposal, all the better. Such reviewers will be inclined to regard your proposal more highly, and they will trust that you understand the proposed work, an essential prerequisite for funding.

If the audience is not clearly defined in the RFP, we recommend a two-tiered approach. When you introduce your ideas and describe the importance of your work, target a general to scientific audience:

- Explain important concepts using analogies that less expert readers will understand.

- Define abbreviations, acronyms, and other terms that some readers will be unfamiliar with.
- Use graphics to illustrate ideas or share preliminary results.

When you describe your proposed methodology, shift to a more expert audience:

- Demonstrate your expertise and ability to do the proposed work.
- Use abbreviated terms, as you would in a journal article.

By following this two-tiered approach, a less expert reviewer will be able to follow the main ideas of your work, while an expert reviewer will be able to judge the intellectual merit of your proposal fully.

Another aspect of audience is readability. Although your science must be top-notch (complete with sophisticated terminology), your proposal must also be highly readable. This apparent conflict can be resolved by using an accessible writing style: keep sentences short, use bulleted lists and other easy-to-read formats, and include graphics to break up the text. Reiterating themes and "telling a story" throughout your proposal will also increase your proposal's readability. As a rule of thumb, a 15-page proposal should take about an hour to read (less time than it typically takes to read a 5-page journal article). If your proposal is too difficult to read in this time frame, it may be written for the wrong audience.

Organizing the Proposal

If you are lucky, the RFP for your targeted proposal will offer organizational guidelines. If it does, by all means, follow them, to the point of mirroring the suggested headings and subheadings in your proposal. This will help reviewers locate pertinent information in your proposal and will improve your proposal's readability (and rating). If, however, your RFP does not provide such guidance, or you are writing a fictitious proposal to improve your writing skills, we suggest these three generically named sections to get you started (Table 10.1). Each section is described briefly here and is discussed in more detail in subsequent chapters.

Table 10.1. Main Sections of a Research Proposal

section	chapter
Goals and Importance	11
Approach and Methods	12
Outcomes and Impacts	13

In the Goals and Importance section, you introduce the overarching goals of your proposed work and justify their importance. Your goals must overlap with the purposes and needs of the funding agency. Applicants are encouraged to visit the funding agency's website to learn more about the agency's mission, strategic plans, or rationale for the current funding initiative. Knowledge of the agency's mission allows you to align your research interests and goals with those of the funding agency.

In the Approach and Methods section, you (1) describe your approach and methods and (2) identify any resources you will need to carry out the proposed work. Some grants require that you already have access to these resources (e.g., instrumentation, computational facilities, and laboratory space); others allow you to request new instrumentation in your proposal (with an accompanying budget). The proposed work plan is often presented chronologically, listing a sequence of steps you will take to reach your ultimate goal. This section also typically includes a summary of your previous (related) works and offers a sneak preview at early findings in the proposed project. When reporting previous works or results, it is important to be clear about who did the work, distinguishing your own efforts from work done by other members of your research team.

In the Outcomes and Impacts section, you share expected outcomes of your proposed work and address the intellectual merit and broader impacts of your work. In many NSF proposals (including the NSF GRFP proposal), specific sections labeled *intellectual merit* and *broader impacts* are required, but even if the labels are not required, your proposal must address these topics. The intellectual merit of your work is evidenced by factors such as (1) the creativity and originality of your ideas, (2) the soundness of your approach, and (3) the new knowledge that your work, if successful, will provide. Broader impacts refer to the potential of your research to benefit and transform others. Minimally, your work should benefit other scientists; optimally, your work will benefit society at large.

Each of these sections plays a critical role in the ultimate success of a proposal. Proposals must include a clearly defined research goal that targets the mission of the funding agency. They must also include a well-researched and realistic plan for achieving this goal. Furthermore, successful proposals demonstrate intellectual merit and address broader impacts of the work. The converse is also true; proposals that fail to demonstrate these qualities are unlikely to be funded. Table 10.2 itemizes common reasons for the rejection of proposals, essentially identifying the major pitfalls to avoid when writing a proposal.

Table 10.2. Common Reasons for the Rejection of Research Proposals[a]

Mechanical Reasons

1. Submission deadline was not met.
2. RFP guidelines for content, format, and length were not followed exactly.
3. The proposal was not absolutely clear in describing one or several elements of the study.
4. The proposal was not complete in describing one or several elements of the study.
5. The author(s) took highly partisan positions on issues and thus became vulnerable to the prejudices of reviewers.
6. The quality of writing was poor (e.g., sweeping and grandiose claims, convoluted reasoning, excessive repetition, unreasonable length).
7. The proposal contained an unreasonable number of mechanical defects that reflected carelessness and the author's unwillingness to attend to detail. The risk that the same attitude might extend to the proposed study was not acceptable to reviewers.

Methodological Reasons

8. The proposed question, design, and method were completely traditional, with nothing that could strike reviewers as unusual, intriguing, or clever.
9. The proposed method was unsuited to the purpose of the research.

Personnel Reasons

10. As revealed in the literature review, the author(s) simply did not know the territory.
11. The proposed study appeared to be beyond the capacity of the author(s) in terms of training, experience, and available resources.

Cost–Benefit Reasons

12. The proposed study was not an agency priority for *this* year.
13. The budget was unrealistic in estimating costs for equipment, supplies, and personnel.
14. The cost of the proposed project appeared to be greater than any possible benefit to be derived from its completion.

[a] Adapted from Locke, Spirduso, & Silverman (2014, p 188).

Exercise 10.3.

Review the list in Table 10.2. Identify one or more reasons for the rejection of a research proposal that links directly to the following. Be prepared to explain your choices.

a. the author's failure to address the need(s) of the funding agency
b. the author's failure to establish intellectual merit
c. the author's failure to address broader impacts

d. the author's poor writing skills and lack of attention to detail
e. the author's limited familiarity with others' works related to the proposed study

Writing a Title and References

The first thing . . . reviewers see is the title. It is therefore very important that your title is strong. A strong title should fulfill three criteria: It needs to be clear, complete, and succinct.

—From Hofmann (2020)

We now briefly address titles and references for a research proposal. Like journal article titles, a proposal title should follow the X of Y by Z pattern (Table 7.1 in Chapter 7) and include keywords that will help the program office decide what discipline and area the proposal should be sent to for review. The title should also include keywords that capture the need(s) targeted by the funding agency and any specific words requested by the agency in the solicitation. For example, NSF requires applicants for the CAREER award to begin their titles with "CAREER:". Abbreviations and acronyms should be avoided in titles. Although explicit restrictions on title length are rare, shorter titles are preferred provided that they accurately depict the contents and emphases of the proposed work.

Exercise 10.4.

Consider the following NSF CAREER proposal titles. Each title follows some form of the X of Y by Z pattern found in journal article titles (see Table 7.1). Which one(s) include all three components (X, Y, and Z)? Which one(s) include only two components (X and Y)? Identify Y in each title.

[*Note:* The original CAREER titles all began with "CAREER:" (as specified in an NSF directive), which has been omitted here.]

a. Combined Experimental and Computational Approach to Controlling Site Selectivity in Cross Coupling Chemical Reactions (Neufeldt, 2019)
b. Defects and Polarons in Complex Materials (Janotti, 2017)
c. Phase-Separating Membrane Materials for Efficient and Specific Molecular Delivery to Cells (Stachowiak, 2014)
d. Harnessing the Reactivity of Neutral Aminyl Radicals for Complex Heterocycle Synthesis (Stockdill, 2016)
e. Ionic and Thermal Transport Properties of Complex Oxides from First Principles (Benedek, 2015)

References, an essential component of a proposal, serve as important indicators of your familiarity with the professional literature, your currency in the field, and your preparedness to complete the proposed work. References also assist reviewers in situating your proposal in the context of previous works. Formats for writing references vary widely among proposals, but the rules are generally specified in the RFP. Nearly all proposals prefer that you use inclusive page numbers, not just the first page. Some proposals also prefer that you include titles of articles, which are useful to reviewers. Often the instructions will tell you to simply use the format most common in your discipline. Importantly, whatever format you use, be sure to follow it consistently in all of your references. In most proposals, references are not included in the page limitations; hence, you are encouraged to cite all relevant works, just as you would in a journal article. Superscripted numbers, as in journal articles, are the most common format for in-line citations (see Chapter 15). Authors prefer this approach as it conserves space.

Chapter Review

As a self-test of what you've learned in this chapter, define each of the following terms and explain its importance, in the context of this chapter, to a friend or chemistry colleague who is new to the field:

broader impacts	RFA	STEM
intellectual merit	RFP	

Also answer the following for a friend or colleague who hasn't yet given much thought to writing a research proposal:

- Who is the audience for a proposal?
- What are common reasons for a proposal to be returned without consideration?
- Why is it important to adhere to RFP guidelines when writing a proposal?
- What are three main sections of a proposal?
- Do you need to include references in a proposal? If so, how are they formatted?

11 *Writing the Goals and Importance Section*

Good proposals stimulate my curiosity. As I read a proposal, I begin to think up new questions about the project and its science. The best proposals answer those questions before I'm done reading. I have a lot of respect for writers who can anticipate the questions that a reader will have and provide answers.
—Alexander Grushow, Rider University

This chapter focuses on writing the first section of a research proposal in which you are asked to provide an overview of your project and describe your **research idea** or **research goal**. The central purpose of this section is to give reviewers a first glimpse at what you propose to do. In so doing, you will also provide relevant background information and highlight the importance of your work. By the end of this chapter, you should be able to do the following:

- Write a sentence that describes your research goal or idea
- Emphasize the importance of your research
- Provide background information in your research area

Research Idea or Research Goal?

Students describe a *research idea* in NSF Graduate Research Fellowship Program (GRFP) proposals; faculty describe a *research goal* in NSF CAREER proposals. In this book, we use these terms interchangeably. In your own writing, we recommend that you mirror the term(s) used in your targeted RFP (Request for Proposals).

Write Like a Chemist. Marin S. Robinson and Fredricka L. Stoller, Oxford University Press. © Oxford University Press 2022.
DOI: 10.1093/oso/9780190098940.003.0011

Writing on Your Own: Goals and Importance Section

Visit the *Write Like a Chemist* (2nd ed.) website for step-by-step guidance in writing the Goals and Importance section of your research proposal.

Reading and Analyzing Writing

We begin with an excerpt from an NSF GRFP proposal for you to read and analyze on your own (excerpt 11A). The excerpt includes the first two paragraphs of Pearson's (2019) Graduate Research Plan Statement. In the first paragraph (with the subheading **Overview**), the author describes her research idea and explains why her research is important. Her idea is to create a nanoparticle structure with a cisplatin shell and an iron oxide core; this structure is important because it has the potential to both improve the targeting ability of cisplatin (a cancer drug) and reduce its side effects. In the second paragraph (labeled **Background**), she shares information about both cisplatin and iron oxide and describes previous efforts to pair these two compounds. She ends this section with her hypothesis. **Hypotheses** are common in the biomedical and social sciences but less common in the physical sciences, including chemistry. Nonetheless, a hypothesis serves as a useful way to clarify the assumption(s) that will be tested in your proposed work.

Hypotheses

Hypotheses are used more extensively in the biomedical and social sciences than in the physical sciences. They are generally not included (or required) in chemistry proposals.

Exercise 11.1.

Read and analyze excerpt 11A and complete these tasks:

a. Based on the first paragraph, explain why the proposed research is important.
b. The author's subheadings are parallel. Explain what makes them parallel. (For guidance, see Chapter 16, "Language Tips," Part 4: Grammar, for tips on parallelism.)
c. What concept does the author illustrate in the graphic (Figure 1) and text? (Note, too, how the graphic is embedded in the text.)
d. In the background section, how do works cited relate to the author's research idea?
e. Why do you think the author decides to bold the last sentences in the Overview and Background sections?

Excerpt 11A (Pearson, 2019)

Overview: The purpose of this study is to determine the effects of combining a cis-diamminedichloroplatinum (cisplatin) shell with an iron oxide core in a core–shell nanoparticle (NP) structure with the ultimate goal of improving cisplatin targeting capabilities and reducing the drug's side effects on the kidneys. Although cisplatin has been in testing and/or treatment use since the early 1980s, it still exhibits significant nephrotoxicity, resulting in side effects throughout the body, not least of which is the kidney, where problems like chronic kidney disease (CKD) can arise and continue to affect patients decades after they finish treatment. Because of the kidney side effects, the maximum dosage is limited, thus limiting the overall effectiveness of the drug.[1] To combat those side effects without reducing drug dosage, the cisplatin-kidney interactions must be significantly reduced. Furthermore, the development of iron oxide NPs has been closely tied to the development of core–shell NPs (Figure 1), with many studies suggesting that iron oxide cores or shells work compatibly with many other compounds, resulting in dual- or triple-functional NPs with specific functionalities.[2] **This study aims to reduce cisplatin-kidney interactions by using the tunable nature of core–shell NPs to combine cisplatin with tumor-targeting, magnetic iron oxide NPs, achieving dual-functional targeted cisplatin drug delivery.**

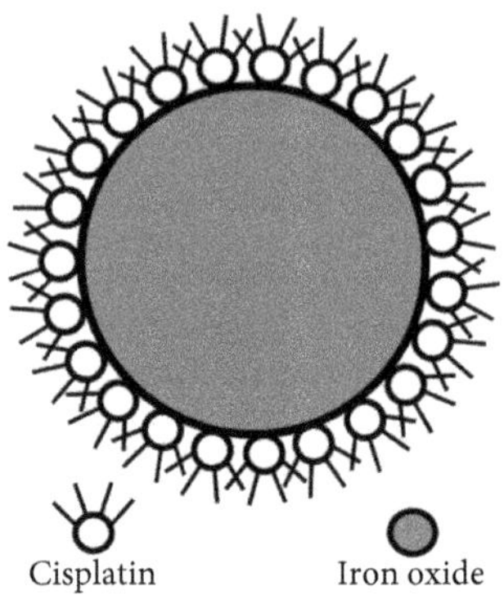

Figure 1. Graphic depiction of core–shell NPs.

Background: The mechanisms of cisplatin tumor destruction are still largely unknown, although it is generally agreed that the drug interacts with cell components including cellular DNA, mitochondria, proteins, and RNA, eventually triggering the cell's death pathway via apoptosis.[3] When applied in sufficient dosage, the drug is verified to be an effective treatment method against many forms of cancer including breast cancer, the focus of this study, and cisplatin has been shown to work cooperatively with other cancer drugs.[1] Iron oxide, a common choice for drug pairing, has been established to be an effective targeting method for tumors due to the magnetic fields inherent to many tumors that attract the magnetic iron oxide NPs.[4] For example, cisplatin and iron oxide, when combined with steroids and applied directly to the cochlea, have been shown to improve inner-ear protection from cisplatin side effects.[5] **Such findings indicate that cisplatin and iron oxide may be effectively combined to improve the targeting characteristics of the drug, resulting in fewer side effects, higher dosages, and higher overall cancer treatment efficacy.**

Hypothesis: *Core–shell iron oxide-cisplatin NPs will achieve higher levels of tumor targeting and lower levels of adverse kidney side effects, resulting in a more effective treatment method overall.*

Affect vs. Effect

See Chapter 16, "Language Tips," Part 6: Word Usage, for tips on differentiating these commonly confused words.

Capitalization

Chemical names (e.g., diamminedichloroplatinum) are written in lowercase except at the start of a sentence. Terms used to indicate spatial orientation [e.g., *cis*-, *trans*-, (*R*)-, (*S*)-] are italicized; *cis*- and *trans*- are written in lowercase, even at the start of a sentence. See Chapter 16, "Language Tips," Part 3: Writing Conventions, for additional tips on capitalization.

DNA and RNA

Some abbreviations, like DNA and RNA, do not need to be spelled out at first use. See Chapter 16, "Language Tips," Part 3: Writing Conventions, for a partial list of abbreviations that do not need to be spelled out for a chemistry audience.

Analyzing Organization

A typical move structure for the Goals and Importance section of a proposal is presented in Figure 11.1. (Of course, if your targeted proposal recommends a different organizational structure, use that one instead.) The overarching purpose of this section is to introduce your proposed research. This task is generally accomplished in three moves. In move 1, the research area is identified, and the importance of continued research in this area is stressed. You also provide readers with a succinct statement of your research idea or goal. In move 2, you explain concepts and introduce relevant works that situate your work within the context of existing knowledge. Move 2 also serves the purpose of convincing reviewers that you have "done your homework" and are knowledgeable about key concepts and previous achievements in your field. In move 3, you introduce your proposed project. This move typically begins with a gap statement, indicating that despite previous accomplishments (highlighted in move 2), continued research (i.e., your proposed

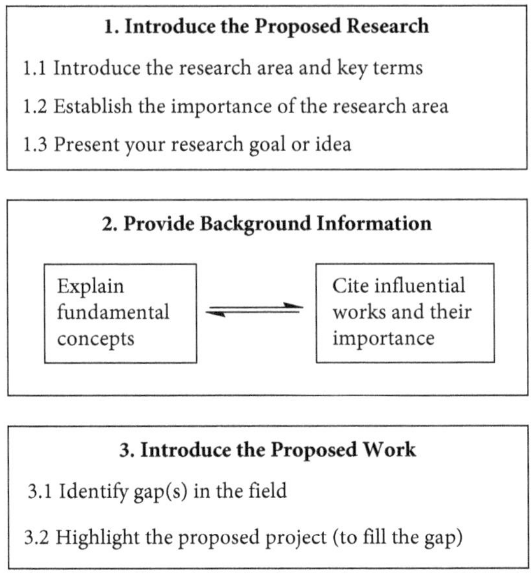

Figure 11.1. A suggested move structure for the Goals and Importance section of a research proposal (unless the RFP specifies different sequencing and/or contents).

work) is still needed. The move ends by highlighting the proposed work, which will fill this gap.

With Figure 11.1 in mind, we are ready to consider these moves in more depth and analyze how authors accomplish these moves in authentic GRFP and CAREER proposals. Later in the chapter, we examine common writing conventions used in this section.

Analyzing Excerpts Move by Move

I no longer expect the reviewer to do a lot of work to understand my ideas or recognize the significance of my proposal. I assume it is my job to convince the reader that the problem I am working on and the solution I have proposed are compelling and have high value. I work to grab the reader's attention immediately and then serve up a well-organized, persuasive, and easily readable presentation of my ideas.

—Joan Curry, University of Arizona

Move 1: Introduce the Proposed Research

The key purpose of move 1 is to introduce your research idea or goal. The move is often accomplished in the opening paragraph of the proposal and is commonly signaled with a heading (e.g., **GOALS AND IMPORTANCE**) or subheading (e.g., **Research Idea**). Although headings are not mandatory, they are useful for quickly alerting reviewers to the contents of a particular section and the general organization of your proposal. If you do include headings and subheadings in your writing, be sure to format them consistently throughout your proposal.

As shown in Figure 11.1, move 1 has three submoves. The first submove introduces the research area, and key terms that readers may be unfamiliar with are defined. The second submove stresses the importance of continued research in this area. The third submove presents the research idea or goal; this idea is typically expressed in a succinct sentence at the end of the opening paragraph. We begin by examining move 1 in successful NSF GRFP proposals (excerpts 11B–11D); the RFP specifically prompts students to state their research idea. For clarity, we have marked the start of the three submoves using brackets in excerpts 11B–11D.

Exercise 11.2.

Glance ahead to excerpts 11B–11D and briefly summarize the research idea expressed in each excerpt.

In excerpt 11B, the paragraph begins by introducing the research area, photosynthesis, and the term *photoexcitation* is defined, describing what is transferred through cell membranes to accomplish the conversion of solar energy to chemical energy (submove 1.1). However, the underlying mechanism of this "remarkably efficient process" is not fully understood. This *gap* underscores the need for the proposed research (submove 1.2). We first encountered gaps and **gap statements** in the Introduction of a journal article (Chapter 6), where they are used to establish the need for the current work. They serve a similar purpose here, establishing the need for the proposed work. They also provide a natural segue to the research idea (*I propose to measure . . .*), which concludes the opening paragraph. We will encounter gap statements again in move 3 (submove 3.1).

Gap Statements

Gap statements are often used to stress the need for additional research by pointing out existing weaknesses in the field. (See also Table 6.1 in Chapter 6.)

Excerpt 11B (Weaver, 2019)

[Submove 1.1] In the first few picoseconds of photosynthesis, a photoexcitation is transferred through a series of antenna complexes to a reaction center where solar energy is converted into chemical energy. This transport occurs with near unit quantum efficiency (QE), [submove 1.2] yet its underlying molecular mechanisms are not fully understood. To uncover what drives this remarkably efficient process, [submove 1.3] I propose to measure ultrafast energy transport in a photosynthetic light harvesting membrane with in situ correlation to the underlying nanoscale protein organization.

in situ and in vivo

In situ and in vivo are generally written without italics, as shown in excerpt 11B.

Excerpt 11C also begins by introducing the research area, which involves the preservation of blood platelets. The term *lyopreservation* is introduced as a way to desiccate (dry) blood platelets so they can survive long-term storage at room temperature (submove 1.1). The proposed research is needed (submove 1.2) because lyopreservation methodologies for blood platelets have not been widely implemented in medicine (the gap statement). The paragraph ends with the research idea (submove 1.3), which will fill the gap. As in excerpt 11B, the sentence begins with a personal pronoun (*I propose to use . . .*). Personal pronouns such as *we* or *I* are common in the research idea statement of NSF GRFP proposals.

Excerpt 11C (Samuels, 2019)

Motivation: [Submove 1.1] Blood platelets function in many biological processes from immune response to liver regeneration,[1] making a near-continuous supply of concentrated blood platelets necessary for medicine. Yet the short shelf-life of blood platelets, only 5 days at room temperature in solution, causes shortages in blood banks and hospitals.[1] Although lyopreservation (i.e., desiccation) has been demonstrated in blood platelets, [submove 1.2] the methodology has not been widely implemented in medicine.[1] A more robust understanding of the fundamental cellular principles causing successful lyopreservation will allow further development of this preservation method. [Submove 1.3] I propose to use nonlinear coherent stimulated Raman scattering (SRS) microscopy and/or coherent anti-Stokes

Raman scattering (CARS) microscopy to image blood platelets as they are being exposed to a lyoprotectant and measure the permeation and diffusion of the lyoprotectant in the cell.

Although vs. While

See Chapter 16, "Language Tips," Part 6: Word Usage, for tips on using these easily confused words.

In excerpt 11D, the research area involves theoretical modeling of chemical reactions. The term *molecular dynamics* (MD)—an important tool for describing a reactive system using classical mechanics—is introduced. However, MD alone does not provide accurate results (a gap). The research idea, signaled by the phrase *we propose a multiscale approach*, will improve accuracy by combining molecular dynamics with quantum mechanics.

Excerpt 11D (Wait, 2019)

Research Idea. [Submove 1.1] Reaction rates depend on the molecular environment of the reactive complex, and this dependence can be predicted with molecular dynamics (MD). [Submove 1.2] However, MD provides the classical description of the reactive system itself, which is not sufficiently accurate. [Submove 1.3] For this purpose, we propose a multiscale approach: activation free energies of the reactive complex will be obtained quantum-mechanically, and the effects of the molecular environment will be accounted for by classical mechanics.

_____, which

See Chapter 16, "Language Tips," Part 5: Punctuation, for rules on comma use; see also Word Usage (Part 6) for the important distinctions between *that* and *which*.

Exercise 11.3.

Re-read excerpt 11A.

a. What gap is suggested in the opening paragraph?
b. Find the research idea or goal statement(s).
c. How will the research idea fill the gap?

Now let's consider move 1 in NSF CAREER proposals. Recall, in NSF CAREER proposals, authors are prompted for their research goal rather than their research idea. We first analyze excerpt 11E. The author begins by introducing the research area (submove 1.1) and its importance (submove 1.2) in the opening paragraph. Because both submoves are addressed continuously throughout the opening paragraph, they are grouped together (i.e., submoves 1.1 and 1.2). The author's research goal (submove 1.3) is presented at the start of the second paragraph, rather than at the end of the first paragraph, and two additional sentences are used to further elucidate this goal.

Note that the research goal in excerpt 11E does not use *I* or *we* but instead begins *The principal goal of the PI's planned research is . . . .* In NSF CAREER proposals, goals are typically written without personal pronouns. A few more examples are shown in passages P1–P3. Each statement includes the word *goal* (which serves as a direct link to the proposal instructions) and is preceded by an adjective (*overall*, *long-term*, or *ultimate*) to denote that the goal is broad in scope and will take multiple years to accomplish.

P1 The overall goal of this research is to explore the use of thermotropic liquid crystal (LC) solvents as new and versatile media for the engineered growth of molecular thin films. (From Patrick, 2000)

P2 The long-term goal of the proposed research is to create functional model complexes of metalloprotein active sites. (From Houser, 2001)

P3 The ultimate goal of the proposed research is the identification and development of fluorescent chemosensors for heavy metal ions and small organic molecules in water. (From Finney, 1999)

Exercise 11.4.

Read excerpt 11E and briefly summarize the research area, the importance of the proposed research, and the research goal.

Excerpt 11E (Aga, 2002)

GOALS AND SIGNIFICANCE OF PROPOSED WORK

[Submoves 1.1 and 1.2] The increased use of agricultural chemicals such as herbicides and animal antibiotics has been mirrored by an increased public concern regarding the impact of xenobiotic compounds on the environment and human health. The pressure to provide new information on the fate of pesticides in the environment has become a monumental task for government and industry. In addition to pesticides, antibiotics are now being recognized as emerging contaminants because of their high potential to enter surface and ground waters from animal confinement operations and manure-treated agricultural fields. The widespread use of antibiotics in animal production is an important issue that has currently attracted attention due to the increased incidence of antimicrobial resistance in food-borne microorganisms. Knowledge of the mechanism for degradation and transport of environmental contaminants is essential in developing guidelines for effective water-quality management actions.

[Submove 1.3] The principal goal of the PI's planned research activities is to develop innovative analytical methods, such as immunochemical techniques, and apply these methods in environmental investigations. Emphasis is placed on the development of immunochemical methods because of their cost effectiveness and general applicability in conducting environmental investigations. In addition, there is an increased interest from regulatory agencies to develop analytical methods that are fast and field-portable for on-site monitoring; these are characteristics of immunochemical methods.

Principal vs. Principle

Principle (noun): a basic truth
Principal (adj.): main, chief

See Chapter 16, "Language Tips," Part 6: Word Usage, for tips on using these easily confused words.

We should note that many NSF CAREER proposals begin (rather than end) the opening paragraph with the research goal (submove 1.3), thus altering the move structure shown in Figure 11.1. In fact, the research goals listed earlier in P1–P3 are all opening sentences, and in each case, the goal statement is followed by a brief overview of project objectives. The research area (submove 1.1) and its importance (submove 1.2) can be deduced from the research goal and objectives. (The importance of the research is also stressed in numerous other ways throughout these proposals.) An example of this approach is shown in excerpt 11F. The paragraph begins with the research goal, which also introduces the research area (*nucleotide chemistry*). Two objectives are shared, and the importance of each objective is stressed (*to understand fundamental electronic reactivity* and *to elucidate the mechanism of conversion*, respectively). Note, too, the use of ordinal language (e.g., *first* and *second*) to enumerate the objectives in excerpt 11F. Although ordinal language is discouraged in journal articles (see Chapter 3), it is often used in proposals.

Excerpt 11F (Lee, 2001)

SIGNIFICANCE AND OBJECTIVES

The overall goal of the proposed work is to broaden our understanding of nucleotide chemistry through gas-phase mass spectrometric and theoretical quantum mechanical studies. The objectives are twofold. First, the gas-phase acidity and nucleophilicity of different sites on nucleobases will be explored to understand their fundamental electronic reactivity. Second, the chemistry of orotic acid will be examined to elucidate the mechanism of conversion of this important nucleic biosynthetic precursor to uracil.

First,

Ordinal language may be used in proposals. Use a comma after ordinal terms in addition to words and phrases used to create linkages, e.g.,

First, . . .	However,
Second, . . .	For example,
. . . , and third, . . .	Accordingly,

See Table 6.6 in Chapter 6 for a more complete list of common phrases used to create linkages, organized by their functions.

Moves 2 and 3: Provide Background Information and Introduce the Proposed Work

Remember that the best proposals are built from the best science. Effective proposals require a sound scientific basis.

—Friedland, Folt, and Mercer (2018)

If I have seen further, it is by standing on the shoulders of giants.

—Isaac Newton

You are likely familiar with the second quote by Isaac Newton. It is a popular one among scientists because it acknowledges that new ideas are not created in a vacuum; new insights build on previous discoveries and existing knowledge. Move 2 is where authors pay tribute to that knowledge base. Fundamental concepts are explained, and background information is shared. Pivotal works are cited and the importance of continued research in this area is stressed. Move 2 transitions naturally into move 3, where gaps in the previous works are identified and the proposed project (which will fill the gap) is introduced. Because moves 2 and 3 are often addressed together in the same paragraph, we consider them together here.

Let's start by examining moves 2 and 3 in successful NSF GRFP proposals. Consider excerpts 11G–11I, which continue excerpts 11B–11D, respectively. Each excerpt has a subtitle (**Background** or **Background Information**) informing the reader what to expect in this section. For clarity, we demark the start of moves 2 and 3 in each excerpt using brackets. In this way, you can easily see where each move begins and ends.

In excerpt 11G, the author begins move 2 by sharing what is currently known about photosynthesis, describing how plants absorb light in a light-harvesting complex (LHC). This light generates an *exciton* that is transferred, in picoseconds, over tens of nanometers to a photochemical reaction center (RC) with remarkable quantum efficiency (QE). How does the exciton find the RC with such great efficiency? Current approaches that might be able to answer this question are cited, but *unfortunately* (signaling the gap statement, submove 3.1) these approaches fall short in their ability to discern important details about the molecular complexes that facilitate this transfer. The author's proposed research (submove 3.2) will fill this gap by elucidating the *underlying energetic and structural factors that guide exciton trajectories through an intricate molecular landscape with unrivaled QE*.

Excerpt 11G (Weaver, 2019)

[*Continues excerpt 11B*]

Background and Motivation: [Move 2] In plants, the light reactions of photosynthesis occur in structured membranes called thylakoids that form folded 'grana' stacks inside the chloroplast. Within each membrane, densely packed specialized proteins and pigments perform the vital function of solar energy conversion. Most light is initially absorbed by chlorophyll molecules associated with light harvesting complexes (LHCs), generating a tightly bound excited electron-hole pair or 'exciton'. This photoexcitation then migrates across an organized network of these pigment-protein complexes until it is harnessed by a photochemical reaction center (RC), a dedicated location for charge separation that triggers downstream biochemistry.

Even though there is no obvious energetic driving force that funnels an exciton toward a reaction center, it finds its target with high fidelity. Because this long-range energy transfer occurs in a matter of picoseconds over tens of nanometers, simultaneously probing its dynamics on native length and time scales is essential. [Move 3] Unfortunately, current approaches such as ultrafast spectroscopy[1] or high-resolution microscopy[2] characterize either ultrafast excited state dynamics or nanoscale structural complexity, obscuring details about intrinsic protein disorder or temporal variations in the membrane, respectively. The goal of this project is, therefore, to measure the dynamics of exciton migration during photosynthesis on characteristic spatiotemporal scales to elucidate the underlying energetic and structural factors that guide exciton trajectories through an intricate molecular landscape with unrivaled QE. By altering the functionality of RCs, I will (1) determine if exciton trajectories are diffusive or directed by the presence of active RCs and (2) characterize the effective attractive potential of a RC.

Because vs. Since

See Chapter 16, "Language Tips," Part 6: Word Usage, for help using these commonly confused words.

Colons

Colons should be used only after a complete sentence, e.g.,

correct	I will (1) determine . . . and (2) characterize . . .
incorrect	I will: (1) determine . . . and (2) characterize . . .

See Chapter 16, "Language Tips," Part 5: Punctuation, for tips on using colons.

In excerpt 11H, the author begins move 2 by sharing known information on lyoprotectants, which are sugars and proteins found in nature that help organisms survive desiccation (dehydration). One such lyoprotectant is trehalose. The field of lyopreservation uses lyoprotectants, such as trehalose, to preserve cells at room temperature by desiccation, without the need for freeze drying or refrigeration. However, cells treated with trehalose do not always survive desiccation. Nonuniform distribution of trehalose in the cell may be the cause of this failure, but this has not been investigated (the gap statement, submove 3.1). The next two paragraphs describe the proposed work that will fill this gap (submove 3.2): blood platelets will be studied using nonlinear coherent Raman microscopy to investigate trehalose distribution in viably preserved cells. This work is important for its potential to extend the shelf life of blood platelets and for its insights into the mechanisms of desiccation.

Excerpt 11H (Samuels, 2019)

[*Continues excerpt 11C*]

Background: [Move 2] There are many organisms in nature that can survive near complete desiccation.[2] These organisms produce high concentrations of various sugars and proteins, referred to as lyoprotectants, to survive harsh conditions.[2] Trehalose is a common disaccharide produced by dry-tolerant organisms and has become a central lyoprotectant in the field of lyopreservation.[1,2] [Move 3] However, the use of trehalose in lyopreservation has not led to universal cell viability after desiccation, perhaps due to its low membrane permeability leading to irregular distribution of trehalose throughout a cell, although this has never been directly investigated.[3] Nonuniform distribution of lyoprotectants in a cell is thought to be one cause of low cell survival after lyopreservation as lyoprotectants likely act as a water substitute in stabilizing the cellular structure after desication.[3]

Blood platelets provide a good system to investigate trehalose distribution in a viably preserved cell because they have demonstrated survival after lyopreservation and are relatively simple sytems.[1] Penetrating cryoprotecting agents (CPAs), such as the commonly used dimethyl sulfoxide (DMSO), are thought to stabilize cellular structure in a similar way to lyoprotectants.[3] Exposing blood platelets to a mixed solution of penetrating CPAs (DMSO) and non-penetrating lyoprotectants (trehalose) before desiccation presents an opportunity to investigate if penetrating CPAs may increase cell survival, perhaps through increasing membrane permeability or by concentrating in parts of the cell non-penetrating lyoprotectants are unable to access.[3]

Nonlinear coherent Raman microscopy allows direct imaging of molecules without the addition of fluorescent tags; deuteration of the molecules of interest will give them a unique spectroscopic signal.[4,5] Thus, this technique offers a superior approach for imaging intracellular kinetics, as the addition of a bulky fluorophore to a lyoprotectant strongly impacts permeation and diffusion of the lyoprotectant, whereas deuteration will have a negligible effect. Nonlinear coherent Raman microscopy also has an excellent spatial resolution and fast response time, only limited by the detector, making it a suitable technique to image the penetration and subsequent distribution of lyoprotectants in a cell.[4,5]

Whereas vs. While

Whereas is a formal, academic word used to indicate a contrast between two facts or ideas. *While* is generally used to indicate time. See Chapter 16, "Language Tips," Part 6: Word Usage, for tips on using *although*, *while*, and *whereas* appropriately.

The third excerpt (excerpt 11I) begins with background information on molecular dynamics and describes previous efforts to use molecular dynamics to estimate free-energy profiles along a reaction coordinate. These energies were used to predict reaction rates under gas-phase and supercritical conditions. However, this theory did not accurately reproduce experimental results (the gap statement, submove 3.1). The author proposes to fill this gap by combining molecular dynamics (to describe the CO_2 environment) with quantum mechanics (to describe the free energies of the reactive complex). Italics were used by the author to call attention to the proposed work. In two subsequent paragraphs (not included), the author goes on to provide background information on transition-state theory and Rice-Ramsperger-Kassel-Marcus (RRKM) theory, both of which will be used in the proposed work.

Excerpt 11I (Wait, 2019)

[*Continues excerpt 11D*]

Background. [Move 2] Molecular dynamics is the method used to describe a molecular system that evolves in time according to classical mechanics (Newtonian equations). Positions of all atoms in the system are calculated at each time step (and collected into trajectories) according to atomic velocities and forces between the atoms. These forces are predicted based on classical energy functions known as force fields. Enhanced MD protocols, such as umbrella sampling or boxed MD, can be used to simulate infrequent events, including reaction barrier crossings. Masunov et al.[1,2] have previously used boxed MD to estimate free-energy profiles along the reaction coordinate for a reactive complex surrounded by 0–90 CO_2 molecules. This allowed the authors to predict reaction rates in gas-phase and supercritical environments, important for studying oxy-fuels where

CO_2 is used to replace nitrogen. [Move 3] However, when compared to experimental measurements, these predictions were found to be inaccurate. *Here, we propose to use classical MD to account for the environmental effects and quantum mechanics to describe the reactive system itself.*

Subject–Verb Agreement

Numerous scientific words end in *s*, but they are singular (e.g., kinetics, quantum mechanics, thermodynamics). In contrast, et al. (which means *and others*) requires a plural verb.

Correct	Molecular dynamics *is* the *method* . . . (dynamics is singular; method is singular, both agree with *is*)
Correct	Masunov *et al.*[1,2] *have* . . .

See Chapter 16, "Language Tips," Part 4: Grammar, for tips on plural and singular scientific words in addition to subject–verb agreement.

We conclude this section by analyzing moves 2 and 3 in NSF CAREER proposals. In move 2 (provide background information), authors must decide how much detail they want to include. Some concepts may be described in considerable detail, approaching a near textbook-like description. **Full-color illustrations** may be used, and **textbooks** may be cited. For example, in excerpt 11J, the author describes quantum mechanical tunneling using a full-color graphic to illustrate the different tunneling behaviors of light (e.g., hydrogen) versus heavy isotopes (e.g., deuterium and tritium). The figure illustrates that the lighter isotope has a higher probability of tunneling than the heavier one, resulting in a kinetic isotope effect (KIE). In the last sentence, the author alludes to the proposed work (move 3), which will use the kinetic isotope effect to investigate tunneling behavior.

Full-Color Illustrations

Proposals often include full-color graphics to elucidate important concepts. Color graphics should also make sense in gray scale, in case reviewers print out a hard copy.

Citing Textbooks?

Authors may cite textbooks (or other authoritative sources) that are used to prepare explanatory remarks in a research proposal. Textbooks are not usually cited in journal articles.

Excerpt 11J (Kohen, 2002)

Quantum mechanical tunneling. [Move 2] Tunneling is the phenomenon by which a particle transfers through a reaction barrier due to its wave-like property.[30,37] Figure 1 graphically illustrates this for a CHC double-well system. . . . Hydrogen is so light that there is a significant uncertainty in its location at a given energy (the Heisenberg uncertainty principle). When it is close to the reaction's barrier, there is some probability of finding it on the product side, if the barrier is narrow enough and the reactant and product energy levels are degenerate. As illustrated in Figure 1, the lighter isotope has a higher tunneling probability than the heavier one, [move 3] suggesting that kinetic isotope effects (KIEs) should be effective tools in studying tunneling.

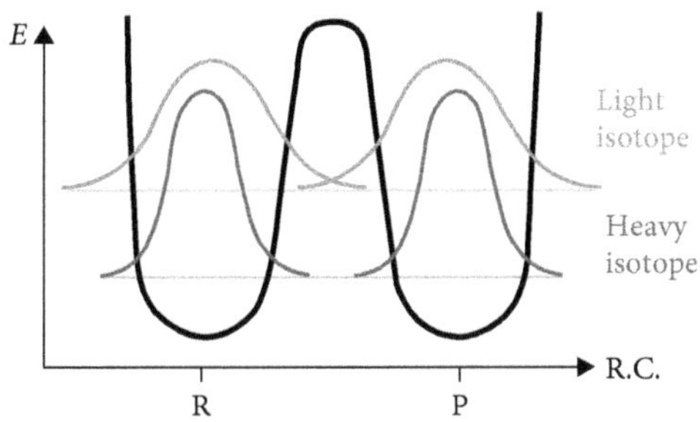

Figure 1. An example of ground state nuclear tunneling along the reaction coordinate (R.C.). The reactant well (R) is on the left side and the product well (P) is on the right. The darker gray (top) and lighter gray (bottom) lines describe a light and a heavy isotope probability function, respectively. The greater the overlap of the R and P probability functions, the higher the tunneling probability.

A second example of moves 2 and 3 in a CAREER proposal is by Vyvyan (excerpt 11K). In the first paragraph, background information on allelopathy is provided (move 2), a field that explores how one plant can harm another by producing chemicals known as allelochemicals. In this example, the concept is explained only briefly, using only a few words embedded within a sentence. In the second paragraph, the author transitions to move 3, where an obstacle (gap) is identified, indicating that allelochemicals are quite difficult to isolate from their natural sources (submove 3.1). In the last paragraph (submove 3.2), the proposed work is introduced; this work will fill the gap by producing allelochemicals synthetically.

Excerpt 11K (Vyvyan, 2001)

BACKGROUND AND SIGNIFICANCE

[Move 2] Maximizing the world's agricultural efficiency depends on controlling unwanted pests—especially weeds. Synthetic herbicides have met the weed control needs of industrialized nations for over 50 years, but mounting ecological and human health concerns are shifting attention to alternative weed control technology based on cues from nature.[1,2] Allelopathy, the chemical interaction between plants and microorganisms, has likely been known for thousands of years. This area, however, has only recently opened up to significant, specific research aimed at determining which natural products are responsible for these interactions.[3] . . .

[Move 3] [An] obstacle to advancing this field is the minute quantities of the active compounds that can be isolated from the natural sources. . . . For example, the heliannuols (**1–11**, Figure 1) are a promising group of phenolic allelochemicals that exhibit activity against dicotyledon plant species.[13] Heliannuols A and D (**1** and **4**, respectively) are the most active members of the family with effective concentrations of 10^{-4} to 10^{-9} M. . . . The recent isolation of heliannuols F–K (**6–11**), in which over 6 kg of sunflower leaves was required to obtain just 1–2 mg of each pure compound, clearly illustrates their scarcity.[16]

Thus, a synthetic source of promising allelochemicals is essential if we are to comprehensively study the agent's mode of activity and establish its basic structure-activity profile. The proposed work addresses this need. We will synthesize alleopathic natural products isolated from the sunflower (the heliannuols) and structurally related compounds, in optically pure form, based on biomimetic phenol-epoxide cyclizations. The bioactivity of the targets and intermediates will be evaluated through laboratory tests on plant germination and growth. Bioassays will be performed on the synthetic intermediates to allow for the development of a preliminary structure-activity profile for these novel natural herbicides.

Perfecting Your Writing

One's research funding is dependent on clear thinking and the expression of ideas in a coherent manner. In other words, you have to convince readers that you know what you are talking about and that you will be able to meet your established goals.

—Richard Malkin, University of California–Berkeley

Excerpts 11A–11K offer excellent models for writing the Goals and Importance section of a research proposal. Here we focus on select aspects of writing that can assist you in writing this section.

Formal Vocabulary

Research proposals, like the other chemistry genres covered in this book, require the use of formal, academic vocabulary. When composing a research idea or goal statement (move 1) or introducing the proposed work (move 3), a novice writer might begin by saying "I will *see*" or "I will *discover*." Although grammatically correct, the first phrase is too nonspecific (what exactly does the writer mean by the word *see*?) and the second is overstated (you cannot know in advance that you will discover something). Another common mistake made by novice writers is to use the word *experiment* (e.g., "In this *experiment*, I will . . ."). Research involves countless experiments, which can span months or years; hence, scientists prefer words such as *work*, *project*, or *study*. These and other words and phrases to avoid are presented in Table 11.1, along with their more appropriate alternatives.

Table 11.1. Words and Phrases to Avoid When Describing Goals and Objectives, and Their More Appropriate Alternatives

Avoid words and phrases that	Instead, use words and phrases that
are informal and nonscientific, e.g.,	are more formal and scientific, e.g.,
"We will (*see, find out, look into, try to figure out if*) . . . "	"We will (*monitor, measure, determine, explore, examine, investigate, analyze*) . . ."
overstate what may be achievable, e.g.,	lead to measurable results, e.g.,
"We will (*discover, prove, create, cure, find, put an end to*) . . ."	"We will (*develop, apply, measure, quantify, explore, compile, examine, characterize, synthesize, modify, adapt*) . . ."
promise success prematurely (before the work has been done), e.g.,	acknowledge assumptions or predictions, e.g.,
"We will (*show, demonstrate, find, prove, discover, create*) *that* . . ."	"We (*hypothesize, expect, predict*) that . . ." "*Preliminary* results *suggest* . . ."
refer to a long-term project as an "experiment" or "lab," e.g.,	convey the long-term nature of the work, e.g.,
"In this *experiment*, we propose to . . ."	"In this *project*, we propose to . . ." "In this *work*, we propose to . . ."

Formal Vocabulary

See Chapter 16, "Language Tips," Part 2: Word Choice, about the formal (and academic) vocabulary that should be used in chemistry writing.

Exercise 11.5.

Consider the following sentences. Identify words that should be avoided and suggest more appropriate alternatives (see Table 11.1).

a. In this experiment, we will prove that polyurethane films can be attached to biosensors and used to detect nerve agents.

b. The goal of this research is to see if cattle injected with antibiotics will have lower carcass mass.

c. We propose to discover new types of sandwich compounds that are composed of polycyclic aromatic hydrocarbons around a single metal atom.

Word Usage, Punctuation, and Subject–Verb Agreement

Here we call your attention to commonly confused words, punctuation, and subject–verb agreement, writing features that are double underlined in excerpts 11A–11I and highlighted in nearby sidebars. These writing features are used correctly in the indicated excerpts (shown in parentheses), but they are often misused in writing.

- Commonly confused words
  - affect vs. effect (11A)
  - although vs. while (11C)
  - principal vs. principle (11E)
  - because vs. since (11G)
  - whereas vs. while (11H)
- Subject–verb agreement (11I)
- Punctuation
  - capitalization of chemical names (11A)
  - *cis* and *trans* (11A)
  - in situ (11B)
  - ______, which (11D)
  - First, (11F)
  - colons (11G)

To test your knowledge of these features, complete exercises 11.6, 11.7, and 11.8. (See also Chapter 16 and its web-based self-study exercises, with answers.)

Exercise 11.6.

Select the correct word(s) in each set of parentheses.

a. The (effects, affects) of the molecular environment are accounted for by classical mechanics.

b. Reactions will involve one-electron reduction of (Nitrite, nitrite) to (Nitric Oxide, nitric oxide).

c. The term (In situ, *in situ*, in situ) means "in place" or "on site."

d. (Because, Since) these ultrafast motions are the essence of every chemical process, detailed knowledge about their nature is of fundamental importance. (Rose-Petruck, 2000)

e. The mechanism is largely unknown, (while, although) it is generally believed to occur through a four-membered ring.
f. The data were analyzed using (principal/principle) component analysis (PCA).

Exercise 11.7.

Are any of these sentences punctuated correctly? Correct sentences with incorrect punctuation; some sentences may have more than one punctuation problem.

a. Our methods will include: (1) extraction, (2) concentration, and (3) purification of liquid and solid samples.
b. For example a polymer's interfacial characteristics determine chemical and physical properties such as these: permeability, wettability, adhesion, and friction. (Adapted from Fairbrother, 2000)
c. The objectives are twofold. First the nucleophilicity of different sites on nucleobases will be explored. (Adapted from Lee, 2001)

Exercise 11.8.

Select the verb that agrees with the subject in these sentences.

a. Burrows et al. (has, have) published an article entitled *Confronting Racism in Chemistry Journals*.
b. The topic of the next conference (is, are) oxy-fuels.
c. Profiling of the most resistant strains (shows, show) promise for identifying the structure–activity relationship.

Verb Tense–Voice Combinations

A few common verb tense–voice combinations for selected functions of the Goals and Importance section are shown in Table 11.2. Other combinations may also be used. As noted in Table 11.2, the research idea or goal is generally written in present tense and active voice. If the research idea or goal is stated in the first sentence of the proposal, we recommend avoiding personal pronouns. Personals pronouns (e.g., I, we) can be used elsewhere in the proposal.

present–active (first sentence)	The goal of this research *is* to investigate . . .
present–active (last sentence)	I *propose* to measure the energy transport of . . .

Table 11.2. Common Verb Tense–Voice Combinations for Key Functions in the Goals and Importance Section of a Research Proposal (other combinations are possible)

function	tense–voice combination	example
a. to introduce research area	present–active	Blood platelets *function* in many . . .
		Reaction rates *depend* on the . . .
b. to emphasize importance	present–active	The use of . . . *is* important because . . .
c. to state research idea or goal	present–active	I *propose* to measure . . . The overall goal of this research *is* to . . .
d. to explain concepts	present–active	Trehalose *is* a common disaccharide . . .
		The reactions of photosynthesis *occur* . . .
e. to cite others' works	present perfect–active	Trehalose *has become* . . . in the field of lyopreservation.[1,2]
		Masunov et al.[1,2] *have* previously *used* . . .
f. to identify a gap	present–active	An obstacle to advancing this field *is* . . .
	past–passive	However, these predictions *were found* to be inaccurate.
g. to highlight proposed work (to fill the gap)	present–active	Here, we *propose* to use . . .
	future–active	To address this need, we *will synthesize* . . .

Exercise 11.9.

Review the research goals in P1–P3. Identify the verb tense and voice used in each sentence.

Exercise 11.10.

Consult Table 11.2 to select the most appropriate tense–voice combination in each sentence. We have indicated the move or submove to assist you (see Figure 11.1).

a. (submove 1.2) The catalytic transformation of organic molecules *is/has been* of tremendous importance on laboratory and industrial scales. (From Warren, 2002)

b. (move 2) Most of the theoretical work regarding the utility of time-resolved photo-electron angular distributions to probe electronic, vibrational, and rotational dynamics *concerns/has concerned* neutral photoionization.[1,2,3] (Adapted from Sanov, 2002)

c. (move 3.1) Despite their importance, the surface science associated with these modifications *is/was* largely unexplored.

Phrases to Emphasize Importance

Careful readers may have noticed that importance is emphasized in both moves 1 and 2 of the Goals and Importance section. Indeed, successful writers find ways to stress importance whenever possible throughout their proposals. Here, we highlight a few ways that you can underscore importance in your own writing.

One way to stress importance is to indicate "widespread interest" in your research area. Common phrases used by researchers in their writing to underscore this interest are shown in Table 11.3. The left-hand column lists phrases for research areas that are well established; the right-hand column is for emerging areas of interest.

Research ideas with the potential to discover new truths or extend current knowledge are also important. Few scientists can resist the appeal of new knowledge or a chance to look inside the "black box." Funding is more likely to be granted if you can highlight the potential of your work to *offer new insights*, *shed new light*, *provide greater detail*, *explore in more depth*, or *study for the first time*. Emphasizing how your research goal will benefit others also establishes importance. Scientific advances are remarkable if they can improve quality, reduce cost, simplify measurements, extend life, increase efficiency, and so on. Complete exercises 11.11 and 11.12 for additional examples of how authors emphasize importance in their writing.

Table 11.3. Phrases that Emphasize Importance Due to Widespread Interest

well-established research areas	emerging research areas
long-term interest	previously unstudied
continued interest	promising new area
sparked renewed interest	remains virtually unexplored
extensively studied because	an area of mounting concern
for decades, chemists have studied	an emerging discipline
attracted widespread attention	attracted recent attention

Exercise 11.11.

Consider the following excerpts from successful NSF CAREER proposals. What words and phrases do the authors use to emphasize importance?

a. More and more examples of kinetic-energy-dependent surface dynamics of reagents on low temperature substrates are appearing in the literature. . . . Collectively, these studies point to the importance of the emerging discipline called "hot atom chemistry."[17,21] (Spain, 1997)

b. Aryl nitrenes have been studied extensively over the last decade and are used in several industrial processes such as microlithography.[1] . . . The pursuit for organic magnetic material has sparked renewed interest in nitrene intermediates, which are ideal candidates for magnetic material because of their high spin properties.[2] (Gudmundsdottir, 2001)

c. Among the most remarkable responsive gels have been those involving drug delivery.[14,35–37] (Lyon, 2000)

d. There are two primary reasons why a fluorescent chemosensor with visible emission would be desirable, in addition to its inherent aesthetic appeal. First, . . . (Finney, 1999)

e. The lack of a fundamental mechanistic understanding of surface modification processes has been highlighted in a recent panel report.[11] (Fairbrother, 2000)

f. Given that interfacial solvation affects chemical transport,[16] surface reactivity and electron transfer,[17–20] and macromolecular self-assembly,[21] predictive models of solvent-solute interactions near surfaces will afford researchers deeper insights into a host of phenomena in biology, physics, and engineering. (Walker, 2001)

Exercise 11.12.

Read this excerpt from an NSF CAREER proposal written by Tuckerman (1999). The author uses three approaches for establishing importance; he demonstrates interest, emphasizes a benefit, and establishes the need for new knowledge. Find a sentence that illustrates each approach.

> Several hundred to several thousand feet beneath the ocean floor in permafrost and continental edge regions lies a potentially vast source of natural gas: in excess of 10^{16} cubic meters of gas hydrates, consisting largely of methane clathrate (53–55). Gas hydrates are nonstoichiometric compounds consisting of hydrogen-bonded water molecules in a cage-like structure, which traps small-diameter gas molecules. The methane clathrate, in particular, is stable in the pressure range of roughly 20–40 MPa and temperature range of 175–250 K (55). Interest in these

vast beds of crystallized gas hydrate is rapidly increasing for several reasons. They constitute perhaps the largest untapped source of hydrocarbon energy available. They also pose a possible threat as an enormous source of greenhouse gas. As global warming continues to cause oceanic temperatures to rise and permafrost regions to recede, the risk of destabilization of the beds increases and with it the risk of releasing large quantities of trapped greenhouse gas into the atmosphere. Given the potential importance of gas hydrates, it is vital that we expand our knowledge of their physical properties, which as yet, are only poorly understood.

Chapter Review

Check your understanding of what you've learned in this chapter by defining each of these terms, in the context of this chapter, for a friend or colleague who is new to the field:

background information	gap statement	research goal
formal vocabulary	ordinal language	research idea

As a review, try explaining the following to a friend or colleague who has not yet given much thought to writing the Goals and Importance section of a research proposal:

- main purpose of the Goals and Importance section
- graphics used to illustrate key concepts
- gap statements used to establish need and importance
- words and phrases that emphasize importance

12 *Writing the Approach and Methods Section*

A research proposal must present an effective argument that will persuade those who read and evaluate proposals . . . that the research proposed is significant, that the methods of pursuing this research are well designed, and that the researcher is in fact capable of carrying out the research.

—Beal and Trimbur (2001)

This chapter focuses on writing the methodology section of a research proposal, the section in which you tell readers how you will conduct the proposed work. It is typically the longest section of the proposal. Unlike the experimental section of a journal article, where you describe work completed in the past, here you describe what you propose to do in the future (if funded). Reviewers, after reading this section, should be convinced that the work is plausible and that the investigator has the background and expertise necessary to carry out the proposed work. By the end of this chapter, you should be able to do the following:

- Distinguish between prior accomplishments and preliminary findings
- Describe your proposed methods in a logical series of tasks or aims
- Develop project aims that pass the SAM test
- Identify obstacles in your work and suggest possible contingency plans

Writing on Your Own: Approach and Methods

Visit the *Write Like a Chemist* (2nd ed.) website for step-by-step guidance in writing your own Approach and Methods section for a research proposal.

Write Like a Chemist. Marin S. Robinson and Fredricka L. Stoller, Oxford University Press. © Oxford University Press 2022.
DOI: 10.1093/oso/9780190098940.003.0012

Reading and Analyzing Writing

We begin with excerpt 12A, an Approach and Methods section that you can read and analyze on your own. Excerpt 12A continues excerpt 11A, in which Pearson (2019) presented her research idea. Recall, she proposed to design nanoparticles (NP) with a cisplatin shell and iron oxide core that would more effectively target cancer cells while reducing unwanted side effects on the kidneys. She labels this section **Experimental Plan** and organizes her work into three phases over three years (36 months).

Exercise 12.1.

As you read and analyze excerpt 12A, complete these tasks:

a. Substantiate or refute the following statement: The author presents a logical progression of projects in her Experimental Plan.

b. The excerpt uses future tense and passive voice extensively (e.g., "The synthesis method . . . will first be developed using . . ."). In addition to "will be developed," list five other verbs that the author uses after the words "will be."

c. The author uses past tense once in this excerpt. Find that sentence and explain why the author shifted from future to past tense.

d. Find examples that illustrate chemistry writing conventions for (1) introducing an abbreviation for the first time, (2) reporting a number with units of time or measure, and (3) formatting an in-line citation.

Excerpt 12A (Pearson, 2019)

Experimental Plan

Months 1–8. The synthesis method for the iron oxide-cisplatin core–shell NPs will first be developed using the aqueous-phase modified polyol synthesis method. After the NPs are synthesized, they will be characterized using dynamic light scattering for size and zeta potential data because size impacts the circulation time of the NPs, and zeta potential impacts the in vivo NP interactions. Characteristic absorption and emission spectra, which will detect any unintentional synthesis side products, will be collected via UV–vis spectroscopy, and FTIR spectroscopy will be used to determine the NP shape.

Months 9–20. The second phase of the study will focus on in vitro testing of the NPs. Because cisplatin and iron oxide have both been used to treat breast cancer, the core–shell NPs will be tested against breast cancer cell line MCF-7 and compared with healthy breast cell line MCF-12A1. These cells were chosen because they require similar culture media and are both epithelial cells.[6] There will be four types of tests for each cell line: cisplatin alone, iron oxide alone, cisplatin-iron oxide core–shell NPs, and no treatment. The cisplatin alone and iron oxide alone tests will be used to determine the inherent cytotoxicity of each as compared with the core–shell NPs. For each test, viability data and confocal images will be collected to determine the overall effects and distributions of the various treatments within each type of cell.

Months 21–36. The third phase of the study will use mice for in vivo testing of the targeting characteristics of the NPs. Female C57BL/6 mice will be used, with half subject to MCF-7 and the other half to MCF-12A cells, which will be injected into the mammary fat pad.[6] After the MCF-7 tumors have grown to 5 mm in diameter, as measured by magnetic resonance imaging (MRI), the two mice groups will each be divided into four groups, again testing cisplatin alone, iron oxide alone, cisplatin-iron oxide core–shell NPs, and no treatment. The mice will be treated weekly for 12 weeks, during which time each mouse will undergo in vivo MRI twice per week to track the NP movements as well as the tumor/healthy cell growth. At the end of the 12 weeks, each mouse will undergo a kidney biopsy to determine the effects of each treatment on kidney health.

UV–vis and FTIR *what?*

Remember to include a noun such as "spectroscopy" or "spectra" after using these well-known abbreviated terms.

Analyzing Organization

Writing a research proposal is a good way for you to learn how chemists work—how they define meaningful problems, design experimental approaches to investigate these problems, and anticipate how their research will contribute reliable scientific knowledge to the community of chemists. Writing research proposals, in other words, asks you to do chemistry.

—Beal and Trimbur (2001)

A move structure for the Approach and Methods section is shown in Figure 12.1. The section is organized around three key moves: highlight prior accomplishments (move 1), share preliminary results (move 2), and describe methodology (move 3). **Prior accomplishments** (move 1) are completed works, typically published, that are building blocks for the proposed project. **Preliminary results** (move 2) are unpublished results that provide promising evidence that the proposed project will be successful. Note that moves 1 and 2 are marked "if applicable" in Figure 12.1. This is because early-career researchers (such as undergraduate or graduate students) may not yet have prior accomplishments or preliminary findings to share. (For example, Pearson omitted both moves 1 and 2 in her proposal shown in excerpt 12A.)

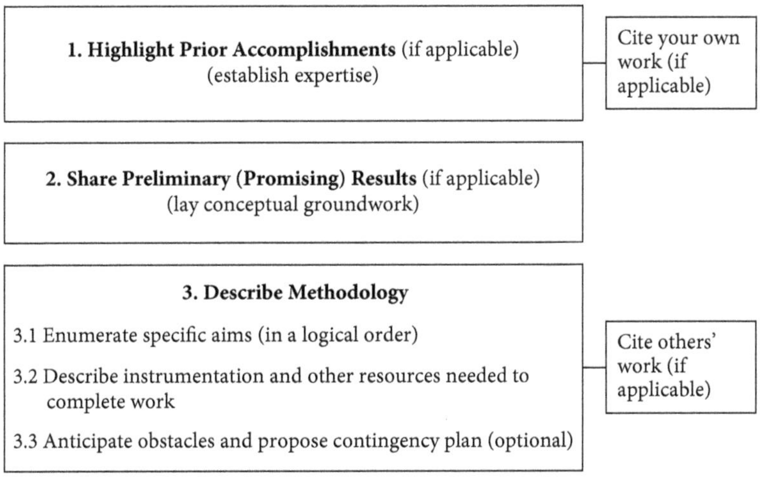

Figure 12.1. A suggested move structure for the Approach and Methods section of a research proposal (unless the RFP specifies different sequencing and/or contents).

Prior Accomplishments vs. Preliminary Results

Prior accomplishments refer to published or disseminated works. *Preliminary results* refer to unpublished findings. Both should be related to the proposed work.

In move 3, authors describe the methodology that they will use to achieve the overarching research goal. In submove 3.1, they enumerate **specific aims** that will be accomplished over the duration of the grant. The term "aims" refers to individual steps or tasks that must be completed to reach the **larger goal(s)**. In submove 3.2, they identify the resources needed to complete the work (e.g., instrumentation, laboratory facilities, supplies). Although some proposals allow you to request new instrumentation, others (like the NSF GRFP proposal) expect you to already have access to needed resources. In submove 3.3, authors anticipate obstacles or challenges that they are likely to encounter in their work. In research, it is understood that not everything will go according to plan; it is best to predict these pitfalls and propose possible alternatives should they arise.

Specific Aims vs. Larger Goal(s)

In proposals, authors must differentiate between the overarching goal of their research (the big goal or research idea, which we addressed in Chapter 11) and the smaller goals that lead up to the larger goal. In this chapter, we use the term *aims* to connote the smaller goals. Words such as *steps* or *tasks* can also be used.

Exercise 12.2.

Review the RFP for the NSF CAREER proposal (excerpt 10B in Chapter 10). Which of the moves in Figure 12.1 are specifically addressed in this RFP?

Analyzing Excerpts Move by Move

With Figure 12.1 in mind, we are ready to consider these moves in more depth and analyze how authors accomplish these moves in authentic GRFP and CAREER proposals. Later in the chapter, we examine common writing conventions used in the Approach and Methods section.

Move 1: Highlight Prior Accomplishments

The overarching goal of move 1 is to establish your expertise and convince your reviewers that you are likely to succeed if awarded funding. Success builds on success, and individuals who have completed a successful project in the past are more likely to succeed again. One of the best ways to demonstrate your expertise is to cite your own publications. It is both appropriate and expected that authors call attention to their authored (or co-authored) works, provided they are related to the proposed work and recent (i.e., published within the past 5 years). The term "prior accomplishments" refers to completed works (e.g., published articles or otherwise disseminated results) and other achievements (e.g., awards, collaborations). (Unrelated publications may be listed in an accompanying biographical statement.) This move should be skipped if there are no prior works, which is often the case for proposals written by undergraduate or graduate students. Subheadings are common in this move (e.g., **Previous Work** or **Prior Accomplishments**).

Let's begin with examples of move 1 from NSF GRFP proposals, which range in length from a few words to an entire paragraph (with a full-color graphic). The length typically reflects how much the proposed work depends on knowledge gained from prior work(s). If dependence is minimal, Cain's approach (P1) is appropriate: she simply states that the proposed work will expand on earlier findings, without sharing any details. Similarly, in Wait's case (P2), she briefly highlights an overlap between earlier work and the proposed work, both of which use RRKM theory to predict rate constants with and without additional CO_2 molecules. The success of the prior work builds confidence that the new work will also succeed.

P1 Expanding upon previous work,[3] the aims of this proposal are. . . (From Cain, 2019)

P2 **Prior studies**. The CO_2 environment of the reactive system contains too many atoms to be treated quantum mechanically with transition state and RRKM theories. However, the few nearest molecules can be included in the extended reactive system. Recently, we applied RRKM to predict the rate constants for the reactions $OH + CO \rightarrow H + CO_2$, and $OH + CH_2O \rightarrow H_2O + CHO$ with and without any additional CO_2 molecules in the reaction complex.[5,6] (From Wait, 2019)

The longest example of move 1 is in Weaver's proposal (excerpt 12B). Her proposed work depends extensively on a previous work that developed a microscope setup (stroboSCAT) to resolve ultrafast anisotropic diffusion in polycrystalline thin films. In her proposal, she illustrates this technique with a graphic (Figure 1, excerpt 12B) that shows (a) the stroboSCAT optical setup and (b) images from the former polycrystalline film study. In her proposed work, this setup will be adapted to investigate ultrafast energy transfer in photosynthesis.

Excerpt 12B (Weaver, 2019)

Prior Work: I plan to extend the capability of an ultrafast microscope developed in the Ginsberg lab to the study of thylakoid membrane samples. The apparatus leverages the sensitivity of interferometric scattering microscopy (iSCAT) as a probe in a time-resolved (stroboscopic) pump–probe technique that we call stroboSCAT (Fig. 1a).[3] A correlative baseline image captures the sample morphology while a pump–probe differential image quantifies energy migration with few-nanometer sensitivity and picosecond resolution. The detection scheme is label free, relying only on the interference between elastically scattered light from the sample and reflected light from the sample–substrate interface. Contrast in the differential image is generated by a change in polarizability between the excited and unpumped populations. We demonstrated the power of this technique to resolve anisotropic diffusion in polycrystalline thin films where grain boundaries act as traps and barriers that halt or redirect charge carriers (Fig. 1b).[3] With improvements in time resolution and sensitivity, this instrument is well suited to investigate energy transfer in photosynthesis.

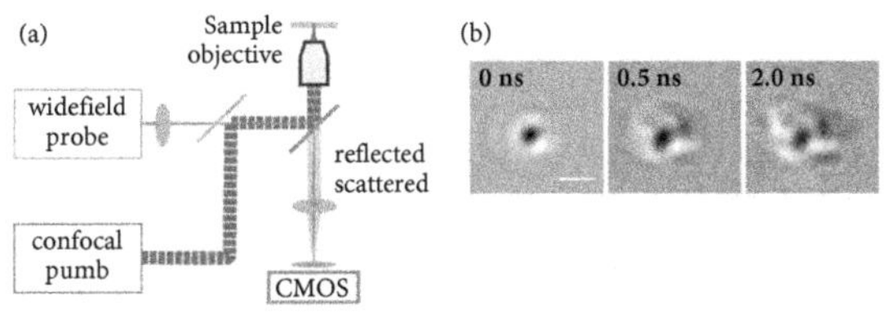

Figure 1. (a) StroboSCAT optical setup and (b) differential stroboSCAT images in $MAPPbI_3(Cl)$ thin film with ~1 μm grains.

Faculty competing for NSF CAREER awards also want to call attention to their previous accomplishments. We consider three examples; in each one, the author cites one or more publications and draws a clear connection between these earlier works and the proposed project. For example, Lee (excerpt 12C) refers to what they "have learned" already about the N1 and N3 acidities of uracil. Lorigan (excerpt 12D) indicates that he has already accomplished the first logical step in his project, as described in several published articles. Similarly, Vyvyan (excerpt 12E) indicates that they recently completed the total synthesis of (±)-helliannuol D; hence, they are now prepared to synthesize additional heliannuols.

Note, too, the different ways in which the authors cite their publications in these excerpts. Lee and Lorigan (excerpts 12C and 12D, respectively) both mention that their papers were published in the *Journal of the American Chemical Society.* Both abbreviate the journal's name correctly, *J. Am. Chem. Soc.*, rather than using the more familiar acronym, *JACS*. This flagship journal for the ACS, like *Nature* and *Science*, is a particularly prestigious one; hence, authors often mention it by name. In contrast, Vyvyan includes only the citation number, thereby conserving space. All three approaches are fine, and which one you choose is at your discretion.

Excerpt 12C (Lee, 2001)

1. Nucleic Base Acidities. *N1 and N3 Acidities of Uracil* (*J. Am Chem. Soc.* **2000**, *122*, 6258–6262). Our progress to date involves calculations and experiments for the determination of nucleobase activities. We have learned that there is an enormous difference in the inherent stabilities of anions at the N1 (**7**) and N3 (**8**) positions in uracil (**5a**, Figure 1).

Excerpt 12D (Lorigan, 2002)

Logically, the first step for developing this method at lower magnetic fields (X-band for EPR studies) is to fully characterize the bicelle system without the added complications induced by protein–lipid interaction. Over the last two years, we have accomplished this goal and published several articles describing our results (*81*, *86–88*). The first paper was published in *J. Am. Chem. Soc.* and demonstrates the effects of magnetic phospholipids bilayer alignment by showing an EPR spectrum of a DMPC/DHPC/Yb^{3+} "bicelle" sample doped with a cholestane spin label (*81*). . . . We followed up our initial report with a second *J. Am. Chem. Soc.* paper that spectroscopically characterizes magnetically oriented phospholipid bilayers with EPR spectroscopy and discusses conditions for optimal bicelle alignment at X-band (*86*).

Excerpt 12E (Vyvyan, 2001)

Our general approach to the synthesis of the heliannuols is illustrated by our recently completed total synthesis of (±)-helliannuol D (**4**).[19]

J. Amer. Chem. Soc. or JACS?

In writing, formal abbreviations should be used for the *Journal of the American Chemical Society* (*J. Amer. Chem. Soc.*, not *JACS*) and *The Journal of Organic Chemistry* (*J. Org. Chem.*, not *JOC*).

Exercise 12.3.

Numbers are used in several different ways in excerpts 12C–12E. Give one example of each of the following (and specify the excerpt in which you found it):

a. a bolded number used as a compound label
b. an italicized number used in a citation
c. a superscripted number used in a citation
d. inclusive page numbers used in a reference

Move 2: Share Preliminary (Promising) Results

In move 2, authors share preliminary results. These are unpublished findings intended to convince reviewers that the proposed work is on the right track and deserves funding. Scientists rarely seek funding for work that has yet to be started. (Indeed, chemists often joke that proposals are only funded if 90% of the work is already done!) Preliminary results should also have a "hook"; that is, they should capture reviewers' attention and help them grasp the value of and need for the proposed investigation. The take-home (implied) message should be "Look at these promising results! It would be a loss to science to not complete this work (so please give me funding to do it!)."

In addition to text, preliminary results are presented in full-color, computer-generated graphics. We first saw computer-generated graphics in excerpts 11A and 11J in Chapter 11, where they were used to introduce fundamental concepts about the research area. Graphics serve a similar purpose here. High-quality graphics help to elucidate important concepts and lay the foundation for the proposed research. The familiar saying "a picture is worth a thousand words" is absolutely true in proposals, and effective writers frequently use illustrations in this section to introduce their research story.

Computer-Generated Graphics

Full-color, high-quality computer-generated graphics (drawings and illustrations) are widely used in proposals to illustrate scientific concepts.

We begin with an excerpt from an NSF GRFP proposal where move 1 (share prior accomplishments) and move 2 (share preliminary results) are accomplished in the same paragraph (excerpt 12F). This is not surprising because most published works become "preliminary findings" for the next project, and the distinction between the two is often subtle. In move 1, the author describes results from a recently published work where a highly hindered olefin (**2**) was successfully arylated using ArBr. He now proposes to vinylate **2** using a halogenated cyclohexene, as illustrated in Figure 1c (a **boxed graphic**). In move 2, he shares preliminary results of the vinylation: 20% of the desired (vinylated) product is produced when ArBr is replaced with 1-bromocyclohexene, even under conditions optimized for arylation. These (promising) results suggest that the proposed transformation is likely to succeed.

Boxed Graphics and Text Wrapping

Graphics in proposals are commonly inserted into the text, with the text wrapping around the graphic. In some cases, the graphic is also boxed. To save space, slightly smaller font is sometimes allowed for figure captions.

Excerpt 12F (Huffman, 2019)

[*Figures 1a and 1b are omitted, which show* ***1****,* ***3****, and* ***5***]

In my first year of graduate school, I developed a carboxylate directed Heck reaction that allowed arylation of highly hindered olefins under oxidatively neutral or reductive conditions (Figure 1b).[5] The directivity of the pendant carboxylate enabled the formation of tertiary and quaternary centers in cyclic systems, giving rise to sp^3–sp^2 attached-ring motifs in several examples (e.g., **3**, **5**). With the goal of accessing more highly saturated systems, I have now turned my attention to a vinylation that could be applied to the synthesis of darwinolide (**1**, Figure 1a). Our conditions were optimized for arylation, but the production of 20% of the desired product using 1-bromocyclohexene suggests that such a transformation is possible (**6**, Figure 1c). This work seeks to address the methodological issues of attached rings apparent in the natural product darwinolide through the two aims described below.

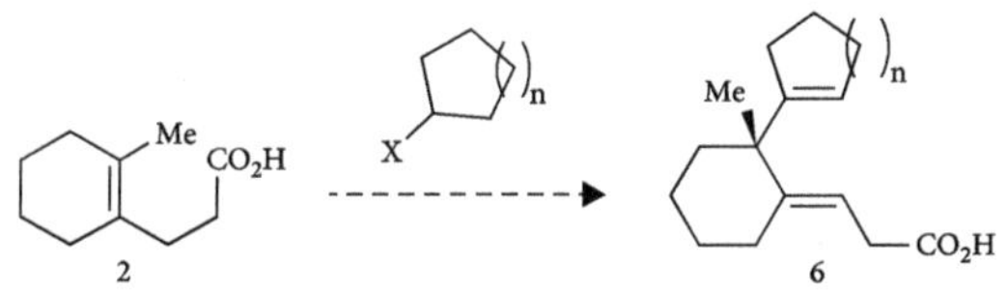

Figure 1c. Proposed vinylation.

"I developed" rather than "I discovered"

Be cautious about claims that you "discovered" something; instead use words such as developed, investigated, or proposed.

That vs. Which

See Chapter 16, "Language Tips," Part 6: Word Usage, for tips on using these commonly confused words.

For more experienced investigators, preliminary results can be extensive and span several pages. Subheadings are common (e.g., **Preliminary Results** or **Preliminary Findings**) to call attention to this section. We consider two examples from NSF CAREER proposals. In both cases, the authors build conceptual understanding of the proposed work as they share preliminary results. In excerpt 12G, Lyon describes his group's progress in using hydrogels to fabricate thin films. As he shares his preliminary findings, he uses text and graphics to help readers better understand his plans to construct multilayered films. In Figure 2, he uses a drawing to illustrate how anionic hydrogel particles are attracted to an amine-coated surface. Three anionic hydrogels are taken up by four amines, which is indicative that a "complete monolayer" does not form. In Figure 3 (*not included*), he presents empirical evidence from atomic force microscopy to support this hypothesis. In Figure 4, he illustrates how a single surface can be extended to multiple layers, formed by alternating layers of hydrogel particles (negatively charged) and poly-allyamines (positively charged). Figure 5 (*not included*) shows "data recently collected in (his) group" that "show conclusively" their ability to construct such multilayered films. After reading this section, his reviewers are better prepared to understand his work and are likely more convinced that it will succeed.

Exercise 12.4.

Read excerpt 12G and then answer these questions:

a. List two concepts that Lyon elaborates upon as he shares his preliminary results. Consider text and graphics.
b. How are Figures 2 and 4 labeled? Are the captions above or below the figures?
c. Would excerpt 12G be as effective without the figures? Explain.

Excerpt 12G (Lyon, 2000)

Preliminary Results

Thin Film Fabrication. The acrylic (AA) or methacrylic (MAA) acid portion of the copolymer also offers a group that will interact electrostatically with amines—a convenient method for immobilization of polymers to surfaces.[55–57] Figure 2 shows the general methods; modification of a solid surface with the appropriate alkylamine renders the surface amine-coated. Exposure of that surface to a suspension of carboxylated hydrogel spheres results in surface immobilization of particle by electrostatic attachment. Figure 3 (*not included*) shows an atomic force microscopy (AFM) image of a representative hydrogel-modified surface recently prepared in our group. Due to interparticle repulsion, a complete monolayer is not formed under these conditions, resulting in a surface decorated with well-defined particles.

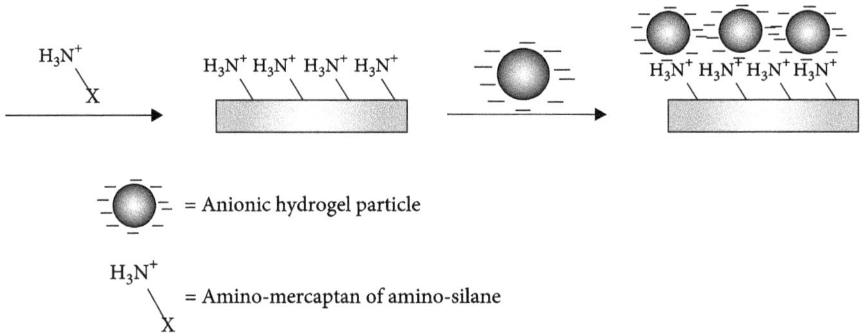

Figure 2. Electrostatic adsorption of colloidal hydrogels to functionalized surfaces.

This result suggests that polyelectrolyte multiplayer assembly methods can be applied to these materials. These methods (developed by Lvov and Decher)[58-60] allow for the construction of stable, reproducible, submicron-thick, polymer films by alternating layer adsorption of polycations and polyanions from aqueous solutions. Accordingly, we have begun investigating the feasibility of this multiplayer film assembly method using hydrogel particle or linear polyelectrolytes (poly-styrenesulfonate) as the anion component and linear polyelectrolytes (poly-allylamine) as the cationic component in an alternate-layer deposition process (Figure 4).

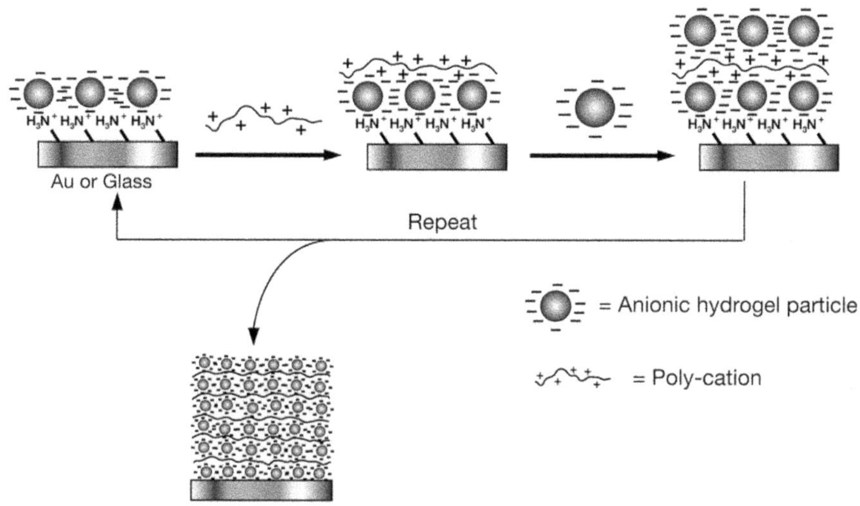

Figure 4. Layer-by-layer polyelectrolyte adsorption.

Figure 5 (*not included*) shows representative quartz crystal microgravimetry data recently collected in our group for the deposition of just such a multilayer film. . . . Despite the need for further investigation of the films, these data show conclusively that we are able to construct multilayered nanoparticulate hydrogel films via simple solution-based assembly methods. Films prepared in this fashion are currently being characterized with respect to their thickness and morphology (via profilometry and atomic force microscopy), porosity (voltammetry), viscoelasticity (crystal impedance), refractive index (surface plasmon resonance), and molecular structure (FTIR spectroscopy). These studies are elaborated upon in the research methodology.

Data/Datum

Data (plural)
Datum (singular, rarely used)

Look over excerpts 12I, 12K, 12L, and 12N for additional uses of the word *data* (plural) in context. See also Chapter 16, "Language Tips," Part 4: Grammar, for tips on other plural and singular scientific words.

Excerpt 12H, also from an NSF CAREER proposal, offers a second example of move 2. The proposed work involves the preparation of thin films at molecular-length scales using strategies of crystal engineering. Patrick proposes to investigate thin film growth on substrates submersed in a liquid crystal (LC) solvent/solute mixture under the influence of an external magnetic field. In a preliminary investigation, they deposited a fatty acid (tetracosanoic acid) from an LC solvent onto a graphite substrate. In the presence of a magnetic field, the deposited film was uniformly oriented (crystalline); in the absence of the field, a randomly oriented (polycrystalline) film was observed. Patrick includes an illustration to depict these phenomena. By combining both text and illustration, Patrick presents his preliminary results in a way that establishes a clear conceptual framework for his proposed work.

Excerpt 12H (Patrick, 2000)

Summary of Preliminary Results

We have conducted preliminary investigations to assess the way LC solvents influence film formation on highly oriented pyrolytic graphite (HOPG) substrates. A number of different molecular solute/LC solvent combinations have been investigated over the last year. In most cases, samples were prepared by deeply immersing a graphite substrate into a reservoir of the LC/solute mixture at ~100 °C, then allowing the system to gradually cool to room temperature (Figure 3). . . . The reservoir was located in a magnetic field, oriented with the field axis in the plane of the substrate surface. After preparation, samples were removed and analyzed with scanning tunneling microscopy (STM). The STM tip penetrates through the thick LC solution to image molecules in the monolayer at the graphite interface. In some systems, we observed films with macroscopic uniformly aligned (crystalline) monolayers formed at the graphite interface in contact with the LC mixture. Results from three representative systems are summarized in Table 1 *(not included)*.

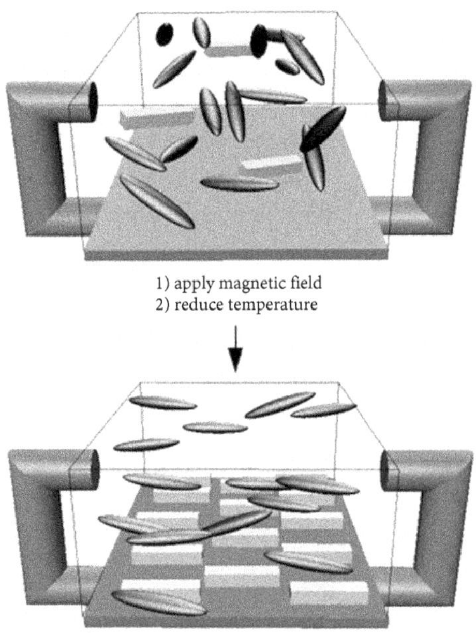

Figure 3. Preparation of oriented films.

100°C

Remember that the degree symbol goes with the C, not the number.

The last example (excerpt 12I) illustrates how authors incorporate words such as *promising* and *recently* in move 2. Phrases such as "early results are *promising*" or "our efforts appear *promising*" are common and can be effective if not overused. The word *recently* (or *recent*) is also common in move 2. Although *recently* should be avoided in journal articles (because *recently* loses its meaning so quickly after a paper is published), its use is less objectionable in proposals, which are short-lived documents. Investigators like the word *recently* because it conjures up the image of a research group actively engaged in the proposed work (e.g., "we have *recently* prepared," "our *recent* NMR data reveal," or "our *recently* completed total synthesis"). Walker also uses the phrase "the *first* evidence of __" in excerpt 12I. The word *first* in this context is used to underscore the originality of the proposed work.

Excerpt 12I (Walker, 2001)

Early results are promising. We have recently completed a preliminary study that examined the solvatochromic behavior of 4-aminobenzophenone (4ABP) adsorbed to hydrophilic substrates from a variety of solvents. Hydrophilic substrates are polished quartz surfaces that demonstrate complete wetting when exposed to water. FTIR spectra reveal that the quartz surfaces are primarily SiOH and adsorbed water with no detectable hydrocarbon contaminants. . . . Figure 5a (*not included*) shows the SHG spectrum of 4ABP adsorbed to a hydrophilic interface from cyclohexane. . . . To our knowledge, these data represent the first evidence of surface-induced solvatochromism at solid–liquid interfaces. Superimposed are UV–vis spectra of 4ABP in bulk cyclohexane and bulk diethyl-ether. As anticipated, data show that the hydrophilic, silanol terminated surface renders the interactial environment more polar than bulk cyclohexane.

Recently

The words *recently* and *recent* should be avoided in journal articles, but they can be used in short-lived research proposals.

Exercise 12.5.

Read the following two passages from an NSF CAREER proposal (Kinsel, 1999). How does the author build the case that his early results are promising without actually using the word "promising"?

> P3 In our initial studies, we have demonstrated, for the first time, that protein binding to surfaces has a significant impact on subsequent protein MALDI MS ion signals. This demonstration of principle was made possible through a number of important collaborations and student interactions between members of our research group and faculty in the Department of Chemistry and the Biomedical Engineering Program at the University of Texas at Arlington.
>
> P4 Our preliminary work strongly suggests that there is an intimate relationship between the MALDI ion signals for surface-adsorbed proteins and the affinity of the surface for binding those proteins. Although this relationship appears well demonstrated in the examples above, numerous questions remain to be answered.

Move 3: Describe Methodology

In scientific papers, you write about events that have already occurred. In a proposal, however, you write a plan for solving a problem. In other words, you write about what you will do, as opposed to what you have done. Describing imagined details is more difficult than describing actual details.

Despite the challenges, the writing of a proposal is a way for you as . . . a scientist to change the world.

—Alley (2018)

In the third (and final) move of the Approach and Methods section, you describe the methodology you will use to achieve your overall research goal. In submove 3.1, investigators present a well-organized and logical progression of project aims. The word "aims" refers to the smaller tasks or steps that collectively lead to the larger research goal. Aims are often presented in a list (numerical or bulleted), organizing the tasks in sequential or chronological order. Lower risk, more familiar projects are generally listed first, followed by higher risk, more exploratory ideas. In submove 3.2, investigators describe the instrumentation and other resources needed to complete the work (laboratory facilities, equipment, computational tools, etc.). In submove 3.3, investigators describe obstacles that they may encounter in their work. Although optional, reviewers are more likely to trust researchers who have considered "what might go wrong" and acknowledge the limitations of their approach.

Project aims should be realistic but ambitious. This balance is not always easy to find. To be successful, you must forge ahead into less familiar territory and approach the "cutting edge" or forefront of your field. At the same time, your aims should be achievable. One way to discern the latter is to apply the SAM test: are your aims *specific* (with a clear road map and timeline), *achievable* (within the projected time frame), and *measurable* (with tangible, measurable outcomes)? Of course, not all project aims will be achieved, and initial assumptions may have to be discarded in the harsh light of empirical evidence. Serendipitous discoveries may also take you in unexpected directions—such is the nature of science. Reviewers understand this process, but they are also unlikely to support a project that is overly ambitious or too speculative.

SAM Test

Effective project aims should pass the **SAM** test by being
Specific
Achievable
Measurable

Let's now examine how authors accomplish move 3 in their research proposals. We begin with three examples from NSF GFRP applications. In each case, the project aims are presented in a logical order. You may not be familiar with much of the chemistry described in these aims, yet it is still apparent that they are organized sequentially and that a great deal of thought has gone into their organization.

The project aims in these excerpts have subheadings that are parallel in terms of punctuation, capitalization, font, and part of speech. For example, in excerpt 12J, each subheading is written in italic font and begins with a verb (*measure*, *measure*, and *image*). In excerpt 12K, the subheadings are bolded and in title case (all key words capitalized), and they end with a period; the specific aims, after the colon, begin with adjectives followed by a nominalization (vinylation, synthesis). In excerpt 12L, the tasks are bolded, end with a period, and begin with the word "Use." The use of parallelism in subheadings and lists makes it easier for readers (i.e., reviewers) to quickly grasp the ideas that are presented.

The project aims in excerpts 12J–12L are also numbered. Alternatively, ordinal language can be used (e.g., first, second, last) to reinforce a logical progression of ideas or outline a road map of planned activities. That said, in sections of a proposal that describe a proposed method, such as in a synthesis, ordinal language is used sparingly. This is particularly true for the word "next," which is easily overused and can often be deleted without loss of meaning.

Parallelism

Headings and subheadings should be parallel in format and begin with the same form of speech. See Chapter 16, "Language Tips," Part 4: Grammar, for tips about parallelism in written work.

Exercise 12.6.

Read excerpts 12J–12L. How many instances of the words *first*, *second*, *then*, and *next* do you find in these excerpts? What purpose do they serve? Can any of these words be deleted without loss of meaning?

Excerpt 12J (Samuels, 2019)

Specific Aims

1. Measure the permeation, concentration, and kinetics of d14-trehalose and d6-DMSO interacting with blood platelets in solution.

I will accomplish Aim 1 by first establishing the location of blood platelets through Raman imaging of their lipid membranes. Selected deuterated lyoprotectant will then be imaged as the cells are exposed to a solution containing 0.1 M d14-trehalose in phosphate buffered saline (PBS) or 10% d6-DMSO in PBS. Switching between imaging lipid membranes and cryoprotectant can be accomplished by changing the wavelength of the pump beam of the instrument. Because the Raman signal is proportional to concentration, the relative concentrations of d14-trehalose and d6-DMSO can be determined, as well as the location of each lyoprotectant. I will measure the diffusion of the molecules by imaging the cells over time and recording how the cellular concentration of each molecule changes with time. I expect the diffusion of trehalose to be much slower than that of d6-DMSO, as trehalose is considered a non-penetrating molecule and DMSO a penetrating molecule.

2. Measure the permeation, concentration, and kinetics of a mixture of trehalose and DMSO interacting with blood platelets in solution.

Similar to Aim 1, I will accomplish Aim 2 by first imaging the lipid membranes of the blood platelets to establish location and then imaging the cells as they are exposed to a mixed solution of 0.1 M d14-trehalose and 10% protonated DMSO in PBS or 0.1 M pronated trehalose and 10% d6-DMSO in PBS. Comparing the concentration of trehalose in cells exposed to a trehalose-DMSO mixture to the concentration of trehalose in cells exposed to a 0.1 M trehalose mixture will indicate if DMSO is increasing the permeability of the trehalose.

3. Image the platelets after they have been desiccated to determine if the concentrations of lyoprotectant molecules change as the cells dry.

Aim 3 will be accomplished through SRS imaging of the lipid membranes of the blood platelets and lyoprotectants contained within the cells after they dry.

Among vs. Between

Use "between" for two items; use "among" for three or more items. See Chapter 16, "Language Tips," Part 6: Word Usage, for more details on these easily confused words.

Excerpt 12K (Huffman, 2019)

This work seeks to address the methodological issues of attached rings apparent in the natural product darwinolide through the two aims described below.

Aim 1: Carboxylate-Directed Heck Vinylation. Extension of our directed Heck arylation to a vinylation is challenged by the unique behavior of vinyl vs. aryl halides; although oxidative addition is faster,[6] the resulting vinyl-PdII intermediate is more prone to decomposition.[7] The reaction conditions will thus need to be tuned to suppress undesired reaction pathways. I plan to use the automated synthesis facility at Scripps Research to screen ligand, solvent, base, halide, temperature, and possible additives. Statistical 'Design of Experiments' (DoE) optimization around this parameter space is expected to identify trends that lead to a productive reaction. These data should both produce optimal reaction conditions and inform about details of the reaction behavior. Exploration of various olefin and vinyl halide partners will provide information about the scope and limitations of the reaction.

Aim 2: Total Synthesis of Darwinolide. With a successful vinylation procedure in hand, I intend to pursue the synthesis of darwinolide as summarized in Figure 2. Vinyl-siloxycyclopropane **7**, of which the analogous ethyl ester is known, can be prepared in a three-step sequence of Wittig olefination, silyl enol ether formation, and Simmons-Smith cyclopropanation. A rhodium catalyzed formal [5+2] cycloaddition, a transformation previously rendered asymmetric using phosphoramidite ligands,[8] is envisaged to establish the key 7-membered carbocycle. Reduction of **8** to **9**, followed by oxidation to the dialdehyde and acidic deprotection-lactonization-acetal formation cascade, should form **10** with stereochemistry arising from steric-enforced thermodynamic equilibrium. After bromination to **11**, the key directed Heck vinylation can be performed using known unsaturated acid **12** under reductive conditions, with the stereochemistry specified by the carboxylate direction. Decarboxylation and metal hydride-catalyzed olefin isomerization of **13** will give the natural product (**1**), with the final stereocenter determined by *syn*-facial hydrogen migration. The ability to couple different olefin fragments late-stage and the synthetic handle present in the carboxylic acid will allow rapid generation of analogues and probes.

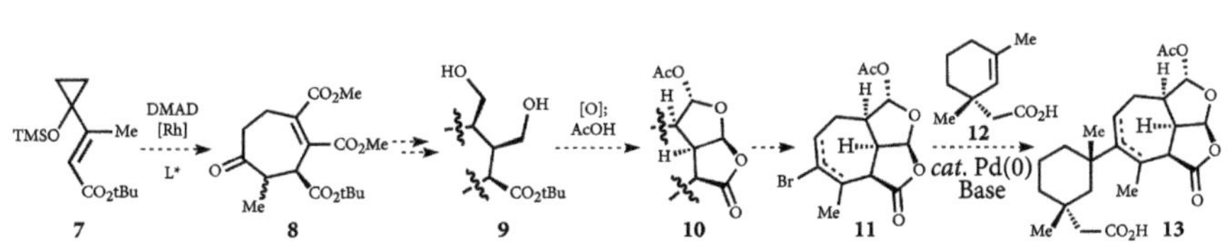

Figure 2. Proposed synthesis of darwinolide.

Excerpt 12L (Wait, 2019)

Tasks to be performed. To implement the proposed approach to predict kinetic effects of supercritical environment, the following tasks will be performed:

Task 1. Use enhanced MD to predict the effects of a supercritical environment on activation free energy. The CHARMM computer program will be used to carry out molecular dynamics simulations. The reactive system will be described with an empirical valence bond force field. It will be surrounded by 0 or 90 CO_2 molecules. Periodic boundary conditions will be applied to keep the density constant. Series of external restraints will be enforced to keep the reaction coordinate to the narrow intervals required for the umbrella sampling protocol. The obtained trajectories will be post-processed using the weighted histogram analysis method (WHAM)[7] so that free energy can be plotted as a function of the reaction coordinate. From this plot, free energy of activation for the reactive system alone (ΔG_{gas}) and the reactive system surrounded by 90 CO_2 molecules (ΔG_{90CO2}) will be calculated. Their difference constitutes the environment effect on activation free energy: $\Delta G_{env} = \Delta G_{90CO2} - \Delta G_{gas}$ and will be used in Tasks 2 and 3.

Task 2. Use TST to estimate high-pressure limiting rate coefficients. We will use the ab initio CCSD level of theory, as implemented in Gaussian 09, to predict electronic energies and vibrational frequencies for reactants and TS of the isolated reactive system. TST methods will use these data to predict k^{∞} for the gas phase. The energies of TS will be adjusted by the value of ΔG_{env}, calculated in Task 1. As a result, density-dependent predictions for k^{∞} in supercritical environments will be obtained.

Task 3. Use RRKM to estimate pressure-dependent rate coefficients. The predictions for the pressure-dependent rate constants k for reactions important in oxycombustion will be performed using RRKM theory as described above. However, electronic energies for transition states and intermediates will be adjusted by the value of ΔG_{env}, calculated in Task 1. In addition, the real gas equation of state will be used to relate pressure and molecular collision frequency.

Ab initio

The term "ab initio" means *from the beginning* or *from first principles*. It is usually not italicized in ACS journals, but it is italicized in other journals. In research proposals, we recommend that you follow the practice that is most common in your area of chemistry.

Project aims are also enumerated in NSF CAREER proposals. For example, in excerpt 12M, Lyon divides his project aims into two stages: the second stage is described as more *complex* and *challenging* than the first, and fewer details are included in stage 2. The lack of specificity in stage 2 is acceptable given the difficulties associated with projecting your work into the future.

Excerpt 12M (Lyon, 2000)

Bioconjugate Synthesis—Stage 1. As described above, we have already demonstrated our ability to produce thermoresponsive copolymer hydrogel particles that possess groups for protein attachment. These syntheses will continue, with the goal of producing gels across a range of NIPA/AA copolymer ratios (protein attachment points). . . .

The bioresponsivity of the hydrogel conjugates will initially be investigated with simple protein–protein and protein–ligand pairs where one of the binding partners will be covalently attached to the hydrogel. . . . First, we will use carboiimide coupling methods[54] to covalently attach one of the binding partners to the carboxylate moieties in the hydrogel. Specifically, we will . . . [*goes on to describe more specific methods*].

Bioconjugate Synthesis—Stage 2. The second stage of the bioconjugate project will involve more complex polymer structures and more challenging biological targets. First, we will explore factors such as protein loading, particle size, polymer crosslinking density, and copolymer ratio/identity in a highly parallel fashion in order to evaluate the biosensitivity of a wide range of hydrogel bioconjugates. This will allow us to rapidly converge upon a highly optimized polymer–protein conjugate based on NIPA hydrogels. Second, we will expand our approach to other coupling chemistries (e.g., maleimide, biotin–streptavidin, photo-crosslinking). By broadening the range of chemistries available, these colloidal bioconjugates will have wider applicability to other biochemistries and film fabrication methods. We will also begin to explore other biological interactions such as ligand–receptor binding, DNA hybridization, enzyme–substrate interactions, or protein–DNA complexation to further evaluate the generality of our approach.

Exercise 12.7.

Dyer (2001), in his CAREER proposal, proposes six key experiments to evaluate polar order and polar stability of organic self-assembled monolayers. Dyer uses ordinal language (first, second, third, fourth, fifth, and sixth) to make the sequence of experiments clear. However, even without this language, the order is predictable because the experiments are arranged in order of increasing complexity.

We have listed the first three of Dyer's six experiments (a–c). Arrange them in order of complexity (and, hence, the order in which they will be performed), identifying the correct ordinal language (first, second, third) where indicated. (SHG = second harmonic generation; SAM = self-assembled monolayer)

a. *First/Second/Third*, the SHG signal will be measured when both surfaces exhibit complementary H-bonding moieties. Ideally, we predict that the SHG signal should be twice that of the (*first, second, third*) experiment because both surfaces will contribute to the bulk polar orientation.

b. *First/Second/Third*, a control experiment will be performed . . . where the terminal functionality on both surfaces is an alkane. . . . In this instance, the SAM surface will not participate in H-bonding with the bulk liquid crystal (LC) and therefore the bulk material will not exhibit polar order; thus, the SHG signal should be very low.

c. *First/Second/Third*, the SHG signal will be measured when only one surface has hydrogen bonding SAM and the other surface is terminated with an alkane. We expect the SHG signal to be significantly larger than for the *first/second/third* experiment where both surfaces were terminated with alkanes.

The excerpts that we have considered thus far have all organized project aims sequentially, but other approaches are possible. One common alternative is to organize proposed work into themes or categories. For example, Harpp (excerpt 12N) proposed to collect samples that would be analyzed for trace metals, major elements, and isotopic ratios. One logical approach would be to organize her planned work based on these data types. However, the data themselves are not as important as the insights they can provide; hence, she organized her work instead by areas of inquiry that the data will be used to elucidate.

Excerpt 12N (Harpp, 1998)

Proposed Plan

[*Several sentences omitted*] The samples will be examined for petrographic variations, trace element and major element concentrations, and isotopic ratios. With these data, we will evaluate the following:

(1) Formation mechanisms. How are the northern Galápagos Islands and seamounts related to the plume and the ridge? Is the WDL a channel to the ridge from the plume, a leaky fault, or related to near-ridge volcanism? How is Genovesa related to the ridge, and the adjacent islands of Pinta and Marchena? We will examine the geochemical evolution of Wolf, Darwin, and Genovesa lavas to address these questions. . . .

(2) Melting dynamics and source composition. Rare earth inversion using the path integral method of White et al. (1992; McKenzie and O'Nions, 1991) will be used to quantify the melting parameters and approximate source compositions (e.g., Harpp, 1995; Mahoney et al., 1993). . . .

Rare earth element data will also serve as the basis for a forward modeling study to better constrain melting systematics in the Galápagos. The melting model will invoke clinopyroxene-rare earth element partition coefficients, which vary with composition (Gallahan and Nielsen, 1992), and a polybaric or column melting process . . .

Isotopic ratios will be used to constrain source compositions and the contributions from the plume and ridge to the island lavas . . . This information will assist in developing a map of the thermal and compositional structure of the present-day Galápagos plume and its interaction with the GSC.

(3) Spatial distribution of geochemical variations. Existing data from Marchena and Pinta Islands, PLUM02 data from the WDL, and the new Wolf, Darwin, and Genovesa data will be compiled into a comprehensive geochemical database for further examination of the formation mechanisms described above. The spatial distribution of geochemical variations relative to the plume and the ridge will be examined in detail.

Author–Year Citations

The author–year format is one of three common ways to cite others' works. (See Chapter 15.)

Exercise 12.8.

It is appropriate to cite others' works when describing project aims. Review excerpts 12J–12N and answer these questions:

a. Which excerpts include citations to others' works?
b. What format is used in each excerpt: superscripted numbers, author–year, or italic numbers in parentheses?

In submove 3.2, instrumentation and other resources needed to complete the work are described. The NSF GRFP RFP asks students to "describe resources needed to accomplish the research goal." Three responses to this prompt are shown in passages P5, P6, and P7. Note that each student describes resources that they will use both at their own institutions and in collaboration with other institutions or national facilities.

P5 The Levinger group plans to build an SRS microscope at CSU in collaboration with Dr. Jesse Wilson. Before completion of the SRS microscope, I have access to a CARS microscope at the Advanced Light Microscopy Core Facility at CU Denver. I have used this instrument to successfully image A431 cells, and plan on continued use of this instrument until the SRS microscope is functional. (Samuels, 2019)

P6 **Unique Resources.** To carry out this work, access to a supercomputing center is required. I am already a user at both NSF's XSEDE and UCF's Advanced Research Computing Center. The high-pressure shock tube setup is operational in Dr. Vasu's lab.

Knowledge and Tools Available. We presently have the necessary software tools (such as Gaussian09, CHARMM, and SSUMES) installed. (Wait, 2019)

P7 I plan to use the automated synthesis facility at Scripps Research to screen ligand, solvent, base, halide, temperature, and possible additives. Statistical 'Design of Experiments' (DoE) optimization around this parameter space is expected to identify trends that lead to productive reaction. (Huffman, 2019)

Instrumentation is also described in submove 3.2. Authors are sometimes unsure how much detail to include when describing instrumentation in research proposals. Unlike journal articles, where the general rule of thumb is to include enough detail so that an expert can repeat your work, in a proposal, your main goal is to convince reviewers that *you* can complete the work. Given that, and the fact that you likely do not yet know specific details regarding instrumental setup or optimization, a more general description of your instrumental needs is appropriate in this submove. For example, in excerpt 12O, the author only briefly describes the instrument she will use, a Fourier transform ion cyclotron resonance mass spectrometer (FTMS). She stresses the high-resolution and high-sensitivity features of the instrument that are essential to her work and introduces acronyms that will be used elsewhere in the proposal.

Excerpt 12O (Lee, 2001)

General Methods. The instrument that will be used to execute the gas-phase experimental portion of the proposed research is a Finnigan 2001 dual-cell Fourier transform ion cyclotron resonance mass spectrometer (FTMS or FTICR), equipped with both electron impact (EI) and electrospray ionization (ESI). FTMS is a high-resolution, high-sensitivity technique that allows the entrapment and detection of gas-phase species.[17,18] Gas-phase ions are trapped in a magnetic field, much like a reactant sits in a flask in solution.

Hyphenated Two-Word Modifiers

Two words that together describe a noun are usually hyphenated, unless the first word ends with -ly (e.g., *gas-phase species* but *highly concentrated acid*). See Chapter 16, "Language Tips," Part 5: Punctuation, for details on hyphenated two-word modifiers.

In the last submove of move 3, authors anticipate obstacles or caveats they are likely to encounter in their work. Unlike lab experiments, where most of the "bugs" have been worked out, research is full of challenges that must be overcome. It is advantageous to identify these challenges in your proposal and suggest how they will be addressed. Moreover, most analytical instruments have inherent limitations (e.g., sensitivity or detection limits), and synthetic procedures are often fraught with competing reactions.

Although shown as a separate submove in Figure 12.1, obstacles can be addressed anywhere in the Approach and Methods section. For example, in excerpt 12P, Cain describes two possible challenges to her work while enumerating aims (submove 3.1): retention time shifts (in aim 1) and ion suppression (in aim 2). "Contingency plans" are offered to solve them should they arise.

Contingency Plans

Effective proposals (1) anticipate obstacles and challenges and (2) include contingency plans to solve them.

Excerpt 12P (Cain, 2019)

Aim 1. *GC × GC–MS*. The advantage of having an additional separation dimension is the inherent increases in chromatographic peak capacity, but the complexity of the data structure requires analytes to be resolved in three dimensions. Use of the chemometric resolution limits can also increase peak capacity because more analytes can be partially separated chromatographically and fully resolved chemometrically. Parallel factor analysis (PARAFAC) can directly analyze trilinear data from multidimensional separations, provided that there are no retention time shifts.[4] Alternatively, if retention time shifts are present, then GC × GC–MS data can be unfolded and recombined into bilinear forms for resolution via MCR-ALS. . . .

Aim 2. *LC–MS and LC × LC–MS*. LC–MS presents another challenge for establishing chemometric resolution limits, ion suppression. Ion suppression is induced by mobile phase additives or coeluting species, which alter the concentration of analyte ions reaching the MS detector.[5] This effect can distort the resolved peak shape and inhibit MCR–ALS from obtaining a pure mass spectral profile, which degrades the performance of the curve resolution technique due to deviations from bilinearity.[6] If the data structure behaves bilinearly, the trilinear behavior of LC × LC–MS data using the PARAFAC model will also be studied. Should deviations from trilinearity arise, then the data structure will be treated bilinearly with MCR–ALS.[6]

Similarly, Weaver also presents challenges in her project aims subsection (excerpt 12Q). Recall that her goal is to map long-range exciton migration during photosynthesis using stroboSCAT images at ultrafast time delays; however, to do this, the instrument must first be modified to improve response time and increase sensitivity. In excerpt 12Q, she describes how this will be accomplished and also how the pump–and–probe laser fluences will be reduced to mitigate sample damage.

Excerpt 12Q (Weaver, 2019)

Proposed Work: I will map long-range exciton migration in spinach thylakoids with varying concentrations of active RCs using sensitive differential stroboSCAT images at ultrafast time delays. By comparing samples with fully chemically inhibited RCs and low-concentration operational (open) RCs, I will characterize the attractive effect of RCs. To enable these measurements, I will replace existing pump–and–probe lasers with available femtosecond sources to reduce the instrument response time. Enclosing the microscope stage in an isolated chamber and implementing probe scanning will reduce noise due to sample vibrations. These improvements will facilitate the necessary time resolution and sensitivity for this measurement, which I estimate to be 20 ps and 0.5 nm, respectively. . . . Low pump fluences (~10 μJ/cm^2) will be used during measurements to mitigate sample damage and light-induced compositional changes.

Respectively

Sentences can be made more concise by using the word "respectively". Note that if units are the same, they are mentioned only once (e.g., The widths of A and B are 10 and 12 nm, respectively.) See Chapter 16, "Language Tips," Part 1: Conciseness, for tips on using the word "respectively" appropriately.

Exercise 12.9.

Identify the sentences that offer contingency plans for the challenges that Cain and Weaver address in excerpts 12P and 12Q, respectively.

In NSF CAREER proposals, obstacles are often identified in a separate section of move 3. Lyon (excerpt 12R) identifies a potential pitfall in his work, specifically that the thermoresponsive polymers may lack sufficient bioresponsivity. As a contingency plan, he will draw on a large array of more complex responsive polymers, if the simpler ones do not work. Similarly, Lee (excerpt 12S) reassures readers that she will address potential complications due to "hot" ions in her instrument by using argon as a cooling gas. She will also improve her ability to measure acidity of less acidic sites by using deuterated reference acids.

Excerpt 12R (Lyon, 2000)

Potential Pitfalls. The major pitfall in this work concerns the ability to induce bioresponsivity in thermoresponsive polymers. It is conceivable that although some thermodynamic perturbations may occur, they may not be large enough for the material to be of high utility or applicability. Fortunately, a tremendous array of more complex responsive polymers have been described, which implement backbones other than those found in the traditional NIPA gel (dextrans, carbohydrates, polyglycols, polystyrenes, polyamines, etc.)[24,29,32,45,100–102] These materials are often multiresponsive (temperature, pH, and/or ionic strength) and display much more complicated phase transition behavior. Although it would be unwise to use these materials in the early stages of the project due to their complexity, we will quickly move to explore these materials as colloidal bioconjugates if our initial choice of hydrogel proves to be unsuccessful. It is expected that we will learn enough about how proteins modulate gel solvation thermodynamics from the early work that our choice of the second-generation hydrogel will be an educated, rational, and successful one.

Excerpt 12S (Lee, 2001)

A few caveats should be mentioned. First, one must worry that one's ions are not "hot"; that is, we are measuring thermochemical quantities and the ions should be at thermal temperature for an accurate determination of acidity. We will use argon to cool our ions.

Second, a drawback with multiple-acidic site bracketing, as we learned with uracil, is that the more basic N_3^- ion can deprotonate a weak acid, which produces a strong base, which can then deprotonate the N_1H of thymine. . . . Our proposed plan to circumvent this problem is to use deuterated acids. The N_3^- ion will deprotonate the deuterated acid; if the conjugate base deprotonates the N1H of thymine, then we will detect the deuterated thymine anionic product with a *m/z* ratio one mass unit higher than deprotonated thymine! In this way, we hope to gain accurate values, even for the less acidic site.

---

m/z

The abbreviation for mass-to-charge (*m/z*) ratio is italicized (not bolded).

Perfecting Your Writing

Excerpts 12A–12S offer excellent models for writing the Approach and Methods section of a research proposal. Here we focus on select aspects of writing that can assist you in writing this section.

Word Choice and Usage, Subject–Verb Agreement, Writing Conventions, and Punctuation

We call your attention to the double underlined words in excerpts throughout the chapter, which are elaborated upon in nearby sidebars. These writing features are used correctly in the indicated excerpts (shown in parentheses), but they are often misused in writing.

- Word choice and usage
  - "I developed" not "I discovered" (12F)
  - that vs. which (12F)
  - recently (12I)
  - among vs. between (12J)
- Subject–verb agreement
  - "data show" not "data shows" (12G; see also 12I, 12K, 12L, 12N for proper uses of "these data")
- Writing conventions
  - Boxed graphics and text wrapping (12F, 12G, 12H)
  - 100 °C not 100° C (12H)
  - ab initio (12L)
  - author–year citations (12N)
  - *m/z* (12S)
- Punctuation
  - hyphenated two-word modifiers (12O)

To test your knowledge of these words and conventions, complete exercises 12.10–12.12. (For additional insights, see Chapter 16 and its web-based self-study exercises, with answers.)

Exercise 12.10.

Which phrases (a–f) include a two-word modifier that should be hyphenated?

a. ligand receptor binding
b. in the gas phase
c. gas phase processes
d. fully resolved spectra
e. long range excitation
f. magnetically oriented phospholipid bilayers

Exercise 12.11.

Select the correct word(s) in each set of parentheses.

a. The instrument will be shared (between, among) three universities.
b. The difference (between, among) the two values will be determined.
c. The data (suggests, suggest) that this potential problem can be circumvented.
d. The results will be calculated using (*ab initio*, ab initio) quantum chemistry.

Exercise 12.12.

Find and correct any mistakes in the italicized sections of these sentences, or indicate that the section is correct as written:

a. The *data confirm* that the mechanism *we discovered* is temperature dependent.
b. The first set of reactions will be conducted at *100° C*.
c. The *data clearly shows* the expected relationship between time and temperature.
d. We recently completed a preliminary study *which* examined the behavior of **2** in acidic conditions.

Verb Tense–Voice Combinations

A few common verb tense–voice combinations for key functions in the Approach and Methods section of a research proposal are shown in Table 12.1. Other combinations may also be used.

Table 12.1. Common Verb Tense–Voice Combinations for Key Functions in the Approach and Methods Section of a Research Proposal (other combinations are possible)

function	tense–voice combination	example
a. to highlight prior accomplishments	past–passive	Our first paper *was published* in . . .
	prt pf[a]–active	Our previous work *has demonstrated* . . .
b. to describe preliminary work	prt pf[a]–active	We *have* recently *completed* . . .
		We *have conducted* initial studies of . . .
c. to share preliminary findings	present–active	Early results *are* promising.
		These data *represent* the first evidence of . . .
d. to enumerate project aims	future–active	We *will begin* with a study of . . .
		Our approach *will examine* . . .
	future–passive	Aim 1 *will be accomplished* by . . .
e. to describe instrumentation or other methods	future–active	The instrument we *will use* is . . .
	future–passive	The CCSD level of theory *will be used* to . . .
f. to anticipate obstacles	present–active	The major pitfall in this work *concerns* . . .
		Another challenge we face *is* . . .
g. to offer a contingency plan	future–active	To circumvent this problem, we *will use* . . .

[a] prt pf = present perfect

Exercise 12.13.

Consider the following sentences used to describe preliminary results. Which are written in active voice? Which are written in passive voice? Convert the sentences that are written in passive voice to active voice to place greater emphasis on the scientist/researcher.

a. A number of different molecular solute/LC solvent combinations have been investigated over the last year. (From Patrick, 2000)
b. Samples were prepared by deeply immersing a graphite substrate into a reservoir of the LC/solute mixture at ~100 °C. (From Patrick, 2000)
c. Our group has investigated the synthesis of hydrogel particles . . . using standard aqueous crosslinking procedures. (From Lyon, 2000)
d. Our initial synthetic efforts focused on lightly crosslinked copolymers of NIPA and acrylic (AA) or methacrylic acid (MAA).[51–53] (From Lyon, 2000)

Fluent and Concise Writing

In all your writing, you should strive to be fluent and concise. See Chapter 16 (Parts 1 and 7) for tips on concise and fluent writing. Then complete exercises 12.14 and 12.15.

Exercise 12.14.

A fluent writer avoids using the same word repeatedly. Suggest at least two alternative choices for the italicized words in a–c. For some ideas, review excerpts 12P–12S and read the additional bulleted excerpts provided.

a. If problems *arise*, we will . . .
b. A possible *problem* is . . .
c. If plan A *proves unsuccessful*, we will . . .

- If problems develop with the interpretation of the magnetically aligned nitroxide spin labeled-peptide bicelle data, we will incorporate the TOAC spin label onto different residues of the peptides and repeat the experiments. (From Lorigan, 2002)
- One drawback of immunoassays is the random occurrence of "false positives" (detection of the analyte when it is not really in the sample) and "false negatives" (analyte is not detected when it is actually present). (From Aga, 2002)
- Should the expected ring opening prove unfruitful, the intermediate could still be converted to the desired product via the pathway shown in Scheme 3.

Exercise 12.15.

Review excerpt 12Q (and the nearby sidebar) to view the way in which the author uses the word *respectively* to achieve conciseness. Then revise these two sentences to make them more concise by using the word *respectively* correctly. Remember to check units and punctuation.

a. Our calculations predict that the gas-phase acidity of thymine N1 should be 331 kcal/mol and thymine N3 should be 343 kcal/mol. (Adapted from Lee, 2001)

b. Protium (H), deuterium (D), and tritium (T) have different relative masses (1 amu, 2 amu, and 3 amu, respectively).

Chapter Review

To judge what you've learned from this chapter, define each of these terms and explain its importance, in the context of this chapter, to a friend or chemistry colleague who is new to the field:

computer-generated graphics	parallelism	project aims
contingency plan	preliminary results	SAM test
obstacles	prior accomplishments	

Also explain the following to a chemistry colleague who has not yet given much thought to writing the Approach and Methods section of a research proposal.

- main purpose of the Approach and Methods section
- ways to cite your previous works when describing prior accomplishments
- ways to organize your project aims in a logical order
- why it's important to anticipate obstacles in your proposed research

13 *Writing the Outcomes and Impacts Section*

A well-developed proposal shows that the investigator has grasped a problem well enough to justify second-party sponsorship.
—Paradis and Zimmerman (2002)

All good proposals must come to an end. Like the earlier sections of the research proposal (Chapters 11 and 12), there is no one right way to accomplish this task. However, proposal guidelines generally instruct authors to conclude with expected outcomes and statements that describe the significance and relevance of these outcomes. Some NSF proposals specifically ask applicants to address the intellectual merit and broader impacts of the proposed work. In this chapter, we consider examples of expected outcomes and related project timelines, in addition to statements on intellectual merit and broader impacts. By the end of this chapter, you should be able to

- Generate a list of expected outcomes
- Identify ways to demonstrate intellectual merit
- Give examples of broader impacts

Writing on Your Own: Outcomes and Impacts Section

Visit the *Write Like a Chemist* (2nd ed.) website for step-by-step guidance in writing your own Outcomes and Impacts section for a research proposal.

Reading and Analyzing Writing

We begin with an excerpt that you can read and analyze on your own (excerpt 13A), specifically, the conclusion to Pearson's (2019) proposal regarding cisplatin–iron

Write Like a Chemist. Marin S. Robinson and Fredricka L. Stoller, Oxford University Press.
DOI: 10.1093/oso/9780190098940.003.0013

oxide core–shell nanoparticles. She includes three subsections: Anticipated Results, Intellectual Merit, and Broader Impacts.

Exercise 13.1.

Read excerpt 13A. (Note: NP is the abbreviation for nanoparticle.) Then answer these questions:

a. Briefly describe Pearson's anticipated results in your own words. How well does the graphic in Figure 2 help explain these results?
b. Based on this excerpt, which is correct: in vitro or *in vitro*?
c. One criterion for judging intellectual merit is if a proposed work will generate new knowledge. What new knowledge will this proposal contribute?
d. One criterion for judging broader impacts is if a proposed work will benefit society. How will this work benefit society? How does the author hedge her claims about the benefits?

Excerpt 13A (Pearson, 2019)

Anticipated Results: NP characterization will determine if the cisplatin–iron oxide core–shell NPs developed in the expected manner, and in vitro analysis will determine if the core–shell NPs had inherent cytotoxicity and 2D targeting characteristics. The in vitro analysis will determine if the hypothesis was correct and if the core–shell NPs resulted in higher cisplatin tumor targeting and lower kidney side effects, as shown in Figure 2.

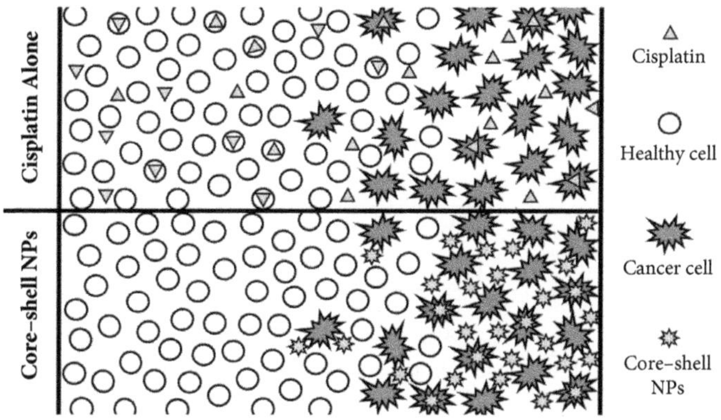

Figure 2. Graphic depiction of anticipated increase in drug-targeting characteristics.

Intellectual Merit: Due to the novel combination of cisplatin and iron oxide for tumor targeting, this study will contribute to our knowledge of cisplatin mechanisms by determining if its bonding characteristics interfere with iron oxide's targeting characteristics. The more that cisplatin bonds to healthy cells, the more likely a trend in cisplatin bonding affinities will result. By identifying which cell components have the greater bonding affinities, the mechanisms of cisplatin will be better understood, enabling more efficient applications of the drug in future treatment schemes. This study will also determine the effectiveness of preventative therapy via core-shell NPs with regard to the kidneys, which can lead to future safe therapy options.

Broader Impacts: Should this study be successful, cancer patients treated with cisplatin–iron oxide NPs will experience ***significantly reduced side effects in the kidneys*** due to increased targeting.

Analyzing Organization

A suggested move structure for the Outcomes and Impacts section of a research proposal is shown in Figure 13.1. There are three moves. The first move highlights expected outcomes, also referred to as anticipated results. Building on the goals and aims presented earlier in the proposal, this move summarizes **deliverables** or tangible accomplishments that will result from the work. Moves 2 and 3, which emphasize intellectual merit and broader impacts, are required explicitly in many NSF proposals, but similar information is expected in nearly all proposals. Together these three moves serve to conclude the proposal and focus on how the research, if successful, will provide new knowledge (intellectual merit) and benefit the larger community (broader impacts). Each of these moves includes a subheading, as shown in excerpt 13A.

1. Highlight Expected Outcomes

2. Demonstrate Intellectual Merit

3. Address Broader Impacts

Figure 13.1 A suggested move structure for the Outcomes and Impacts section of a research proposal (unless the RFP specifies different sequencing and/or contents).

Deliverables

Deliverables are the more tangible products of a research project, such as a new assay, device, instrument, synthesis, or measurement. Deliverables also include the dissemination of findings through publications, patents, and conference presentations.

Analyzing Excerpts Move by Move

With Figure 13.1 in mind, we are ready to consider these moves in more depth and to analyze how authors accomplish these moves in authentic excerpts and passages. Note that we have double underlined occasional words and phrases in these excerpts and passages to illustrate the correct use of writing features, some of them previously addressed in Chapters 10–12. Later in the chapter, we examine these common writing conventions, among others, used in the Outcomes and Impacts section.

Move 1: Highlight Expected Outcomes

A realistic time schedule becomes an effective argument. It suggests to readers that you understand how much time your project will take and that you are not promising miracles just to win approval.

—Adapted from Houp et al. (2006)

Expected outcomes are measurable achievements that link back to the initial goal statement at the start of the proposal (Chapter 11) and the project aims described in the methodology section (Chapter 12). Move 1 is often demarked with a subheading such as **Expected Outcomes** or **Anticipated Results**. The outcomes can be presented in a narrative format or a timeline. A timeline includes a graphic that demarks specific milestones in time, through the duration of the grant. Combinations of the two approaches can also be used, as will be seen in the excerpts that follow.

Excerpt 13B, like excerpt 13A, is from an NSF GRFP application. The author's anticipated results link directly to her goal statement in excerpt 11G, where she stated that her goal was to "elucidate the underlying energetic and structural factors that guide exciton trajectories through an intricate molecular landscape with unrivaled QE" (quantum efficiency). Recall that in photosynthesis, light is initially absorbed by a light harvesting complex (LHC), generating an exciton. This exciton migrates across a network of LHCs until it is harnessed by a reaction center (RC). Weaver illustrates this process in Figure 2 of her proposal (excerpt 13B). When the RCs are closed (inactive), excitons diffuse randomly, with no preference for any direction (isotropic diffusion), resulting in a circular Gaussian profile (Figure 2, top row). When the RCs are open (active), the migration pathway is influenced by the RCs, and the exciton is harnessed by an RC. Diffusion is anisotropic (varies with direction) and results in an anisotropic differential image (Figure 2, bottom row). By varying the concentration of RCs, Weaver can determine how close an exciton must be to an RC to "feel" it. These data will allow her to accomplish her goal of elucidating the exciton's migration pathway.

Excerpt 13B (Weaver, 2019)

Anticipated Results: By obtaining a functional image at ultrafast time delays, I will directly track exciton trajectories with in situ correlation to the underlying RC distribution. I expect that exciton trajectories will terminate on open RCs, which will serve as effective exciton sinks. A relatively flat energetic landscape (i.e., when all RCs are closed) will manifest as an isotropically advancing Gaussian profile in the differential stroboSCAT images, consistent with a random walk model (Figure 2). From these characterizations, I will directly determine the intrinsic diffusion length. With a low concentration of active RCs, anisotropy in the differential image would indicate the location of active RCs and quantify the extent to which they energetically redirect excitons. Varying the concentration of active RCs will reveal how close an exciton must be to a RC to "feel" its influence, establishing the nature of the effective potential well of a RC. This versatile platform could also permit studies of photoprotective mechanisms in real time and comparisons across other mutants and organisms like cyanobacteria.

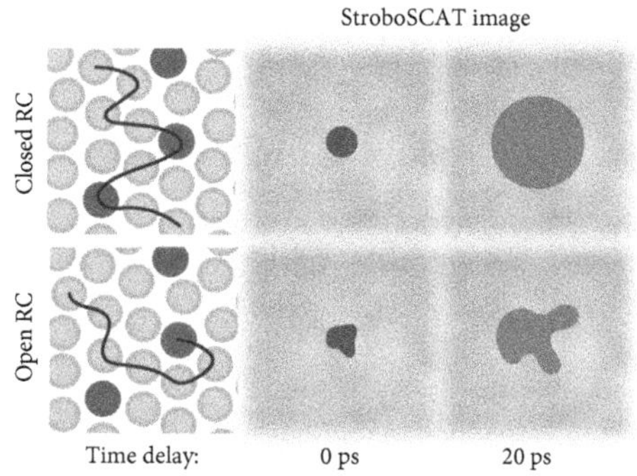

Figure 2. Anticipated stroboSCAT data in a photosynthetic membrane with different functional RCs. Light gray: LHC; Dark gray: RC. Scale bar represents 1 μm.

We next consider two examples of move 1 from NSF CAREER proposals. In excerpt 13C, Johnson enumerates the expected outcomes of his proposed Lewis acid–promoted [3+2] cycloaddition reactions using a numbered list. Three outcomes are presented, and they are written in phrases rather than complete sentences. When read together, the three phrases form a complete sentence, beginning with the lead-in phrase "The expected outcomes of the proposed work will include." Semicolons (rather than commas) are used to separate the phrases because the first phrase already used commas. Note, too, that the opening phrases in the numbered entries are grammatically parallel (noun phrases): a variety of strategies, a diverse range of heterocycles, and a collection of focused approaches.

Excerpt 13C (Johnson, 2003)

Lewis acid-promoted [3+2] cycloadditions of aziridines and epoxides proceeding via carbon–carbon bond cleavage of three-membered ring heterocycles were demonstrated for the first time. This proposal details plans for extending these initial results into a general synthetic method for the enantioselective synthesis of structurally diverse pyrrolidine- and tetrahydrofuran-containing organic compounds. Expected outcomes of the proposed work include

1. a variety of strategies for effecting dipolar cycloadditions from stable, neutral precursors, including carbon–carbon bond heterolysis induced by Lewis acidic metal complexes and chiral organic catalysts;
2. a diverse range of heterocycles and other chiral building blocks that can be generated by this cycloaddition strategy; and
3. a collection of focused approaches for rapidly evaluating metal complexes that are candidates to catalyze the cycloadditions.

Colons and Semicolons

Use *colons* only after full sentences; hence, no colon appears after the phrase "Expected outcomes of the proposed work will include" in excerpt 13C. Use *semicolons* (instead of commas) to separate a series of three or more items when a comma is used in one or more item(s). See Chapter 16, "Language Tips," Part 5: Punctuation, for more tips on colons, semicolons, and commas.

A common variation of move 1 is to report outcomes using a timeline, which includes a schedule that predicts (often tentatively) when various outcomes will be achieved. In excerpt 13D, Sanov presents a timeline in narrative form. The section (labeled TIMELINE) begins by acknowledging how ambitious the project is and points out that progress will be guided by currently unknown intermediate results. Nevertheless, the researcher does his best to identify "strategic landmarks" and describes his hope to start tackling his most ambitious goal (bimolecular reactions in cluster) in the second half of the award period.

Excerpt 13D (Sanov, 2002)

TIMELINE
It is recognized that the range of the experiments described in this proposal is very ambitious. But this is a long-term program at the frontier of science, and I must be guided by intermediate results to decide which directions merit the greatest attention. Nonetheless, I envision the following strategic landmarks. Within several months, we will complete the diagnostic experiments on I^- and other atomic anions. By the end of 2001, we will have carried out the studies of O_2^- and S_2^- and should soon have the results on the OCS^- hydration. In 2002, we will begin the time-resolved experiments on $(OCS)_2^-$ and continue with reaction dynamics and alignment studies in other systems. Bimolecular reactions in clusters are our most ambitious goal, which I hope to start tackling in the second half of the award period.

An example of a timeline in a figure or graphic is shown in excerpt 13E. Kohen presents a 5-year timeline across the top of the figure and enumerates expected outcomes, numbered 1–8, down the left-hand axis. Figures and charts are excellent ways to illustrate how smaller, individual tasks contribute to larger research goals and how smaller projects complement one another and overlap in time.

Excerpt 13E (Kohen, 2002)

[Kohen uses the following abbreviations in his text and timeline, each defined previously in the proposal: hydrogen (H), tritium (T), deuterium (D), nicotinamide adenine dinucleotide phosphate (NADPH), kinetic isotope effect (KIE), dihydrofolate reductase (DHFR), and wild type (WT).]

Timetable for work plan. The work plan spans five years and involves eight related projects (Figure 3). Enzyme preparation, syntheses of H/T and D/T labeled NADPH, and preparation of high-resolution crystals have been started, or will be started, in the early stages of the project. Studies related to KIEs, the temperature dependence of KIEs, and DHFR dynamics will take place in years 1–5; modeling and interpretation of the results will take place in years 2–5.

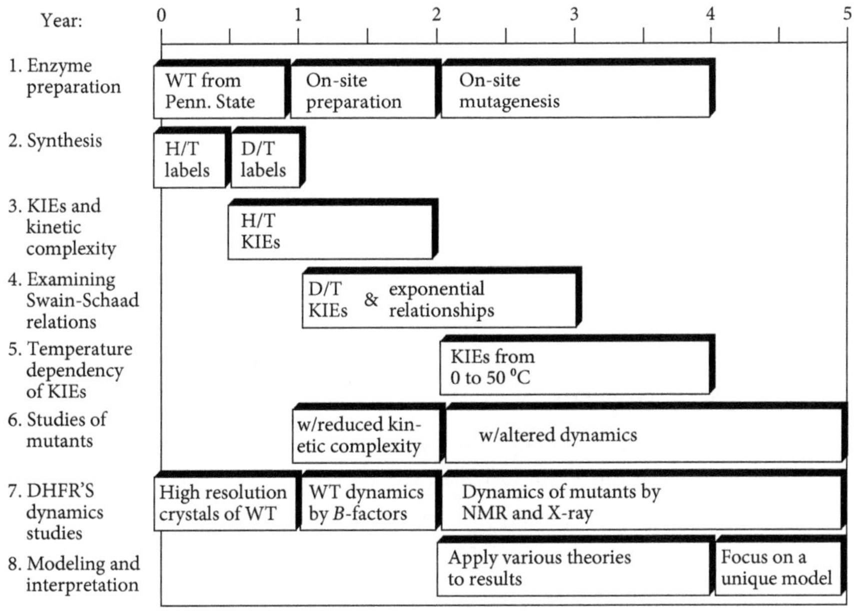

Figure 3. Timetable for work plan.

Five or 5?

In most instances, chemistry writers use numerals (e.g., 5) rather than words (e.g., five) with units of measure and time. Yet, in a single sentence with more than one number, numbers (as either numerals or words) should be used in the same way. For example, in excerpt 13E, "five years" was used instead of "5 years" to be consistent with "eight projects." See Chapter 16, "Language Tips," Part 3: Writing Conventions, for more details on numbers and units.

> *Hedging reduces the certainty with which researchers present their claims: A "modesty" strategy in science writing that dates back to Robert Boyle [well known for Boyle's Law].*
>
> —Adapted from Hyland and Salager–Meyer (2008)

When you are writing your expected outcomes, be careful not to promise something you cannot deliver. In other words, pay careful attention to your word choices. Hedging language can be quite useful in some instances. For example, in excerpt 13A, the author hedges when explaining her goal to deliver a more effective cancer drug with the phrase "Should this study be successful. . . ." Similarly, as in project goals statements (Chapter 11), avoid forms of the word "discover" when sharing outcomes. Rather than saying "X will be *discovered*," use more objective words like "X will be *studied*, *investigated*, *tested*, *developed*, *evaluated*, or *piloted*." The same caution applies to timelines. In excerpt 13E, Kohen gives himself a broad time span (years 1–5 or 2–5) for his studies to take place. Similarly, Sanov (excerpt 13D) hedges his target dates with the sentence "I envision the following strategic landmarks." Other phrases that can be used to hedge or soften timelines are "We *expect to* complete X within 6 months" or "We *propose to* develop Y within the first 2 years."

Hedge Your Outcomes

Although most scientists are intelligent, they are not clairvoyant! Hence, it is best to avoid saying that you will "discover" or "create" something in your expected outcomes. Soften your claims with hedging and careful word choice. See Chapter 16, "Language Tips," Part 2: Word Choice, for more details on hedging and formal vocabulary. See also Tables 5.1 and 5.2 in Chapter 5.

Exercise 13.2.

Suggest ways to hedge (or soften) the following outcomes and timelines.

- In year 1, compound **3** will be synthesized.
- In year 3, 50 related compounds will be isolated.
- In year 5, we will discover the mechanism of the reaction.

Move 2: Demonstrate Intellectual Merit

When I was Program Officer at the National Science Foundation, I witnessed without a doubt that simply and clearly written proposals sailed through more easily than ones that were not so. Surely there is a lesson there.

—Robert Damrauer, University of Colorado–Denver

For a proposal to be funded, the ideas presented must be intellectually sound. Move 2 emphasizes this point and allows authors to make it crystal clear that their ideas represent "good science." Reviewers typically judge the intellectual merit of a proposal based on "three Cs":

- *creative* (are the ideas original?)
- *credible* (is the work plan realistic?)
- *contributory* (will the work contribute to new knowledge?)

Successful authors strive to emphasize one or more of these Cs throughout their proposals, especially in move 2 (Figure 13.1).

Three Cs

Is your proposal *creative*? Is your work plan *credible*? Will your findings be *contributory*? The intellectual merit of a proposal is often judged by these three Cs.

Proposals are awarded high points by reviewers if the proposal writer presents a *creative* approach or an innovative idea. A good idea that has already been explored, by you or by others, is seldom funded. Your research idea must have the potential to generate new knowledge and answer questions that have yet to be addressed.

Intellectual merit is also based on the *credibility* of your work plan, including both proposed research methods and projected timeline. Understanding the literature and demonstrating knowledge of what is currently known must be evident in your writing, even if you are proposing a novel approach. Credibility is also enhanced by describing the inherent limitations of your methods (e.g., by including error bars, detections limits, or the range over which a calibration is accurate). Timelines should be ambitious but realistic. If you are an undergraduate student, you are likely to underestimate how long your project will take. Consult your research mentor for ideas about realistic timelines.

The *contributory* potential of your work is also a factor in determining its intellectual merit. Clearly not all grant-supported works will result in a Nobel Prize, but your work should advance knowledge, shed light on unresolved questions, provide new insights, or delve a bit deeper into areas that remain poorly understood. In short, others need to learn something from your work.

Finally, intellectual merit is also judged by your writing skills (hence the reason for including this module in our book). The value of a well-written proposal cannot be overstated. An articulate and clearly organized proposal promotes the original qualities of your work and convinces readers that you can carry out the work described. It is not an exaggeration to say that your competence as a scientist will be judged by your writing. A competent scientist pays attention to detail; a carelessly written proposal that includes typos, poor organization, and confusing sentence structure leaves a negative impression and undermines any confidence readers might have had in the author. Thus, although writing skills cannot replace good science, writing skills are needed to convey good science.

Writing and Science

Writing skills cannot replace good science, but writing skills are needed to convey good science. Your competence as a scientist will be judged, in large part, by your writing.

Five examples of intellectual merit from NSF GRFP applications are shown in passages P1–P5. All include (as required) the subheading **Intellectual Merit**, which is followed by either a colon or a period. Either punctuation mark is fine, but remember to be consistent throughout your proposal.

P1 **Intellectual Merit:** This project will expand our fundamental knowledge of photosynthesis, an essential process for life. Not only will exciton migration be directly visualized on ultrafast time scales, but it will also be correlated to underlying functional structures. Through our collaboration with the Niyogi group and other groups in the STROBE Science and Technology Center, we will use this platform to construct a more complete picture of what drives efficient energy transfer across different photosynthetic organisms and with different imaging platforms. (Weaver, 2019)

P2 **Intellectual Merit.** The work proposed here is founded on two well-validated approaches to calculate the free energy change during a chemical reaction: (1) quantum treatment of the reactive system with TST/RRKM and (2) classical MD technique with Umbrella sampling/WHAM protocol. (Wait, 2019)

P3 ***Intellectual Merit.*** Enumeration of chemometric resolution limits can demonstrate the capabilities of coupling instrumentation and advanced multivariate data analysis. Chemometric resolution can increase peak capacity for both one- and two-dimensional chromatography because it allows for analytes partially resolved during the separation to be fully resolved mathematically. Also, the effects of ion suppression in LC–MS and LC × LC–MS separations on chemometric model behavior have not been described in the current literature. Therefore, communicating these results through conference presentations and publications will help close these known knowledge gaps. (Cain, 2019)

P4 **Intellectual merit:** Previous lyopreservation studies have not determined where lyoprotectants concentrate in the cell. Blood platelets are simple cells that have been successfully lyopreserved in the past and provide a close to model system to investigate lyoprotectant distribution in a cell. This research will provide quantitative data of the distribution of trehalose and DMSO in hydrated blood platelets and desiccated blood platelets, information that can be used to improve the current lyopreservation methodology enough to be useful for medicine. To my knowledge, direct imaging of trehalose diffusing into blood platelets has not been performed, and the data generated through my research will be a novel contribution to this field. (Samuels, 2019)

P5 **Intellectual Merit**: The challenge presented by attached ring motifs has yet to be adequately addressed by the synthetic community. If successful, this approach will provide access to a large subset of underexplored chemical space. Although the problem is challenging, my successes in this area give me experience that will aid in tackling any issues that arise in the process. Extension of the developed methodology to a complex system as in darwinolide will demonstrate the utility of the approach and give access to a product that existing methods cannot produce. Several related attached ring spongian diterpenes have been isolated,[4] and the proposed route could enable a unified synthetic solution to the entire subclass. (Huffman, 2019)

Exercise 13.3.

Re-read Intellectual Merit statements P1–P5 and re-read the second subsection in excerpt 13A (Pearson, 2019). Then complete these tasks:

a. Which one of the three Cs (creative, credible, or contributory) is best accomplished in each statement? Briefly justify your choice.
b. Imagine that you are the reviewer who is judging the intellectual merit of these proposals. Which one would you award the highest score? Briefly justify your choice.

Move 3: Address Broader Impacts

Proposals that can potentially benefit a large segment of humankind (i.e., have a "broad impact") are more likely to receive funding. Indeed, the most successful proposals posit ideas that can potentially transform society. Transformative discoveries have already occurred in countless areas of science (e.g., medicine, technology, atmospheric science, materials science, and computer science) and are possible in nearly all areas of scientific investigation.

A scientifically literate society and a sustainable scientific workforce are also "broader impacts" of research. Indeed, proposals are viewed more favorably if they include ways to train and educate various segments of society such as (a) existing scientists (in peer-reviewed publications and conference presentations); (b) new scientists (in elementary, secondary, and postsecondary schools and universities); or (c) the general public (through news media and non-scientific journals). Trained scientists are national (and international) resources, and the opportunity to participate in a research group assures that a "pipeline" of individuals—undergraduate, graduate, and postdoctoral researchers—will continually be renewed. To ensure that this pipeline reflects the ever-changing composition of a diverse society, research that encourages the participation of underrepresented populations is particularly valued.

Broader Impacts

In this book, the term *broader impacts* refers to how your proposed work will benefit other scientists, society at large, and the training of future scientists.

Gender-Neutral Language

Use gender-neutral language when identifying broader impacts (and elsewhere in the proposal). For example, refer to the general public, people, or humankind, rather than mankind. Gender-neutral language is expected in scientific writing. Refer to the *ACS Guide to Scholarly Communication* and chemistry-journal Author Guidelines for gender-neutral language expectations.

With that background in mind, we consider broader impact statements written for NSF GRFP applications (P6–P10), each following the corresponding Intellectual Merit statement shown in P1–P5, respectively. Note the parallel formatting in the subheadings (e.g., P1 and P6 both use a colon, P3 and P8 both use italic font, and P4 and P9 both use lowercase for the second word of the subheading).

P6 **Broader Impact:** By uncovering the design principles at work in natural photosynthetic systems, this work will inform the next generation of synthetic light harvesting materials, potentially bringing electricity to more parts of the world in a sustainable way. A detailed mechanistic understanding of the initial steps in photosynthesis will also direct the design of engineered crops to feed the world's growing population. I will leverage the intuitive connection between photosynthesis and imaging to communicate my findings in a way that is relatable to the general public. Motivated by the excitement of this experiment, I will also develop a new BASIS lesson for elementary students to inspire young future scientists. (Weaver, 2019)

P7 **Broader Impact.** To mitigate the negative effects of climate change, it is necessary to reduce the amount of CO_2 released into the atmosphere. The technology of carbon capture and sequestration offers a clean energy, zero-emission alternative.[8] This technology can be simplified by removing nitrogen from the oxidant during combustion-based power generation, a process known as oxy-combustion. To control the combustion temperature, pure oxygen can be diluted with recovered CO_2.[9] In addition, CO_2 can be allowed to reach a supercritical state, increasing its thermal expansion coefficient and thereby improving the efficiency of power generation. Knowledge of how CO_2 affects the reaction rates and mechanisms of oxy-combustion is required for the design of high-pressure, oxy-combustion turbines. Presently, these processes are not well understood. In this proposal, theoretical chemistry will be used to increase our understanding of these reactions. (Wait, 2019)

P8 ***Broader Impact—Outreach to Chemists.*** The traditional practice of limiting the number of analytes within a sample to increase peak capacity is impractical for today's environmental and biomedical samples. Chemometric resolution allows for utilization of the full separation space without limiting the analytes present in a sample. However, many chemists and other scientists do not have experience with chemometrics. An interactive video tutorial will be developed to educate undergraduates, graduate students, professors, and industrial chemists on chemometric principles in chromatography. Along with fostering interdisciplinary connections, chemists at different stages in their research careers will be united through this video tutorial. Building an international workforce capable of using chemometrics to increase resolution and peak capacity when analyzing samples with over 100 components is integral to continued progress in data-driven fields that rely on chromatography. (Cain, 2019)

P9 **Broader impact:** Understanding the distribution of lyoprotectants in blood platelets and the interaction of different protectants in platelets will provide information useful for the development of new and better preservation methods. With better platelet preservation methods, blood banks and hospitals will be able to maintain room temperature supplies of platelets for longer than 5 days, relieving the pressure of platelet shortages. Applying what is learned from this research to other types of cells will help the development of lyopreservation as an alternative form of biologics storage. (Samuels, 2019)

P10 **Broader Impact**: Carboxylate directing groups seem to occupy a "Goldilocks" region in Pd-directing strength for achieving regioselective palladation and β-H elimination, and this study will improve understanding of this directing effect throughout the Pd catalysis community. Discussions among groups at Scripps on this topic have already generated novel ideas, and new results will be presented at an ACS national conference this coming year to further communicate this research. Access to darwinolide will enable mechanism-of-action identification through a collaborative partnership, an effort that will likely be aided by my undergraduate experience in identifying the targets of the natural products withaferin A[9] and parthenolide.[10] In addition to furthering the fundamental biological understanding of biofilm behavior, identifying antibiofilm targets in MRSA will have broad implications for addressing the threat of antibiotic-resistant bacteria. Darwinolide and related analogues can serve both as leads for therapeutic development and as tool molecules. (Huffman, 2019)

Exercise 13.4.

Re-read Broader Impact statements P6–P10 and re-read the one in excerpt 13A (Pearson, 2019). Then complete these tasks:

a. Briefly describe the potential benefits of each proposed work to society through possible scientific breakthroughs and/or educational contributions.

b. Imagine you are the reviewer who is judging the broader impacts of these proposals. Which one would you award the highest score? Briefly justify your choice.

Perfecting Your Writing

Excerpts 13A–13E and passages P1–P10 offer excellent models for writing your own Outcomes and Impacts section. Here we focus on select aspects of writing that can assist you in writing this section.

Word Choice and Usage, Subject–Verb Agreement, Punctuation, and Number-Unit Writing Conventions

The writing conventions itemized here are highlighted throughout the chapter with double underlining and are elaborated upon in nearby side bars. These writing features are used correctly in the indicated excerpts (shown in parentheses), but they are often misused in writing.

- Word choice and usage
  - hedging (13A, 13D, 13E)
  - because vs. since (P3)
  - affect vs. effect (P3, P7)
  - that vs. which (P4, P6, P10)
  - although vs. while (P5)
  - between vs. among (P6)
- Subject–verb agreement (P7, P10)
- Punctuation
  - colons and semicolons (13C, P2)
  - comma usage before "but" (P1) and before "and" (P8, P9, P10)
  - hyphenated two-word modifiers (P2)
- Writing conventions
  - numbers and units (5 or five years?) (13E)

Review the excerpts indicated and read over the nearby sidebars. Also refer to Chapter 16 (and its web-based self-study exercises and answers) for tips on the proper use of these important writing features. To test your knowledge of these features, complete exercise 13.5.

Exercise 13.5.

Sentences a–d were adapted from project timelines. Correct the errors in each sentence. Focus on hedging, use of colons or semicolons, numbers (written out or numerical), and subject–verb agreement.

a. In the first year, we will discover at least five new compounds.

b. The initial project will take 2 years and will result in the synthesis of six compounds.

c. In year 2, samples will be collected at the Grand Canyon site, in year 3, samples will be collected at the Yosemite site, and in year 4, samples will be digested and analyzed.

d. Time-resolved experiments includes: (1) alignment studies, (2) kinetic studies, and (3) thermodynamic analysis.

Verb Tense–Voice Combinations

Typical verb tense-voice combinations for key functions of the Outcomes and Impacts section of a research proposal are highlighted in Table 13.1. Other combinations are also used.

Table 13.1. Common Verb Tense–Voice Combinations for Key Functions in the Outcomes and Impacts Section of a Research Proposal (other combinations are possible)

function	tense–voice combination	example
a. to highlight expected outcomes	future–active	NP characterization *will determine* if . . .
	present–active	Expected outcomes of this work *include* . . .
		I *expect* that trajectories *will terminate* on . . . (present active, future active)
b. to demonstrate intellectual merit	future–active	This project *will expand* our knowledge of . . .
	prt pf[a]–passive	These effects *have not yet been described.*
	prt pf[a]–active	Previous studies *have not yet determined* if . . .
c. to address broader impacts	future–active	This work *will inform* the next generation . . .
		Identifying antibiofilm targets in MRSA *will have* broad implications for . . .
	future–passive	An interactive video tutorial *will be developed* to educate undergraduates.
		New results *will be presented* at an ACS national conference.
	present–active	Carbon capture and sequestration *offer* clean energy alternatives to . . .

[a] prt pf = present perfect

Exercise 13.6.

Consider the following sentences taken from one of the three moves of the Outcomes and Impacts section of a research proposal. Consult Table 13.1 and select the most appropriate verb tense and voice in each sentence. Name your tense–voice choice.

a. (move 1) Within several months, we *have completed/will complete* the diagnostic experiments on I^- and other atomic anions.
b. (move 1) Nonetheless, I *envision/will envision* the following strategic landmarks.
c. (move 2) If successful, this approach *provides/will provide* access to . . .
d. (move 2) Several related attached ring spongian diterpenes *are isolated/have been isolated*, . . .
e. (move 3) The proposed synthesis *benefits/has benefited/will benefit* biochemists in both academia and industry.
f. (move 3) Blood banks and hospitals *have/will have* the capability to maintain room temperature supplies of platelets for longer than 5 days, . . .

Personal Pronouns

Personal pronouns (i.e., I, we, my, our) can be used, at the author's discretion, throughout the Outcomes and Impacts section. Consider these examples:

a. *I* will directly determine the intrinsic diffusion length. (excerpt 13B, move 1)
b. By the end of 2001, *we* will have carried out the studies of . . . (excerpt 13D, move 1)
c. Bimolecular reactions in clusters are *our* most ambitious goal . . . (excerpt 13D, move 1)
d. *We* will use this platform to construct a more complete picture of . . . (excerpt P1, move 2)
e. The data generated through *my* research will be a novel contribution to this field. (excerpt P4, move 2)
f. *I* will leverage the intuitive connection between photosynthesis and imaging . . . (excerpt P6, move 3)

Exercise 13.7.

Why do you think personal pronouns are used more often in a research proposal than in a journal article or conference abstract?

Finalizing on Your Own: Research Proposal

Visit the *Write Like a Chemist* (2nd ed.) website for tips on finalizing your research proposal.

Chapter Review

Check your understanding of what you've learned in this chapter by defining each of these terms and explaining its importance, in the context of the chapter, to a friend or chemistry colleague who is new to the field:

broader impacts	gender-neutral language	the three "Cs"
deliverables	intellectual merit	timeline
expected outcomes		

Also try explaining the following to a chemistry colleague who has not yet given much thought to writing the Outcomes and Impacts section of a research proposal:

- the main purpose(s) of the Outcomes and Impacts section
- the role of hedging in the section
- different ways to present a timeline
- different ways to demonstrate intellectual merit
- the purpose of the broader impacts section

Section 2

Writing Conventions

14 *Figures and Tables*

Some visual forms [tables, charts, graphs] suit particular data and messages better than others. . . . Choose the graphic form that best helps readers both grasp your data and understand how they support your argument.

—Booth, Colomb, Williams, Bizup, and FitzGerald (2016)

This chapter provides general formatting guidelines for figures and tables (collectively referred to as graphics). Figures and tables, in combination with the text, allow authors to communicate information efficiently. Graphics and text work together, reinforcing each other without duplicating information. Effective graphics help readers comprehend complex relationships between and among sets of data, relationships that would be difficult, if not impossible, to explain in words alone. Graphics also allow readers to view results independently, without being influenced by authors' interpretations.

Calling out Graphics

Graphics should be "called out" (i.e., mentioned) by name and number in the text (e.g., Figure 1, Table 2). The callout should precede the graphic. The most concise way to call out a graphic is to place its name in parentheses, e.g., (Table 2). The phrase "as shown in ___" is also commonly used.

The **formatting guidelines** presented in this chapter are general in nature and are designed to help you prepare graphics appropriate for most chemistry genres. However, specific formatting requirements (such as resolution, color, font, size, line weights, and layout) vary across genres; hence, you also will need to consult the appropriate resource for specifics (e.g., Author Guidelines for journal articles or submission guidelines for conference posters and research proposals). Author

Write Like a Chemist. Marin S. Robinson and Fredricka L. Stoller, Oxford University Press. © Oxford University Press 2022.
DOI: 10.1093/oso/9780190098940.003.0014

Guidelines for journal articles will also specify if the graphics should be attached to the end of the manuscript or inserted directly into the text.

Formatting Guidelines

The technical formatting and submission requirements for figures and tables differ from genre to genre. It is the author's responsibility to consult relevant guidelines for specific requirements.

Formatting Figures

Prepare your graphs well. Doing so will help make your findings and their value clear and will help show the care with which you do your work.

—Gastel and Day (2016)

We first consider formatting guidelines for figures. Figures come in various forms (e.g., graphs, spectra, illustrations, diagrams, maps, photographs), but regardless of form, each figure is numbered sequentially in the text in order of presentation (Figure 1, Figure 2, Figure 3, etc.). One of the most common figure types is the graph (e.g., bar graphs, scatter plots, contour plots, two-and three-dimensional line graphs). Most graphs follow similar formatting guidelines; the examples shown in this section illustrate common formatting practices.

We begin by examining the two graphs shown in Figure 14.1: a scatter plot is shown in Figure 1 (top), and a bar graph is shown in Figure 2 (bottom). The first feature to notice is that both figures have captions, not titles. The caption is placed *below* the figure; the identifier is bolded (e.g., **Figure 1.**) and is followed by a short descriptor (a phrase or sentence) written in sentence case. The second feature to notice is that neither figure is "boxed in," that is, there is no additional box surrounding the entire figure; lines are solely used for the graphs' axes. Including figure titles and adding extra boxes are common errors made by novice writers, in part because most graphing programs (including Excel) add these unwanted features automatically; hence, both boxes and figure titles must be removed manually.

Several additional formatting conventions are illustrated in Figure 14.1. First, both graphs have white backgrounds. Although color may be used in posters and proposals to add visual appeal, a white background is best for graphs in journal articles. Second, there are no vertical or horizontal lines (gridlines) in the graphing area. Light gray horizontal lines are sometimes used to guide the reader, but dark gridlines should be avoided. Third, the scaling of the *y* axis is appropriate for the data presented in each graph. Scaling should not be misleading; large differences should not be made to look small, or vice versa. Moreover, in Figure 2, the units on the y axis are correctly formatted (μg/m^3), and the legend is placed inside the graphing area (rather than to the right of graph). In Figure 1, the legend is omitted entirely, and the symbols are defined instead in the figure caption. We have summarized these and other formatting conventions in Table 14.1. Exercises 14.1 and 14.2 are designed to strengthen your knowledge of these conventions.

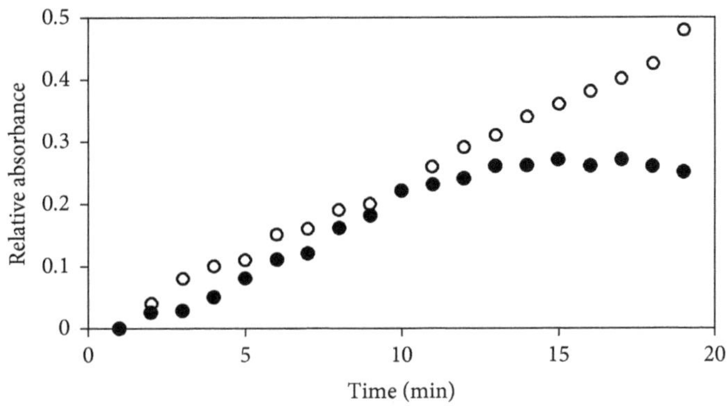

Figure 1. Relative absorbance over time for products in acid-catalyzed (•) and base-catalyzed (◦) reactions.

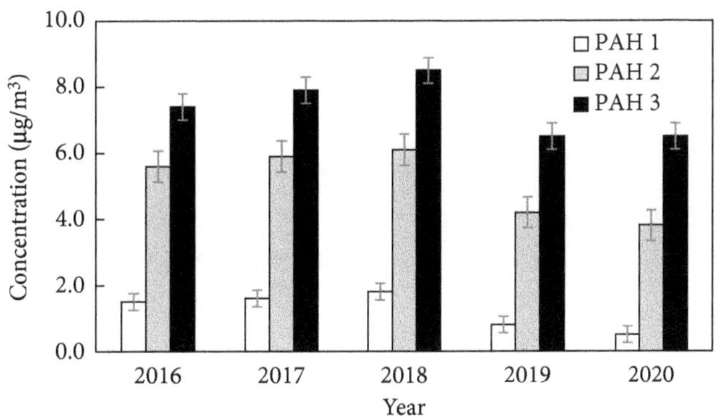

Figure 2. PAH concentrations between 2016 and 2020.

Figure 14.1 A scatter plot (Figure 1) and a bar graph (Figure 2) that follow formatting conventions in most chemistry journals.

Exercise 14.1.

Compare the incorrectly formatted graphs in this exercise with their corresponding and correctly formatted graphs in Figure 14.1. Identify eight formatting errors in each and describe how you would correct them. (Consult Table 14.1, as needed.)

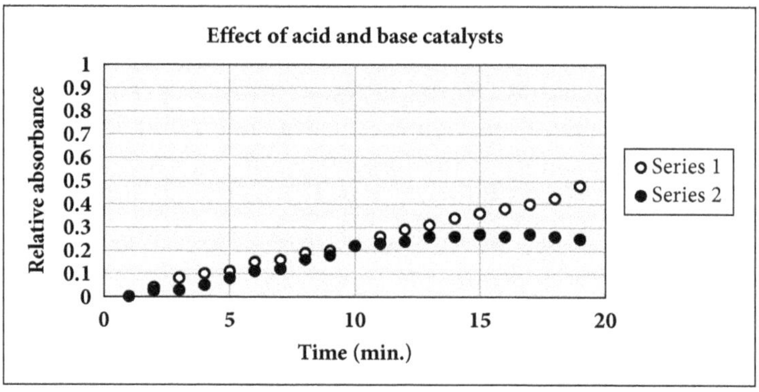

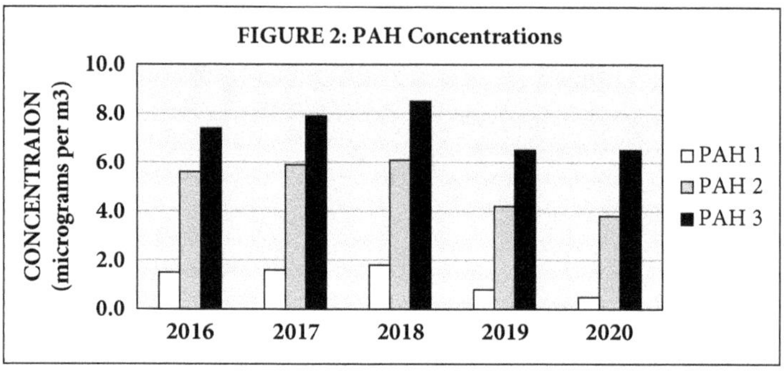

Exercise 14.2.

Log on to the website of your favorite chemistry journal and search recent issues of the journal for examples of graphs. Compare those graphs to the ones shown in Figure 14.1 and consult the guidelines presented in Table 14.1. What similarities and differences do you notice? In particular, comment on these formatting conventions:

a. Are the graphs boxed in? Do they include gridlines?
b. Is there a figure title or a figure caption?
c. Is there a legend? If so, is it placed inside the graphing area?
d. Is the font used for axis labels bolded or a larger size than the text?
e. Are all axes labeled appropriately with correctly formatted units?

Table 14.1. Generally Accepted Guidelines for Formatting Graphs

attribute	description
axis labels	Label all axes. A common format is to use unbolded font and to capitalize only the first letter of words (e.g., Reaction Time, not REACTION TIME). Place units in parentheses, using abbreviations (e.g., min, s), superscripts and subscripts (e.g., m^3, Cl_2), and special characters (e.g., α, μ) appropriately.
axis scaling	Choose axis scales that accurately represent the data; do not mislead readers by using scaling that makes small differences appear large or vice versa. When possible, minimize empty space.
background color and gridlines	Use a white background, not a gray or colored background. (Colored backgrounds may be used in posters and proposals to add visual appeal.) Remove horizontal and vertical gridlines unless they are needed to convey scientific information, then use a light gray color.
boxes	In general, do not "box" graphs. Boxes are sometimes used in research proposals, to help the graph stand out, and in posters, to add visual appeal.
captions	Place a caption below the graph, aligned with the left-hand margin (not above the graph, except in posters). Begin with the figure name and number in bolded font (e.g., **Figure 2.**). Continue with a short, informative descriptor that can be understood apart from the text. The descriptor can be a fragment, single sentence, or several sentences; it is usually written in sentence case and ends with a period (e.g., **Figure 2.** Arrival times of isomeric peptides.). In general, treat captions as text, and use the same font and font size as in the text. (A slightly smaller font size may be used in posters and proposals.)
color	Use color to convey information and increase clarity. Use color consistently in all graphics (e.g., red for reactive and blue for unreactive). Minimize combinations of red and green, which may be hard for colorblind individuals to distinguish. Use light and dark shades that will also be visible in a grayscale version of the graph.
curves	Avoid including more than five curves per graph. Label all curves clearly. Leave sufficient space between curves so that they are easily distinguishable.
font sizes	Choose font sizes that are roughly the same size as the text. Avoid using more than two font sizes.
legends	Use legends with two or more data sets. If possible, make the legend part of the caption (e.g., **Figure 2.** Yields with (•) and without (○) heating.). Otherwise, place the legend inside the graph (without a box), or, if it overlaps data, place it to the right of the graph.

Despite the widespread use of graphs, they are not appropriate in all instances. Sometimes graphs contain too much information and can be overwhelming. If a graph is too crowded, consider graphing a representative sample of the data instead and describe similar or related trends in the text. Graphs are also inappropriate if they contain too little information. Sometimes words alone, without graphs, are preferred:

- Use words to report a linear relationship between variables rather than graphing a single straight line, for example, "A linear relationship was observed between A and B ($r = 0.97$)."
- Use words to describe the results of calibration procedures rather than graphing calibration curves, for example, "All calibrations were accurate to ±0.05%."
- Use words to describe simple relative abundances rather than graphing such data in a pie or bar graph, for example, "Compounds **2**, **3**, **5**, and **7** were produced in 25, 30, 36, and 40% yield, respectively."
- Use words to report spectral data that verify compound identity, rather than include the spectrum (e.g., see how NMR data are reported in excerpt 3S in Chapter 3).
- Use words to describe related or similar results rather than plotting multiple sets of data that convey the same trend, for example, "Reaction 2 followed a decay curve like the one shown in Figure 2."

Formatting Tables

Let's now consider formatting guidelines for tables. Tables are used to summarize, group, and highlight numerical data so that they can be viewed collectively. Tables are numbered consecutively throughout the text (Table 1, Table 2, etc.), and, like figures, tables are designed to conserve space, not waste it. A general rule of thumb is to use a table only if you have enough data to fill at least three rows and three columns; otherwise, describe the data in the text. Also, be careful not to report the same data in both a figure and a table; typically, data should be reported in either a table or a figure, not both. Exceptions do occur; for example, a table could be used to present a comprehensive set of data and then one or more graphs could be used to highlight important features or trends in these data. But do not simply repeat the data in two formats.

Tables

In general, tables summarize, group, and highlight numerical data so that the data can be viewed collectively. A bare minimum for a table is three rows and three columns. Be sure to call out all tables in the text.

In Tables 14.2 and 14.3, we illustrate how tables are formatted in most chemistry journals. We recommend that you follow these examples (or similar ones) whenever you prepare a table for a chemistry audience (e.g., in a journal article, research proposal, thesis or dissertation, or even a lab report in a chemistry course). The tables are uncluttered, easy to read, and adhere to formatting practices used by most chemists. As you view these tables, pay particular attention to the following details (which address four common errors made by novice writers when preparing their first tables):

- Column headings are not in all caps or bolded.
- Units are included in the column headings (not with column entries).
- Tables have only three horizontal lines (no gridlines).
- Words in columns are left-justified; numbers in columns are aligned by decimal.

Table 14.2. Solvent Screening for Formation of Product 3[a]

entry	solvent	T (°C)	time (min)	yield of **3** (%)	*er*
1	H_2O	25	5	42	75:25
2	H_2O	100	10	93	72:28
3	CH_2Cl_2	25	5	80	69:31
4	CH_2Cl_2	100	10	53	60:40
5	Me-THF	25	30	29	75:25
6	Me-THF	100	100	78	73:27
7	DMC	25	30	50	55:45
8	DMC	100	100	32	42:48

[a]All reactions used 0.5 equiv **1** and 0.1 equiv **2**.

Table 14.3. Photometric Characterization of Each LED in the Detector (from Hemida et al., 2020)

LED @ 20 mA	typical power (μW)	observed power (μW)	coupling (%)	transmission (%T)
LED 240	70.0	62.0	3.3	3.2
LED 255	300.0	421.0	3.6	3.1
LED 275	800.0	1032.0	3.1	2.8

Some table-formatting conventions vary among genres and from journal to journal. For example, in Tables 14.2 and 14.3, both table titles are bolded and written in title case, but some journals expect titles to be written in sentence case (without bolding) and end with a period. Similarly, although bolded font is almost never observed in column headings, some journals expect the first letter of a column heading to be capitalized (e.g., "Yield" rather than "yield"). The placement of footnotes also varies; sometimes footnotes are placed just below (rather than above) the horizontal line at the bottom of the table. We have summarized these and other common table-formatting conventions in Table 14.4. To strengthen your knowledge of table-formatting conventions, complete exercises 14.3 and 14.4.

Table 14.4. Generally Accepted Guidelines for Formatting Tables

attribute	description
column headings	Give every column a brief heading. In general, headings are centered and unbolded, except for normally bolded items (e.g., **2** as a compound label). Some journals use lowercase for headings (e.g., reaction time); others use title case (e.g., Reaction Time) or sentence case (e.g., Reaction time). Title case is common in posters. Capitalized abbreviations (e.g., T for temperature) are capitalized even in lowercase headings.
column numbers	Align measured numerical values on the decimal. Include a leading zero for numbers <1 (e.g., 0.5). Include zeroes to the right of the decimal point only if they are known (e.g., 10.0 when the tenths place is known).
column order	Place independent variables (i.e., what you controlled) in the leftmost columns, in a logical order. Place dependent variables (i.e., what you measured) in the rightmost columns. Hence, tables are read from left to right, from controlled variables to measured values.
column repeating values	Do not include a column that repeats the same value for each row. Instead, state the value in the table title or in a footnote (e.g., The temperature was 25 °C.).
column units	Include units in column headings, in parentheses, not with the values in the columns. (The unit alone as a column heading is not acceptable.) Format units and symbols properly (e.g., °C, μg, m^3). A second line may be used for the unit. In this case, drop single-line headings to the second line.
column words	Left-justify words in columns; use lowercase (e.g., ethanol not Ethanol) unless the term is normally capitalized (e.g., EtOH); use unbolded font unless the entry is normally bolded (e.g., **2** as a compound label).

Table 14.4. Continued

attribute	description
entry column	Include an entry column (leftmost column) to list sample numbers or trials. Include this column *only* if you refer to specific entries in the text. Order entries according to their presentation in the text, and number them sequentially (1, 2, 3, etc.). Center entry numbers under the heading "entry." Do not use labels that were used in your research (e.g., S01) because they are meaningless to readers.
table font	Strive to use the same font in the table as in the text. If necessary, reduce the table font by one size. Use the same font and font size in the entire table (i.e., for titles, headings, entries, and footnotes).
table footnotes	Use footnotes, if necessary, to share information about the whole table, explain column headings, define abbreviations, or provide additional information. Call out footnotes using italicized, superscripted letters. Place the footnote below the last row of the table and above the bottom horizonal line. (Some journals place footnotes below the bottom horizontal line.) Use the same font size in the footnote as in the rest of the table.
table gridlines	Avoid gridlines, boxes, and vertical lines. (Such lines may be visible on your computer but should not be visible in the printed document.) Use only horizontal lines; three are most common. In the three-line format, horizontal lines are placed below the table title, below column headings, and below the table.
table size, color, and orientation	Try to fit your table to the width of one page. If necessary, exchange rows and columns, or break a long table into smaller ones. Avoid horizontal tables (landscape orientation). Use black and white, unless the table is in a poster. Single-space table titles and table entries to minimize unused space. Similarly, plan your table so that there are no empty cells (or very few).
table title	Place a title above the table. Begin with an identifier (e.g., **Table 2.**). Continue with a brief, informative descriptor that can be understood apart from the text (e.g., Reaction Conditions for Synthesis of **3**). The descriptor is usually not a complete sentence. Consult guidelines for formatting titles because rules vary. The identifier and descriptor are usually bolded and written in title case without a period at the end (e.g., **Table 2. Reaction Conditions for Aldol Synthesis**). Some titles are left-justified; others are centered.

Exercise 14.3.

Compare the following incorrectly formatted table with Table 14.2. Identify eight formatting errors and explain how you can correct them. (Consult Table 14.4, as needed.)

TABLE 1. Reaction conditions for formation of product **3**.[a]

ENTRY	SOLVENT	T (°C)	TIME (min.)	YIELD OF 3
1	H_2O	0	75	42%
2	H_2O	25	75	93%
3	H_2O	25	150	80%
4	H_2O	100	75	53%
5	CH_3CN	0	75	29%
6	CH_3CN	25	75	78%
7	CH_3CN	25	150	50%
8	CH_3CN	100	75	32%

[a]All reactions used 0.5 equiv **1** and 0.1 equiv **2**.

Exercise 14.4.

Log on to the website of your favorite chemistry journal and search recent issues of the journal for examples of tables. Compare those tables to the ones shown in Tables 14.2 and 14.3, while consulting the guidelines in Table 14.4. What similarities and differences do you notice? In particular, consider these formatting conventions:

a. Is the table title bolded? Is it centered or left-justified?
b. Do the column headings use bolded font or all caps? What formatting is used?
c. Are table entries separated by gridlines?
d. Are units included with the table entries or in the column headings?
e. For tables with footnotes, where are the footnotes placed? How are they signaled?

15 *Citations and References*

An author should cite those publications that have been influential in determining the nature of the reported work and that will guide the reader quickly to the earlier work that is essential for understanding the present investigation.

—American Chemical Society, Ethical Guidelines to Publication in Chemical Research

In this chapter, we consider formatting conventions for **citations** and **references**. Writers are obligated to cite others' works that have significantly influenced or are relevant to their own work. It goes without saying that to include another's original ideas without proper acknowledgment is plagiarism. Consequently, this stage of the writing process is critical both professionally and ethically.

Citations and References

Citation	An authoritative source referred to *in the text* to indicate work that has influenced the current work or that provides essential background information.
Reference	Bibliographical information for in-text citations.

Citations

Many chemistry students . . . are intimidated by the polished writing styles of some renowned chemists. Rather than paraphrasing what the experts have said, many students are tempted to use quotations to retain the

Write Like a Chemist. Marin S. Robinson and Fredricka L. Stoller, Oxford University Press. © Oxford University Press 2022.
DOI: 10.1093/oso/9780190098940.003.0015

excellent writing style. However, because my goal is to teach our students to become better writers, I essentially forbid the use of quotations in their papers. I tell them that "one quote is two too many."

—Donald Paulson, California State University, Los Angeles

Citations—inserted into the text of journal articles, posters, and research proposals—identify publications that influenced the authors' knowledge and understanding of the presented work. Although citations are formatted in different ways, they serve the same purposes, including acknowledging others' works, enhancing the writer's credibility, placing the current work into a broader context, and making it easy for readers to locate cited materials.

"How many citations should I include?" is often the first question that novice writers ask. The answer to this question depends on the genre. In conference abstracts, it is exceedingly rare to include citations. Posters have a few citations, often only two or three, limited to the most essential works. The number increases for journal articles and proposals, which typically have between 25 and 75 citations. In journal articles, most citations appear in Introductions (to provide background information) and Discussion sections (to corroborate findings). In posters and proposals, citations are common in Introductions and Methods sections, to cite precedent approaches and instrumentation. In general, authors are cautioned against citing their own works excessively and publications that have not been peer-reviewed. Citations to textbooks are uncommon because textbooks contain general knowledge; general knowledge can be summarized in your own words, without a citation.

Citing Textbooks?

Information from textbooks is generally not cited. Authors may include such information in their writing without a citation provided that the information is rewritten in their own words (i.e., do not plagiarize).

You have likely already encountered citations in term papers or book reports; however, chemistry genres use citations somewhat differently than these more familiar genres. For example, chemists avoid using direct quotations. Although commonplace in many genres, direct quotations—where you quote verbatim the words of others—are exceedingly rare in chemistry writing. They are virtually nonexistent in chemistry journal articles and conference abstracts; they are uncommon in posters and research proposals. Instead of direct quotations, chemists learn to summarize others' works succinctly, in only a few words, and cite sources appropriately.

Direct Quotes

Direct Quotes

Although common in other genres, direct quotes are exceedingly rare in chemistry journal articles and conference abstracts; they are uncommon in posters and research proposals.

A second difference in how citations are used in chemistry writing is that, in general, authors are not referred to by name in sentences that cite their works. Including authors' names adds unnecessary words to a sentence and is redundant because the authors' names can always be found in the references. Consider the following examples:

Not preferred	Jung et al.[1] have shown that ozonolysis can be used to control the taste and odor of drinking water. (19 words)
Preferred	Ozonolysis has been used to control the taste and odor of drinking water.[1] (13 words)

In the preferred sentence, emphasis is placed on the science (ozonolysis) rather than the scientists (Jung et al.), making it easier to read and more concise. On occasion, names are used in formal science writing (last names only), but this practice should be kept to a minimum. If authors' names are included, a superscripted number is placed directly after the last author's name (e.g., Kim and Jeon[1]) or after "et al." (e.g., Jung et al.[1]).

Citation formats used by chemists typically involve numbers; superscripted numbers are the most common format, followed by italic numbers in parentheses. In each case, the citations are numbered in order of appearance (1, 2, 3, . . .) and are placed at appropriate locations in the text. A third citation format is the author–year approach. This format is relatively uncommon in chemistry journals (particularly ACS journals) but is frequently used in journals of other scientific disciplines such as geology and physics. Each of these formats is described in more detail in the sections that follow.

Citation Formats

The two most common citation formats used in chemistry writing are

- superscripted numbers ______[1,2]
- italic numbers in parentheses ______(*1, 2*)

The author–year format is rarely used in chemistry journals, but is often used in other disciplines.

- author–year ______ (Garcia et al., 2021; Roberts et al., 2022)

Superscripted Numbers

The most common citation format used by chemists in journal articles, proposals, and posters is superscripted numbers. The popularity of this format can be attributed to its conciseness. Let's examine this format more closely in excerpt 15A, the opening paragraph of a journal article. First, note how succinctly and concisely the authors cite others' works. No direct quotes are used, and no authors' names are mentioned. Others' works are summarized in a single word or phrase, allowing multiple works to be cited within a single sentence. Also, as you browse through this excerpt, note the following citation-formatting conventions:

- Citations begin with 1 and continue in numerical order.
- A comma (without a space) is used to separate two citations (e.g., [1,2]); an en dash is used for a consecutive sequence of three or more citations (e.g., [3–5]).
- Superscripted numbers are placed immediately (without a space) after punctuation: after the period at the end of the sentence or after a comma (or semicolon) within a sentence.
- Superscripted numbers are placed at appropriate locations in the text, not always at the end of a sentence. Thus, in excerpt 15A, it is clear that citations 7 and 8 refer to neurodegenerative disorders and citation 9 refers to cardiomyopathies.

Excerpt 15A (adapted from Song et al., 2020)

[*First paragraph of the journal-article Introduction*]

Mitochondria are organelles that perform vital cellular functions in eukaryotic cells.[1,2] These organelles produce biochemical energy in the form of ATP for cell growth and biological activities. They also play an important role in biosynthesis, cellular redox homeostasis, intracellular signaling, and apoptosis regulation.[3–5] Mitochondrial dysfunction is associated with a variety of different diseases,[6] such as neurodegenerative disorders,[7,8] cardiomyopathies,[9] and cancer.[10] Mitochondria are highly dynamic and continuously move and undergo fusion and fission, which are essential for the maintenance, function, distribution, and transmission of mitochondria.[11,12]

Did you also notice in excerpt 15A that all but one sentence (the second sentence) included citations? In general, authors cite information that appears in the published literature (e.g., journal articles and technical reports); general knowledge, however, need not be cited. In excerpt 15A, the fact that "organelles produce biochemical energy in the form of ATP for cell growth and biological activities" is an example of general knowledge.

A second example of superscripted numbers is shown in excerpt 15B. Here, the excerpt is from the second paragraph of the Introduction. The first numbers that appear are 9 and 10; hence, numbers 1–8 were cited sequentially in the first paragraph. Again, no direct quotes are used, and no authors' names are included in the sentences that cite their works. All sentences include citations except the final sentence, which is used to emphasize the importance of continued research in this area. Note, too, additional conventions that are illustrated in excerpt 15B:

- Citations are repeated each time a work is mentioned. For example, citations 1–4 and 4 (alone) are repeated in the second paragraph after being introduced in the first paragraph.
- Multiple citations are separated by commas without spaces (e.g., [1–4,11,12]).
- Citations can also be placed directly after a word (without a space) in a sentence (e.g., . . . heteroatom substitution[1–4,11,12] as well as preparing . . .). There is a space between the citation and the next word in the sentence.

Excerpt 15B (adapted from Balijepalli et al., 2020)

[*Beginning of second paragraph in Introduction*]

For example, β-lactams fused to a sulfur-substituted five-membered ring (i.e., penicillin derivatives) exhibit significantly divergent antibiotic properties than when fused to an unsaturated oxygen-substituted six-membered ring (i.e., oxacephem derivatives): the aminopenicillin amoxicillin is more efficacious against *Staphylococcus aureus* than *Pseudomonas aeruginosa*, whereas the converse is observed for the slightly modified carboxypenicillin carbenicillin.[9,10] Current strategies to circumvent antibiotic resistance include synthesizing β-lactam antibiotics with different ring sizes and heteroatom substitution[1–4,11,12] as well as preparing β-lactamase inhibitors such as the oxabicyclic clavulanic acid.[4,13–15] This avenue of research remains a high priority in light of the inevitable obsolescence of traditional antibiotics such as penicillin and amoxicillin.

Exercise 15.1.

Consider this passage from the opening paragraph of a journal article (Lee et al., 2020). Explain why the second sentence includes a citation, but the first sentence does not.

> Antibody drug conjugates (ADCs) are monoclonal antibodies covalently bound to a cytotoxic drug through a chemical linker. Conventionally, ADCs are generated either through nonspecific conjugation to lysines throughout the antibody or through reduced interchain disulfides.[1]

Exercise 15.2.

Rewrite these two sentences (adapted from Monforte et al., 2020) more concisely by omitting authors' names; the citations should still be included.

Rizzi[3] has established that polyphenols are able to degrade methionine and phenylalanine to form their respective aldehydes. Delgado et al.[4] have further demonstrated that a hydroxy group in the *para* or *ortho* position promotes phenylacetaldehyde formation.

Exercise 15.3.

Examine the superscript citation formatting in this passage and correct formatting errors.

Carbon nanotubes (CNTs) have potential applications in fields such as molecular electronics[8], conductive polymers[8, 9, 10], and energy storage.[2–3, 5, 7]

Italic Numbers in Parentheses

The second numerical citation format is italic numbers in parentheses. This approach is less common in chemistry journals than superscripted numbers, but you will still encounter it, particularly in older journals. For example, the *Journal of Agricultural and Food Chemistry* used this format until 2010, after which it switched to superscripted numbers. We show one example in excerpt 15C. Note these formatting conventions:

- Numbers inside the parentheses are in italic font: (*6*).
- Two numbers are separated by a comma (not italicized) and a space: (*1*, *2*).
- En dashes are used to indicate a consecutive sequence of three or more numbers: (*7–10*).
- Multiple citations can be placed inside parentheses, following conventions just stated: (*10*, *14–17*).
- Citations are placed at appropriate locations in the text, including inside sentences.
- Citations are placed *before* punctuation (i.e., commas, semicolons, or periods): . . . and other fungal species (*6*).

Excerpt 15C (adapted from Megadi et al., 2010)

The phthalimide fungicides such as captan, captafol, and folpet are the important surface protectants used to control plant pathogenic fungi (*1*, *2*). Captan (*N*-trichloromethylthio-4-cyclohexene-1,2-dicarboximide) is one of the most widely used surface fungicides to inhibit the growth of fungi on seeds, fruits, vegetables, and ornamental plants (*3–5*). Captan is used as an agricultural fungicide to control diseases caused by

Botrytis, *Fusarium*, *Fusicoccum*, *Pythium*, and other fungal species (*6*). Captan is also used as an industrial fungicide in paints, plastics, leather, paper, cosmetics, textiles, and plaster (*7–10*) to prevent fungal growth.

Author–Year

The author–year format lists the author's (or authors') last name(s), a comma, a space, and the year of publication in parentheses (or brackets in some journals): (Zhao, 2020). This format is uncommon in chemistry journals (which prefer to omit authors' names entirely by using numbers), but it is common in journals of other disciplines (e.g., geology and physics). Common conventions used with the author–year approach are as follows:

- For publications with two authors, the two last names are separated by the word "and" or the symbol "&." For publications with three or more authors, only the first author's last name is listed, and it is followed by "et al." Multiple citations are separated by semicolons and arranged in alphabetical order:

One author	(Pell, 2021)
Two authors	(Liu and Garcia, 2018) or (Liu & Garcia, 2018)
Three or more authors	(Wait et al., 2020)
Multiple citations	(Liu & Garcia, 2018; Pell, 2021; Wait et al., 2020)

- As with italic numbers in parentheses, the author–year citation is placed *before* punctuation (i.e., commas, semicolons, periods).

  . . . as seen in humid climates (Creus et al., 2019).

  . . . less significant in shallow waters (Bernstein, 2017), running rivers (Ewert et al., 2016), and deep lakes (Chamberlin and Nguyen, 2018).

- Authors' names appear in sentences more often in journals that use the author–year format. In such a case, the year is placed in parentheses immediately after the name(s):

  Giadrossich et al. (2015) found that . . .

- Every time a work is cited, the complete and correct citation must be used. Thus, Liu et al. (2020) or (Liu et al., 2020) must be used every time the work by Liu and others is cited.

The correct use of "et al." is particularly important when using the author–year format. The acronym (from the Latin *et alia*) means "and others" and is used to refer to works with three or more authors. In the past, Latin terms were typically italicized in written works, and you will still see "et al." italicized in some journals (i.e., *et al.*). However, most ACS journals no longer follow this practice.

Because "et al." means "and others," the verb following "et al." must agree with the plural subject. Consider these incorrect and correct examples using the author–year and superscripted number formats:

Incorrect	Liu et al. (2020) shows . . .	Incorrect	Liu et al.[1] shows . . .
Correct	Liu et al. (2020) show . . .	Correct	Liu et al.[1] show . . .

et al.—Dos and Don'ts

Do use	**Don't use**
a period after al.	capital letters
a space between et and al.	commas
the plural form of the verb after et al.	quotation marks
	italics (in most journals)
	a period after "et"

Exercise 15.4.

Consider in-text citations a–e. Identify what is wrong with each one and rewrite it correctly. Assume that the article being cited was written by three authors (Snow, Craig, and Kretzer).

a. Snow, Craig, and Kretzer[2] show . . .
b. Snow et al.[2] shows . . .
c. Snow and Craig et al.[2] show . . .
d. Snow *et al.*[2] have shown that . . .
e. Snow et. al[2] have shown that . . .

References

The last required section in most written works is the References (sometimes called Literature Cited). Whatever the section heading is, its purpose is to offer more detailed information about the publications cited within the text, making it easy for readers to retrieve each cited source, if so desired. As the headings imply, only the literature cited in the text should be included in the references. The format used for your references should parallel the format used for your citations. References for numerical citations (superscripted numbers or italicized numbers in parentheses) are listed in numerical order (1, 2, 3, etc.); references for author–year citations are listed in alphabetical order by the last name of the first author.

References must include a minimal amount of information: the authors' surnames and initials, journal name, year of publication, journal volume, and page numbers (inclusive page numbers are preferred). Many journals also require that references include the title of the article, the journal issue number, and the article's digital object identifier (DOI). Journal names should be abbreviated according to the *Chemical Abstracts Service Source Index* (*CASSI*). Table 15.1 lists some *CASSI* abbreviations; note that only journal titles with at least two words are abbreviated. One-word journal names (e.g., *Science*) are not abbreviated. It is the author's responsibility to be sure this information is complete and accurate.

Table 15.1. Sample Journal-Title Abbreviations According to *CASSI*

Acc. Chem. Res.	*Chem. Res. Toxicol.*	*J. Phys. Chem. A*
AIChE J.	*Environ. Sci. Technol.*	*Langmuir*
Anal. Chem.	*Inorg. Chem.*	*Nano Lett.*
Biochem. J.	*J. Agric. Food Chem.*	*Nature*
Biochemistry	*J. Am. Chem. Soc.*	*Org. Lett.*
Can. J. Chem.	*J. Org. Chem.*	*Science*

In the remainder of this chapter, we focus on formatting references in journal articles. References in scientific posters are normally truncated versions of journal article references to conserve space (see Chapter 9), whereas references in research proposals typically mirror the formats used by journals in the author's field. We include a few illustrative examples of numerical references and alphabetical references in the sections that follow. Consult your targeted journal for examples of numerical or alphabetical formats for books in print, ebooks, edited volumes, chapters in edited volumes, technical reports, government documents, conference proceedings, etc.

Numerical References

Numerical references accompany numerical citation formats (i.e., superscripted numbers or italicized numbers in parentheses). The references are numbered and listed in citation order (1, 2, 3, . . .) in the References section, typically the last section of the journal article (after Acknowledgments). The numbers in the reference list must match the in-text citation numbers exactly. Although there can be multiple in-text citations for a single article, the reference to that article appears only once in the References section.

Nearly every journal has a preferred format for references; hence, the first step in preparing your references is to consult the Author Guidelines for your targeted journal. We also recommend downloading a References section from a recent article in that journal to use as a template. Be sure to download the printable PDF version of the references, not the HTML version, because HTML versions do not mirror the correct format for submitted manuscripts. As you analyze these references, pay careful attention to the information contained in each reference and the sequencing and formatting of that information.

PDF or HTML?

When viewing references for formatting guidance, use PDF rather than HTML files.

To help you learn to recognize what details to look for in your references template, we include a few sample journal-article references in Table 15.2. These examples are formatted according to guidelines followed by many (but not all) ACS journals. To make it clear that these are numerical references, we have numbered them arbitrarily: 10–13. After Table 15.2 is a bulleted list that highlights the guidelines we followed in the table. As you study these examples, pay particular attention to (1) what information is included (or excluded), (2) how the information is sequenced, (3) what punctuation is used, and (4) what fonts are used. Although often quite subtle, such details must be observed so that your references match the preferred format for your targeted journal. Strict adherence to these details is expected by editors, reviewers, and readers.

Table 15.2. Examples of a Numerical Reference Format Used in Many ACS Journals

(10) Vecitis, C. D.; Park, H.; Cheng, J.; Mader, B. T.; Hoffmann, M. R. Enhancement of Perfluorooctanoate and Perfluorooctanesulfonate Activity at Acoustic Cavitation Bubble Interfaces. *J. Phys. Chem. C* **2008**, *112* (43), 16850–16857. DOI: org/10.1021/jp804050p

(11) Zhang, T.; Hansen, K.; Politis, A.; Müller. M. M. An Unusually Rapid Protein Backbone Modification Stabilizes the Essential Bacterial Enzyme MurA. *Biochemistry* **2020**, *59* (39), 3683–3695. DOI: org/10.1021/acs.biochem.0c00502

(12) Hemida, M.; Coates, L. J.; Lam, S.; Gupta, V.; Macka, M.; Wirth, H.-J.; Gooley, A. A.; Haddad, P. R.; Paull, B. Miniature Multiwavelength Deep UV-LED-Based Absorption Detection System for Capillary LC. *Anal. Chem.* **2020**, *92* (20), 13688–13693. DOI: 10.1021/acs.analchem.0c03460

(13) Mousavi, E. S.; Kananizadeh, N.; Martinello, R. A.; Sherman, J. D. COVID-19 Outbreak and Hospital Air Quality: A Systematic Review of Evidence on Air Filtration and Recirculation. *Environ. Sci. Technol.* **2020**, Article ASAP. DOI: 10.1021/acs.est.0c03247 (accessed 2020-12-29).

The journal article references in Table 15.2 (entries 10–13) are formatted according to the following guidelines:

- **Author list**. Authors' names are listed in the same order as in the journal article. The last name is written first, followed by initials (no first names). Each initial is followed by a period and a space (e.g., Scott, G. K.). In hyphenated first names and last names, a hyphen is included (e.g., Wirth, H.-J.; Foster-Cohen, S.). Authors' names are separated with semicolons. The word "and" is not used before the last author's name.
- **Article title**. Titles (when required) are written in title case (capitalizing all major words). The article title is followed by a period and space.
- **Journal name**. Journal names are written in italic font using approved *CASSI* abbreviations. (If the name is not listed in *CASSI*, the full name is used.) Periods are used only after abbreviated words in the journal name (e.g., *Anal. Chem.* but *Biochemistry*). The journal name is followed by a space (but no punctuation).
- **Year of Publication**. The year is written in bolded font followed by a comma and a space (e.g., **2021**,).
- **Volume and issue numbers**. The volume number is written in italic font followed by a space; the issue number (not in italic font) is placed inside parentheses and followed by a comma, and a space (e.g., *11* (4),).
- **Page numbers**. Inclusive page numbers are preferred. An en dash (–) is used to separate the first and last page. The last page number is followed by a period. (Page numbers are not included for ASAP articles.)
- **Digital Object Identifier (DOI)**. The DOI is included at the end of the reference entry, if required. DOI is followed by a colon and space; the full identifier follows without a period at the end: DOI: org/10.1021/acs.biochem.0c00502
- **ASAP (as soon as publishable) articles**. The reference for an ASAP article (see entry 13 in Table 15.2) omits the journal volume, journal issue, and page numbers (all of which are unknown). In ASAP articles, the DOI is followed by parentheses, in which the date (year-month-date) on which the article was accessed is documented, followed by a period: (accessed 2021-07-28).

Bolded and Italicized Commas?

Prior to 2021, commas in references that followed years (e.g., **2020,**) and volume numbers (e.g., *44,*) were written in the same font as the preceding number (bolded and italicized, respectively). This practice is no longer recommended by the *ACS Guide to Scholarly Communication* (Banik et al., 2020), but it is still observed in some ACS journals.

Exercise 15.5.

Log on to the ACS website and select a recent issue of an ACS journal of your choice. Compare the generic ACS reference formats for journal articles shown in Table 15.2, with the formats used in your selected journal. What similarities and differences do you notice? In particular, consider these formatting conventions:

a. In the author list, what punctuation separates authors' names? Are first names included? Is there an "and" preceding the last author's name?
b. Is the article title included? If so, how is it formatted?
c. What formatting is used for the journal title, year, volume number, and issue number? What formatting is used for the punctuation following these items?
d. Are page numbers inclusive? If so, is an en dash used to separate the first and last page?
e. Is the DOI included? How is it formatted?

Exercise 15.6.

Repeat exercise 15.5 (a–e) for a non-ACS journal such as *Tetrahedron* or *Science*. What differences do you notice?

Exercise 15.7.

Numerical reference formats differ widely among journals (and even within journals). To see this for yourself, consider the following reference, written first for *The Journal of Organic Chemistry* (published by ACS) and second for *Tetrahedron* (published by Elsevier).

J. Org. Chem.	(1) Alliati, M.; Donaghy, D.; Tu, X.; Bradley, J. W. Ionic Species in a Naphthalene Plasma: Understanding Fragmentation Patterns and Growth of PAHs. *J. Phys. Chem. A* **2019**, *123* (10), 2107–2113. DOI: 10.1021/acs.jpca.9b00100
Tetrahedron	[1] M. Alliati, D. Donaghy, X. Tu, J. W. Bradley, J. Phys. Chem. A 123 (2019) 2107–2113.

Using these formats as templates, convert the following information into a properly formatted reference for *J. Org. Chem* and *Tetrahedron*.

Title	Calibration Transfer of Partial Least Squares Regression Models Between Desktop Nuclear Magnetic Resonance Spectrometers

Authors	Diego Galvan, Evandro Bona, Dionisio Borsato, Ernesto Danieli, and Mário Henrique Montazzolli Killner
Journal	Analytical Chemistry
Year	2020
Volume (issue)	92 (19)
Pages	12,809 to 12,816
DOI	10.1021/acs.analchem.0c00902

Alphabetical References

We conclude this chapter with a brief consideration of alphabetical references, which accompany the author–year citation format. Alphabetical references are rare in chemistry journals, which prefer numerical references, but many other disciplines (both scientific and non-scientific) favor this approach. When the author–year citation format is used in the body of the work, references are listed in alphabetical order (based on the last name of first author). Just like numbered references, the correct format for alphabetical references varies among journals. Two examples follow, where we have formatted the same reference for two journals: (1) *Journal of Geophysical Research: Atmospheres* and (2) *Ecology*.

J. Geophys. Res.: Atmos.	Alliati, M., Donaghy, D., Tu, X., & Bradley, J. W. (2019). Ionic species in a naphthalene plasma: Understanding fragmentation patterns and growth of PAHs. The Journal of Physical Chemistry A, 123(10), 2107–2113. https://doi.org/10.1021/acs.jpca.9b00100
Ecology	Alliati, M., D. Donaghy, X. Tu, and J. W. Bradley. 2019. Ionic species in a naphthalene plasma: Understanding fragmentation patterns and growth of PAHs. The Journal of Physical Chemistry A 123:2107–2113.

Note these details in the two formats:

- Both journals use a hanging indent, with the first line left-justified and subsequent lines indented.
- Both journals use a form of "and" ("and" or "&") before the last author's name.
- In *Ecology*, the second author and all subsequent authors are written with initials first followed by the last name.
- Both journals write out journal names in full.
- No special fonts (italics or bolded) are used in the references.

- Page numbers are inclusive (with en dashes).
- References end with a period.
- DOIs, required in references for the *J. Geophys. Res.: Atmos.*, follow the final period.

Exercise 15.8.

Using the preceding example as a guide, convert the following hastily compiled numerical reference list to a list appropriate for the *Journal of Geophysical Research: Atmospheres*. (Recall that author–year references are sequenced in alphabetical order, by last name of first author.)

(1) Shoeib, M.; Harner, T.; Vlahos, P. Perfluorinated chemicals in the Arctic atmosphere. *Environ. Sci. Technol.* **2006,** *40 (4),* 7577–7583. https://doi.org/10.1021/es0618999

(2) Gilles-Gonzalez, M. A., and Gonzalez, G. (2005) Heme-based sensors: defining characteristics, recent developments, and regulatory hypotheses, *J. Inorg. Biochem. 99 (14),* 1–22. https://doi:10.1016/j.jinorgbio.2004.11.006

(3) Evans, D. C., Watt, A. P., Nicoll-Griffith, D. A., and Baillie, T. A. (2004) Drug-protein adducts: an industry perspective on minimizing the potential for drug bioactivation in drug discovery and development. *Chem. Res. Toxicol. 17 (6),* 3–16. https://doi.org/10.1021/tx034170b

16 *Language Tips*

Writers make linguistic choices that connect [their] texts with their disciplines, which is why physicists don't write like philosophers, nor lawyers like linguists [and, by extension, why chemists don't write like biologists].

—Adapted from Hyland (2019)

This chapter provides details about many of the language features and chemistry writing conventions highlighted throughout *Write Like a Chemist*. Each Language Tip includes helpful hints for communicating effectively in writing. The Language Tips are organized as follows:

Language Tips Contents

Write Like a Chemist. Marin S. Robinson and Fredricka L. Stoller, Oxford University Press. © Oxford University Press 2022.
DOI: 10.1093/oso/9780190098940.003.0016

Self-Study Exercises, Answer Keys, and Proofreading Tips

Visit the *Write Like a Chemist* (2nd ed.) website for self-study exercises, with answers, and proofreading recommendations for each Language Tip in this chapter.

Part 1: Conciseness

One of the hallmarks of writing in chemistry is conciseness. Unnecessary words clutter your writing and hinder effective communication. The three Language Tips here—on concise writing, generally, and on the uses of nominalizations and the word *respectively*, specifically—will help you achieve conciseness in your writing.

Concise Writing

Expert and scientific readers of your work expect conciseness. Four rules will help you write concisely, thereby meeting their expectations.

Rule 1: Eliminate unnecessary words or phrases, including references to scientists.

1. ~~As stated earlier in this paper,~~ TCE demonstrated specific antiproliferative activity against the more progressive and metastatic SW620 cells. (Adapted from Seeram et al., 2004)
2. ~~The cancer cell lines~~ SW480 and SW620 are two colon cancer cell lines ~~that were obtained~~ from the same patient; ~~the~~ SW480 ~~cell line~~ was ~~obtained~~ from the primary tumor, and ~~the~~ SW620 ~~cell line~~ was ~~obtained~~ from a metastatic site.
3. ~~As shown in a paper by Conderoy et al. (2008), there is an~~ increasing ~~amount of~~ evidence ~~that~~ suggests an association between cancer and the cyclooxygenase (COX) enzyme (5); hence, the COX-expressing cell line HT-29 was studied ~~in this present work~~.

Rule 2: Replace wordy phrases with more concise words or phrases that have the same meaning.

1. ~~Similar to this~~ [Similarly], fractions enriched in organic acids also inhibited cell growth.
2. ~~As an example of what we mean~~ [For example], the detection limit increased 3-fold.
3. The rate increased ~~due to the fact that~~ [because] the temperature increased.

Rule 3: Use parentheses to report information such as vendors, reaction conditions, amounts, or product yield.

1. All reagents were research grade ~~and purchased from Aldrich~~ (Aldrich).
2. The highest yield ~~of 92%~~ was observed in THF ~~after 1 h of reaction time~~ (1 h, 92%).

Rule 4: Eliminate redundancy, often found in locations like these:

- immediately after headings and subheadings;
- in sentences that transition the reader from one section to another (e.g., at the start of a Results section to remind readers about methods and at the start of the Discussion section to remind readers about results);
- in sections of a work that have overlapping purposes (e.g., in the Abstract and Discussion sections of a journal article, which both share overlapping purposes with the Introduction). In sections with overlapping purposes, resist the temptation to merely copy sentences from one section to another.

Consider these examples:

Sampling Procedure. ~~The sampling procedure was as follows:~~ A sample (10 mL) was placed in a septum-sealed glass vial (30 mL).

Site of Investigation. ~~The site of investigation was the Coconino National Forest~~. The Coconino National Forest is the largest contiguous ponderosa pine forest in North America.

Conclusions. ~~In conclusion, t~~[T]he concentrations of PAHs in . . .

[Note that the phrase *In conclusion* (followed by a comma) is redundant only if it follows the heading *Conclusions*. Without the heading, the phrase is appropriate and serves as a useful signal to the reader.]

Nominalizations

Nominalizations are nouns that are formed from other parts of speech, usually by adding an ending such as -ment, -tion, or -ity to a verb or an adjective.

Other Parts of Speech	⇨	***Nominalization***
react (verb)		reaction (noun)
concentrate (verb)		concentration (noun)
measure (verb)		measurement (noun)
intense (adj.)		intensity (noun)

Nominalizations are common in scientific writing because, in most instances, they make your writing more concise.

Without nominalization	The heterogeneous property of ricin is generally not considered in studies of how toxic it is. (16 words)
With nominalization	The *heterogeneity* of ricin is generally not considered in *toxicity* studies. (11 words) (Adapted from Fredriksson et al., 2005)

Nominalizations can also minimize awkwardness, adding to the fluidity and clarity of your writing.

Awkward	The ability to deprotect the carboxylic acid and to simultaneously cleave from the resin will provide compounds with the desired side chain.
Less awkward	The *deprotection* of the carboxylic acid and the simultaneous *cleavage* from the resin will provide compounds with the desired side chain. (Adapted from Hergenrother, 2002)

Respectively

The word *respectively* clarifies relationships between two or more sequences in the same sentence. When used correctly, your writing will become more concise and professional.

Less concise	The injector temperature was 260 °C. The ECD temperature was 280 °C.
More concise	Injector and ECD temperatures were 260 and 280 °C, respectively. (From Llompart et al., 2001)

Less concise	Arsenite was dissolved in water; arsenobetaine was dissolved in methanol; arsenocholine was dissolved in chloroform.
More concise	Arsenite, arsenobetaine, and arsenocholine were dissolved in water, methanol, and chloroform, respectively.

Notice how *respectively* is used in the preceding examples:

- *Respectively* is preceded by a comma in all instances.
- The ordering of the sequenced items linked by *respectively* match. That is, the first item in the first series is linked to the first item in the second series, the second item in the first series is linked to the second item in the second series, and so forth.
- *Respectively* most often appears at the end of a sentence, though it can appear within a sentence (e.g., A and B are points above and below the surface, *respectively*, and d represents distance).
- When units for items in a series are the same, the unit is repeated only once at the end of the series (e.g., The aldol reactions required 30 and 60 min, respectively.) This is also true for "%" (e.g., Yields for **1** and **2** were 86 and 92%, respectively.)
- *Respectively* is only used to relate two or more series (or sequences) in the same sentence. *Respectively* is not used in a simple list, as in this example:

  > Flavonoids are of particular interest because of their potential anti-inflammatory, antiallergic, and antiviral activities.

  Because there is only one series in this example (anti-inflammatory, antiallergic, and antiviral activities), the use of *respectively* is inappropriate.

Part 2: Word Choice

Different genres (including journal articles, posters, email messages, and blogs) call for different word choices. In the Tips that follow, we focus on the use of formal academic vocabulary and hedging words that are appropriate for chemistry journal articles, conference abstracts, posters, and proposals.

Formal Vocabulary

Writing in chemistry requires that authors use vocabulary that is more formal and academic than they would use in speaking. For example, you could *say* "we cut back on the solvent," but you would *write* "the solvent was decreased (or reduced)." Consider these examples. Also see Table 16.1.

Less formal	Mutant forms were studied to *find out* what changes, if any, occurred in carbohydrate-binding specificity.
More formal	Mutant forms were studied to *determine* what changes, if any, occurred in carbohydrate-binding specificity.

Table 16.1. Formal (Academic) Verbs Commonly Used in Chemistry Writing

accumulate	decrease	examine	originate
analyze	determine	extract	reduce
comprise	eliminate	imply	remove
conduct	enable	include	report
conserve	ensure	investigate	suggest
consider	enter	limit	utilize
decide	establish	obtain	withdraw

Hedging

Hedging is a technique used by effective writers to indicate some degree of caution, tentativeness, or restraint. Hedging prevents writers from overstating a claim or over-interpreting results. Without hedging, your writing may sound exaggerated or non-scientific, resulting in a loss of credibility. Imagine a weather forecaster uttering these statements:

Without hedging	Sunday will be windy with 60 mile per hour winds between 2:00–4:00 p.m.
With hedging	Sunday is *likely* to be windy; gusts up to 60 miles per hour *might* be observed in the afternoon.

Forecasters who fail to hedge may lose listeners' trust because they will be wrong over and over again. More credible forecasters predict the weather with caution and restraint, using hedging words.

See Tables 5.1 and 5.2 in Chapter 5 for hedging words commonly used by chemists to soften their interpretations and temper their claims.

Part 3: Writing Conventions

The term *writing conventions* is used in *Write Like a Chemist* to refer to generally accepted discipline-specific practices. Chemistry is no different from other fields; its many writing conventions are documented in the *ACS Guide to Scholarly Communication* (Banik et al., 2020) and in Author Guidelines for different ACS journals. Here, we consider writing conventions for abbreviations and acronyms, active and passive voice, capitalization, compound labels, and numbers and units.

Abbreviations and Acronyms

Abbreviations and acronyms are short forms of individual words (e.g., Et for ethanol) and multiword terms (e.g., SPME for solid-phase microextraction). In abbreviations, the individual letters are pronounced (e.g., N-M-R), whereas in acronyms, the letters form a pronounceable word (e.g., NASA). The use of abbreviated terms is encouraged for long names or terms that appear multiple times in a text. Some rules for using abbreviations in your writing follow. [Note: italic font has been added for instructional purposes.]

Rule 1: Place the abbreviated term in parentheses, immediately following the first use of the full term in the text:

> SF_6 has the largest *global warming potential (GWP)*, which is 23,500 times greater than that of the naturally occurring greenhouse gases (e.g., CO_2). (From Yu et al., 2018)

[Note: In the preceding example, the abbreviated term is in all caps (without periods), but the full term is in lowercase.]

Rule 2: Define the abbreviated term only once. Once defined, the abbreviation can be used without definition (and without parentheses) in the remainder of the work:

> The experimental desorption isotherms for *randomly methylated β-cyclodextrin (RAMEB)* and *RAMEB*-enriched minerals are presented in Figure 1. The isotherms were measured for the *RAMEB* in the forms of powder and crystals. (From Jozefaciuk et al., 2001)

Rule 3: After an abbreviated term has been defined, you may use either the full term or the unabbreviated term, whichever seems more appropriate.

Rule 4: Use standardized abbreviations when possible. Many commonly used terms have standardized abbreviations; be sure to use the approved abbreviation (i.e., do not invent your own). For example, abbreviations for elements (e.g., Fe), chemical formulas (e.g., CH_3CH_3), and standard units (e.g., mL, g, μm, °C), in addition to those listed in the *ACS Guide to Scholarly Communication* (Petronella, 2020, sections 5.37 and 5.38), can be used without definition in ACS journals. See also Table 16.2.

Rule 5: Avoid abbreviations and acronyms in titles.

Table 16.2. Abbreviations That Do Not Need to Be Defined for a Chemistry Audience[a]

at. wt	atomic weight	NMR	nuclear magnetic resonance
bp	boiling point	RNA	ribonucleic acid
DNA	deoxyribonucleic acid	U.S.	United States (as adjective)[b]
equiv	equivalent(s)	UV	ultraviolet
fp	freezing point	v/v	volume per volume
GC	gas chromatography	w/v	weight per volume
IR	infrared	w/w	weight per weight
mp	melting point	wt	weight

[a] See Petronella (2020, sections 5.3.7 and 5.3.8) for abbreviations that can be used without definition in ACS journals but may need to be defined in other journals.

[b] U.S. (with periods) and US (without periods) are both used as abbreviations for the United States.

Rule 6: Abbreviations defined in abstracts should be defined again in the text.

Rule 7: Use pronunciation to select which article (*a* or *an*) to use before an abbreviated term. Select *a* or *an* according to the pronunciation (not the letter) of the first sound of the abbreviated form. With a consonant sound, use *a*; with a vowel sound, use *an*:

a nuclear magnetic resonance spectrometer

an NMR spectrometer

Rule 8: Abbreviate units of measure that follow a number. Without a number, spell them out:

9 V/s

measured in volts per second

Rule 9: Use a lowercase *s* to form plurals of abbreviations. Form the plural of multiple-letter, all-capital abbreviations and abbreviations ending in a capital letter by adding a lowercase *s*. Do not put an apostrophe before the *s*; do not add an *s* to units of measure:

PCBs PAHs CFCs pHs 10 mL 10 g

Active and Passive Voice

Active and passive voice are both widely used in chemistry writing. Examples of the two forms in various verb tenses are illustrated in Table 16.3.

Table 16.3. Examples of Active and Passive Voice in Different Verb Tenses

active voice			passive voice	
verb tense	subject (We) + verb	object of sentence	object of active sentence	form of "to be" + past particle form of verb[a]
present	We stir	the mixture.	The mixture	is stirred.
past	We stirred	the mixture.	The mixture	was stirred.
past	We stirred	the mixtures.	The mixtures	were stirred.
future	We will stir	the mixtures.	The mixtures	will be stirred.
prt pf[b]	We have stirred	the mixture.	The mixture	has been stirred.
prt pf[b]	We have stirred	the mixtures.	The mixtures	have been stirred.

[a] Passive voice constructions require the use of the past participle (e.g., shown, become), which is sometimes different from the past tense verb (e.g., showed, became). See footnote in Table 6.2 for more details.

[b] prt pf = present perfect.

Passive voice is preferred when summarizing others' works. Passive voice shifts the emphasis from the researcher to the research, that is, from the scientist to the science, thereby adding objectivity. In many cases, passive voice also makes the sentence more concise.

Active (not preferred)	Anderson et al.[1] have shown that chromium (VI) compounds are carcinogenic.
Passive (preferred)	Chromium (VI) compounds *have been shown* to be carcinogenic.[1]

Passive voice is also preferred when describing procedures in a Methods section (again, because passive voice adds objectivity).

Active (not preferred)	We distilled DMF before use.
Passive (preferred)	DMF *was distilled* before use.

Writers typically have less experience using passive voice than active voice; hence, we provide a few tips to help you use this construction correctly:

Tip 1: To convert active voice to passive voice, follow these steps:

- Identify the object of the active sentence. The object becomes the subject of the passive sentence.
- Follow the "new" subject with a form of *be* (*is*, *was*, *were*, *will be*, *has been*, *have been*) and the past particle form of the verb (often the same as the past tense verb form; see footnote to Table 6.2 in Chapter 6 for more on past participles).
- Check that the *be* verb matches the tense (e.g., *is*, *was*, or *will be*) and the singular or plural form of the "new" subject (e.g., *is* or *are*).

Active sentence (singular)	We stirred the mixture.
Passive sentence (singular)	The mixture *was stirred*.

Active sentence (plural)	We show both spectra in Table 1.
Passive sentence (plural)	Both spectra *are shown* in Table 1.

Tip 2: Do not omit the form of *be* before the verb in the passive sentence.

Incorrect	Raman shifts calibrated with indene.
Correct	Raman shifts *were calibrated* with indene.

Tip 3: Do not repeat the subject and/or form of *be* when the sentence has two passive verbs with the same subject.

Incorrect	The product was decanted, and it was recrystallized.
Incorrect	The product was decanted and was recrystallized.
Correct	The product *was decanted* and *recrystallized*.

Tip 4: Include a form of *be* when two sentences, joined with *and*, have different subjects and two passive verbs.

Incorrect	The solvent was evaporated, and the residue dissolved.
Correct	The solvent *was evaporated*, and the residue *was dissolved*.

Tip 5: Begin passive sentences with the word "To" when using the verb *added*. This construction avoids the awkwardness of starting the sentence with a number.

Awkward	Ten milliliters of ethanol was added to the mixture.
Correct	To the mixture was added 10 mL of ethanol.
Correct	To the mixture was added ethanol (10 mL).

When the amount added is not specified, the more standard passive sentence pattern can be used.

Correct	*cis*-2-Bromo-3-methylcyclohexane *was added* to (*R*)-lactic acid.

Capitalization

Chemists are particular about what they do and do not capitalize. The following tips will help you capitalize correctly, following chemistry conventions.

Tip 1: Use lowercase for compounds and elements except at the start of a sentence:

- Methane was prepared by reacting hydrogen and carbon.
- A synthesis was proposed for 3-methylbutane.
- Hydrogen is the lightest gas; nitrogen and oxygen are the most common gases.

Tip 2: Capitalize the following letters:

- unit K (for kelvin), following a number (e.g., 100 K)
- unit C (for Celsius), following a number (e.g., 100 °C)
- only H in pH
- only first letter in two-letter symbols for elements (Na, Fe, Cl, $FeCl_3$)
- FTIR spectra (*but* Fourier transform infrared spectra)
- NMR spectra (*but* nuclear magnetic resonance spectra)
- first letter of a molecule at the beginning of a sentence (e.g., 3-Butanol was reacted with ethane.)

- letters, except the plural *s* in multiple-letter, all-capital letter abbreviations and abbreviations ending in a capital letter (e.g., PCBs, CFCs, pHs)
- "T" and "F" in Table and Figure when referring to specific graphics (e.g., As shown in Table 1, Figure 2)

Tip 3: Use lowercase for unit surnames not preceded by a number.

Correct	. . . was measured in kelvin.
Incorrect	. . . was measured in Kelvin.

Tip 4: Use lowercase for seasons of the year (fall, spring, summer, winter).

At all observation sites, the concentrations of DHOPA from fall to spring were significantly higher than those in summertime. (From Ding et al., 2017)

Tip 5: Capitalize and italicize genus names; italicize, but do not capitalize, species names. Genus names are spelled out in full at first mention and abbreviated thereafter.

Imazalil is efficient for the control of fungi of the *Penicillium* genus, such as *P. digitatum* and *P. italicum* in citrus fruits,[12] and *P. expansum* in pome fruits.[13] (From Díaz-Mula et al., 2019)

Tip 6: Capitalize main words in ACS journal-article titles and section headings including nouns, pronouns, verbs, adjectives, and adverbs.

Matrix-Assisted Laser Desorption Ionization Imaging and Laser Ablation Sampling for Analysis of Fungicide Distribution in Apples (From Pereira et al., 2019)

Tip 7: Do not capitalize Latin terms except in titles, headings, and at the start of a sentence.

Within a sentence	in vitro, in vivo, in situ
At the start of a sentence	In vitro, In vivo, In situ
In titles and headings	In Vitro, In Vivo, In Situ

Tip 8: In geographic names, capitalize regions but not general adjectives.

The Southwest is a region of the West.
We traveled to southwestern California.

Compound Labels

Compound labels are often assigned to chemical species to prevent the repetition of long chemical names in the text. Some generally accepted rules for compounds labels are as follows:

Rule 1: Use bolded numbers and/or numbers and letters as compound labels (e.g., **1**, **2**, **3** or **2a**, **2b**, **2c**).

Rule 2: Introduce compound labels in the text or graphics (e.g., a table or scheme).

- In text, introduce the compound label *after* the first use of the spelled-out name of the compound, inside parentheses (like most abbreviations).
- In graphics, place the label directly below the structure of the compound.

> The asymmetric Strecker reaction of (*R*)-phenylglycine amide (**1**), pyvaldehyde (**2**), and HCN generated in situ from NaCN and AcOH was studied. (Adapted from Boesten et al., 2001)

Rule 3: Once defined, the bolded compound label may be used without parentheses.

> Unfortunately, the oxidation failed for **14**.
>
> Compounds **5–8** are shown in Scheme 2. The synthesis of **5** was accomplished by . . .

Rule 4: Number compounds consecutively.

Rule 5: In schemes, tables, or figures, assign compound labels only to compounds that are referred to in the text, not all compounds.

Rule 6: For synthesized compounds that are named in bolded subheadings of Experimental Sections, introduce (or reintroduce) the compound label after the name.

> **Preparation of 2,6-di-tert-butyl-4-((4-fluorobenzylidene)-amino)phenol (3a).**

Numbers and Units

Learning how chemists format numbers and units—for example, 15%, 15 mL, three samples (not 15 %, 15mL, 3 samples)—will help you write more like a chemist. Consider these nine rules (and two exceptions) to get you started. (See also Henderson, 2020, section 5.4.4.)

Rule 1: Abbreviate units of measure and time when they follow a number.

Examples of units of measure and time are mL, cm, g, K, s, h, and min. Most abbreviated units do not include periods, unless the abbreviation could be confused with a word (e.g., "in." for inches, "at." for atomic).

Inappropriate	5 seconds	16 milliliters	13.5 in
Appropriate	5 s	16 mL	13.5 in.

Rule 2: Spell out units of measure if they do not follow a number.

Inappropriate	several mg	a few min
Appropriate	several milligrams	a few minutes

Rule 3: Use the numerical form of numbers (e.g., 7) rather than the word form (e.g., seven) with units of measure and time. One exception is at the beginning of a sentence (e.g., Seven hours later, . . .). Include a space between the number and unit, unless you are using the % sign.

Inappropriate	seven mL	five cm	thirty percent	11 hours
Appropriate	7 mL	5 cm	30%	11 h

Rule 4: Place the degree symbol before the C (e.g., 100 °C). Include a space after the number, but no space between the degree symbol and the C. For K (kelvin), include a space between the number and the K; do not use the degree symbol with kelvin.

Inappropriate	10°C	10° C	10K	10 °K
Appropriate	10 °C	10 K		

Rule 5: Do not add a plural *s* to an abbreviated unit.

Inappropriate	4 mgs	3 mols
Appropriate	4 mg	4 mol

Rule 6: Use the word form for numbers less than 10, except with units of measure.

Inappropriate	4 sites	five mg	thirteen sites	8 samples
Appropriate	four sites	5 mg	13 sites	eight samples

Rule 7: In general, use abbreviated forms for units. One exception is when a number and unit are used at the start of a sentence.

Inappropriate	3 minutes	300 milliliters	Seven mL
Appropriate	3 min	300 mL	Seven milliliters

Rule 8: Use all numbers, no words, in series that include the number 10 or greater.

Inappropriate	one, five, 10, and 15 mg
Appropriate	1, 5, 10, and 15 mg

Rule 9: In a series or range of numbers, include the unit only once. Use an en dash (–) to mean "to" or "through."

Inappropriate	3 mL–10 mL	10%, 20%, and 30%
Appropriate	3–10 mL	10, 20, and 30%

Exception 1 to Rule 9: When one or both of the numbers are negative or include a symbol that modifies the number, use the words "to" or "through" rather than the en dash.

Inappropriate	–3–10 mL	<2–4 kg
Appropriate	–3 to 10 mL	<2 to 4 kg

Exception 2 to Rule 9: Do not use the en dash (–) when the words "from" or "between" are used.

Inappropriate	from 200–400 mL	between 1–3 h
Appropriate	from 200 to 400 mL	between 1 and 3 h

Part 4: Grammar

Possessing a mastery of the mechanics of writing (e.g., grammar, punctuation) is not trivial, but rather it indicates an attention to detail that reflects well on the seriousness of the writer.

—Joseph H. Aldstadt, University of Wisconsin, Milwaukee

Unlike chemistry-specific writing conventions, English grammar rules apply across the disciplines. Here we focus on parallelism, plural and singular scientific words, split infinitives, and subject–verb agreement.

Parallelism

Using parallel constructions is important in words and phrases, lists, headings and subheadings in text and tables, and figure captions.

Grammatical and Conceptual Parallelism

Words and phrases joined by *and* or *or* should be parallel; they should have the same grammatical form (grammatical parallelism), and the concepts should be of similar scale and level of importance (conceptual parallelism), as seen in these examples.

Set A (Grammatical parallelism)

Not grammatically parallel	The hydrolysis was suppressed by *a reduction in* the amount of water and *lowering* the reaction temperature.
Grammatically parallel	The hydrolysis was suppressed by *reducing* the amount of water and *lowering* the reaction temperature. (Adapted from Senda et al., 2001)

Set B (Grammatical parallelism)

Not grammatically parallel	The next step was *to collect* the eluting fraction and *subjecting* it to prolonged aerial oxidation.
Grammatically parallel	The next step was *collecting* the eluting fraction and *subjecting* it to prolonged aerial oxidation.
Grammatically parallel	The next step was *to collect* the eluting fraction and (*to*) *subject* it to prolonged aerial oxidation. (Adapted from Napolitano et al., 2001)

Set C (Grammatical parallelism)

Not grammatically parallel	The 15-HLO screen is both *a robust tool* and *reliable for reporting* lipoxygenase inhibition.
Grammatically parallel	The 15-HLO screen is both *a robust tool* and *a reliable reporter* of lipoxygenase inhibition. (Adapted from Carroll et al., 2001)

Set D (Conceptual parallelism)

Not conceptually parallel	This reaction has many applications in *industry*, *synthesis*, and *the creation of amino acid derivatives*.

Explanation: The three items in this series are not conceptually parallel. Both "industry" and "synthesis" are broad terms, while the "creation of amino acid derivatives" is a much narrower, specific task.

Conceptually parallel	This reaction, which can be used to create amino acid derivatives, has many other applications in *industry* and *synthesis*.

Explanation: The entire sentence had to be revised to retain the same content but express it in a parallel manner.

Headings and Subheadings

Headings and subheadings—in journal articles, posters, and proposals—should be parallel to one another in terms of grammar, punctuation, and capitalization.

not parallel	parallel
Summarizing Project.	Project Summary
Description of Project:	Project Description
(1) Giving Background and Stating Objectives	Background and Objectives
2. Design of Research and Methods	Research Design and Methods
3) scheduling of project	Project Schedule
4. Conclusions	Conclusions
references	References

Lists

Lists are commonly used in research proposals and posters but not in journal articles. In proposals, numbered or bulleted lists are often used to describe proposed methods and highlight expected outcomes. In posters, lists can be used to describe methods and procedures, summarize results, interpret findings, and offer concluding remarks. In both genres, lists should use parallel constructions in the following areas:

- Numbering schemes: In continuous text, numbers are enclosed in parentheses: (1) (2) (3). In vertical lists, numbers are enclosed in parentheses or followed by a period:

  1.

  2.

  3.

- Capitalization: Listed items should start with either lowercase letters or uppercase letters, not both.
- Punctuation: Listed items should be punctuated similarly.
- Language: Listed items should begin with the same form of speech (e.g., nouns, verbs).

Plural and Singular Scientific Words

Several words, used often in chemistry, have tricky plural forms. Those plural nouns must be accompanied by plural verb forms. Table 16.4 itemizes some irregular plural/singular word pairs. Other plural/singular scientific words are explained in this section.

Table 16.4. Irregular Plural/Singular Word Pairs

plural form	singular form	example sentences
data[a]	datum	Although the IR *data are* limited, the absorption at 3300 cm^{-1} suggests that it is **5F**.
		The *data suggest* that electron-withdrawing groups increase the cyclization barrier.
spectra	spectrum	The photoelectron *spectrum is* shown on the *y*–axis.
		The NMR *spectrum confirms* this assignment.
		The IR absorption *spectra* of **4** and **5** *were* found to shift slightly upon addition of a metal ion.
		Similar absorption *spectra have* been reported for Co(II) derivatives.
analyses	analysis	Our *analyses confirm* the formation of an ice Ih layer.
		Further *analysis suggests* that the organic layer contains the anions.
syntheses	synthesis	The *syntheses were* most successful when the solvent was polar.
		In this work, a *synthesis* for the formation of alkyl nitriles with a hydroxy group at the α-position *is* presented.

[a] If the term *data set* is used instead of *data*, the singular verb form is used. For examples of *data* in use, consult Table 4.3 in Chapter 4.

Scientific Disciplines That End in *s*

The names of numerous scientific disciplines end in *s*, even though they are singular words (e.g., ballistics, chemometrics, dynamics, genetics, genomics, kinetics, mathematics, physics, quantum mechanics, thermodynamics). These words should be followed by singular verb forms.

> Nonequilibrium *thermodynamics*—with applications to physical, chemical, and biological systems—*has received* much attention in recent years.
>
> The *kinetics* of reactions in the atmosphere *depends* on the concentration of the hydroxyl radical.

et al. (for Multiple Authors)

The abbreviation "et al." (from the Latin *et alia*) means "and others." It is used when citing a work with three or more authors. Instead of writing out all of the authors' names, "et al." is used after the first author's surname. The "surname + et al." combination is followed by a plural verb form.

> Felmatti et al.[4] *have shown* that a compound in smoke promotes seed germination.

Split Infinitives

Infinitives comprise the word *to* plus the base form of a verb (e.g., to contribute, to transfer). A split infinitive occurs when a word (or words) is placed between the two parts of the infinitive (e.g., to fully investigate, to further complicate).

Although split infinitives are common in spoken language, they are less frequent in formal written language. A common belief is that infinitives should not be split. Yet, according to Raunikar (2020a, section 5.1.1), it is acceptable to split infinitives if doing so will prevent awkwardness or ambiguity.

> Ambiguous — The goal of this research is *to develop completely* new classes of bioresponsive materials.

Explanation: What is the source of ambiguity? Is the author's goal completely new classes of bioresponsive materials *or* development that is thorough and complete?

> Unambiguous — The goal of this research is *to completely develop* new classes of bioresponsive materials. (Adapted from Lyon, 2000)

Subject–Verb Agreement

Subject–verb agreement occurs when the subject and verb of a sentence agree in number; that is, a singular subject is accompanied by a singular verb form, and a plural subject is accompanied by a plural verb form. With simple sentences, we usually do not make subject–verb agreement mistakes. However, with more complex sentences, such as those in scientific writing, errors are more common. Six tips will help you avoid common subject–verb agreement errors.

Tip 1: Be careful when the subject and verb are separated. The most common subject–verb agreement mistake is made when the subject of a sentence is separated from its verb by a phrase or clause.

> *The ratio* of the intensities of these two doublets *is* ~5:1. (From Cantrill et al., 2001)

As more words come between the subject and the verb, it becomes even more difficult to check for agreement:

> *The importance* of nonadditive contributions for the accurate description of the intermolecular interactions *is* well documented. (Adapted from Tongraar et al., 2001)
>
> GRACE implements a cutoff distance d_c such that *any lattice mismatch* between a pair of overlayer and substrate lattice points exceeding d_c *is ignored*. (From Mitchell et al., 2001)

Tip 2: Use different verb forms for subjects joined by *and* or *or*. When two or more subjects are joined with the word *and*, use a plural verb form. When two or more subjects are joined by *or*, the verb should take the number (i.e., singular or plural) of the subject (noun) that is closest to the verb.

> Foam *and* flavor stability *are* important considerations for a brewer. (Adapted from Cooper et al., 2002)
>
> Cesium iodide *or* polyethylene glycol *was* employed as a reference compound. (From Cantrill et al., 2001)
>
> The rate constants *or* the appropriate metal ion concentration *was* used.
>
> The appropriate metal ion concentration *or* the rate constants *were* used.

Tip 3: Units of measure take singular verb forms.

> Approximately *2 mg* of the sample *was* placed in the vial.
>
> To this mixture *was* added *25 mL* of water.

Tip 4: *Each* and *every* are followed by singular verb forms.

Each SPE column *was* preconditioned with 10 mL of methanol.

Each point in the series *corresponds* to a new position.

Every compound *is* represented by a single letter.

Tip 5: *All*, *none*, *some*, *most*, and *any* can be followed by singular or plural verb forms, depending on the noun or object of the preposition following it.

All chemicals were used without further purification.

All of the solutions contain the phosphate buffer.

All of the precipitate was transferred to the flask.

None of the analyzed aldehydes exceed their flavor threshold in beer. (Adapted from Vesely et al., 2003)

Some esters play a central role in flavor changes during beer aging. (Adapted from Vesely et al., 2003)

Most aldehydes, except formaldehyde, *form* two geometrical isomers. (Adapted from Vesely et al., 2003)

Tip 6: Use plural verb forms with *data* and *spectra*. See the Plural and Singular Scientific Words Language Tip for more details.

Part 5: Punctuation

Correct punctuation contributes substantially to writing clarity. Here we highlight several forms of punctuation that are commonly used in chemistry writing.

Colons and Semicolons

Colons (:) and semicolons (;) are often confused even though they serve different purposes.

Tip 1: Colons follow complete statements to introduce examples that directly relate to the statement. Whatever comes before the colon must be able to stand alone as a complete sentence.

Incorrect	The relative Ag^+ selectivities to Hg_2^+ were: 34, 36, and 41.
Correct	Three mechanisms are explored: a hydride shift, a carbocation rearrangement, and a nucleophilic substitution.

Explanation: Why is the first statement incorrect? What comes before the colon is not a complete sentence. The incorrect statement could be modified by simply changing it to read as follows: "The relative Ag^+ selectivities to Hg_2^+ were as follows: 34, 36, and 41%." But this correction is wordy. The original sentence is fine without the colon: "The relative Ag^+ selectivities to Hg_2^+ were 34, 36, and 41%."

Tip 2: Semicolons are used to connect two related independent clauses.

The clauses that come *before* and *after* the semicolon must each be able to stand alone as a complete sentence. Notice that the sentence that comes *after* the semicolon begins with a lowercase letter:

> Both first and fourth harmonics gave identical results; however, the ACV spectra have different appearances. (From Yu et al., 2001)
>
> Mechanistic studies imply that zinc is not simply acting as a Lewis acid; a number of other Lewis acids were tested and caused little to no acceleration of the reaction.[23] (Adapted from Demko and Sharpless, 2001)

Tip 3: The semicolon rule does not apply to sentences joined with *and*, *but*, or *or*:

Incorrect	Each diagram contains points; and each point has associated with it a single phase point variable and time.
Correct	Each diagram contains points, and each point has associated with it a single phase point variable and time.
Correct	Each diagram contains points; each point has associated with it a single phase point variable and time. (Adapted from Andersen, 2003)

(See the Commas Language Tip, Rule 3, for more information on sentences that are joined by words like *and*, *but*, *or*.)

Tip 4: Semicolons are used in series that contain commas.

Semicolons, not commas, are used to separate items in a series when one or more of the series items use a comma. The last item in the series is usually preceded by the word *and*, except in a series that reports physical properties and spectral data (see last example).

Correct	In each structure, 01 is coordinated by two M1, one M2, and one Si; 02 by one M1, one M2, and one Si; and 03 by one M2 and two Si. (Adapted from Ashbrook et al., 2002)
Correct	**Figure 1.** EPR spectra at room temperature for $PM_{2.5}$ collected at five cities in the U.S. (trace A, Rubidoux, CA; trace B, Phoenix, AZ; trace C, Philadelphia, PA; trace D, Durham, NC; and trace E, Baton Rouge, LA). (From Dellinger et al., 2001)

Correct The mixture was cooled to room temperature; the resulting solid was collected, washed with ethanol (2 × 50 mL), and dried to give 54.6 g (90%) of pure **2a**: mp 126.0–127.0 °C; IR 2951, 2906, 1676, 1599, 1572 cm^{-1}. (Adapted from Swenson et al., 2002)

Commas

How and where to use commas can be confusing because of the different rules governing their placement. Use rules 1–8 to help you.

Rule 1: Use a comma after most introductory words and phrases. (See Table 6.6 in Chapter 6 for a list of commonly used introductory phrases and their functions.)

Therefore, the GWP for these hybrid molecules is in the range 0–2000, which is considerably lower than that of SF_6. (From Yu et al., 2018)

However, the presence of gallocatechins (GC and EGC) is only relevant in some products, including broad beans, lentils, grapes, wine, beer, coffee, tea, and most of the berries. (From Díaz-Mula et al., 2019)

As previously described, the bridging iodine atom is located at the midpoint between two diplatinum units. (Adapted from Mitsumi et al., 2001)

Rule 2: Use commas to set off words and phrases such as *for example*, *however*, or *therefore* when mid-sentence.

The band at 3300 cm^{-1}, *however*, disappears in solution.

This may occur, *for example*, when the overlayer unit cells are significantly larger than the substrate unit cells. (From Mitchell et al., 2001)

Rule 3: When two sentences are joined by a linking word (e.g., *and*, *but*, *or*, *so*, *yet*), put a comma before the linking word.

Only a minor fraction of this N1 alkylated derivative (~5%) hydrolyzed under acidic conditions, *and* all loss generated an equivalent quantity of dA. (From Veldhuyzen et al., 2001)

A weak band is observed at 114 cm^{-1}, *but* the energy of this band shows little temperature dependence. (Adapted from Mitsumi et al., 2001)

Rule 4: Include a comma before *and* in a list of three or more items.

When a sentence contains a list or series of three or more items, place a comma before the *and* (or *or*) that comes before the last item. This comma is called a *serial comma*.

Samples passivated with PVB, MA, *and* PVA exhibit a substantial increase in quantum efficiency upon UV irradiation. (From Bol and Meijerink, 2001)

Rule 5: Use a comma between adjectives only if you can reverse their order.

With commas A *novel, inexpensive biosensor* was used to measure calcium.

Rule 6: Use commas to set off *extra* information.

If a clause or phrase adds *extra* information about the preceding word or phrase (that is, information that could be deleted without changing the sentence's main message), set the clause or phrase off from the rest of the sentence with commas.

> Concurrently, the α-chloroamine **9a**, *in a solution of THF*, was added to the NaH/Celite column. (From Hafez et al., 2001)

Rule 7: In most cases, include a comma before "which."

If the word *which* is used mid-sentence to introduce a non-restrictive clause (i.e., extra information that does not limit the meaning of the sentence's main message), place a comma before *which* and after the additional information. (See the That/Which Language Tip for more details.)

> This transformation, *which was verified by Raman spectroscopy*, was apparently induced by the mechanical pressure exerted by the AFM tip in contact mode. (From Mitchell et al., 2001)

Rule 8: Include a comma with "i.e." (that is) and "e.g." (for example) inside parentheses.

> Except for a few molecules in the CAS registered reference databases (i.e., SF_3N, SF_4NF, CF_3SO_2F, CF_3OCOF, and CF_2NCF_3), the other fully fluorinated compounds involving multiple hybrid bonds (e.g., S=C and S≡C) are experimentally unknown. (From Yu et al., 2018)

Hyphenated Two-Word Modifiers

Chemistry writing includes many two-word modifiers, two words that together describe a noun:

Two-Word Modifier	*Noun*
air-dried	flask
solid-state	material
radiation-sensitive	compound
above-average	results
high-frequency	sound

Many types of two-word modifiers are used (noun + adjective, adjective + noun, noun + noun); almost all are hyphenated when used to modify a noun (see Table 16.5).

Table 16.5. Different Types of Two-Word Modifiers

type	examples
noun + adjective	time-dependent changes, mass-selective detector
adjective + noun	long-term storage, high-density lipoprotein
adjective + participle	fast-growing bacteria, far-reaching implications
noun + participle	time-consuming calculation, phospholipid-coated capillaries
adverb + adjective	below-average results
noun + noun	solvent-solute results, room-temperature solution

Two-word (and sometimes three-word) modifiers can be used with compound nouns (i.e., two or more terms used to express a single idea). Some compound nouns are hyphenated (e.g., half-life), whereas others are not (e.g., redox couple). The compound noun should not be connected to the hyphenated modifier with an additional hyphen.

Hyphenated Modifier	*Compound Noun*
temperature-dependent	absorption cross section
moisture-sensitive	iodine monochloride
three-dimensional	NMR spectra
light-induced	charge redistribution
light-catalyzed	miniaturized solar batteries
high-resolution	crystal structures

Some exceptions to hyphenation rules for two-word modifiers are explained here:

Exception 1: Do not hyphenate if the first word ends in "ly".

Incorrect	accurately-measured dosage commonly-accepted practice
Correct	accurately measured dosage commonly accepted practice

Exception 2: Do not hyphenate chemical names (e.g., amino acid level, barium sulfate precipitate).

Exception 3: Do not hyphenate if one of the words is a proper name (e.g., Lewis acid catalysis).

Exception 4: For modifiers with the same base, include a hyphen after each modifier.

When multiple two-word modifiers have the same base (the second word in the modifier) and are used to modify the same noun, use a hyphen after each modifier, without repeating the base.

Correct	medium- and long-range forces
Correct	first- and second-order reactions

Exception 5: Hyphenate three- (or more) word modifiers unless other hyphenation rules are broken by doing so.

Correct	general-acid-catalyzed reaction
Correct	Lewis acid catalyzed reaction

Part 6: Word Usage (Commonly Confused Words)

Numerous word pairs are commonly confused by writers. The confusion often stems from similar pronunciations or spellings. Other times the confusion is caused by word meanings that are similar, but distinct. Here we introduce 12 pairs of commonly confused words.

Affect/Effect

The words *affect* and *effect* sound similar, so they are often used incorrectly. In chemistry writing, *affect* and *effect* are typically used as follows:

Affect (verb): to influence, modify, change

Effect (noun): result, consequence, outcome

Effect (verb): to bring about, cause to happen

Consider these examples:

affect (verb)	Of some concern is whether the vacuum drying step . . . could *affect* the measurement accuracy. . . (From Pereira et al., 2019)
	Saponification of fats . . . facilitates the release of PCBs from fatty matrixes . . . without *affecting* the PCBs. (From Llompart et al., 2001)
effect (noun)	Genotoxicity of Trivalent Chromium in Bacterial Cells: Possible *Effects* on DNA Topology (Adapted from Plaper et al., 2002)
effect (verb)	The red signal can be read from the positive sample with the naked eye due to the localized surface plasmon resonance *effect* of Au NPs. (From Hu et al., 2017)
	Increasing the temperature *effected* a rate increase of 20%.

Although/While

Refer to the Because/Since Language Tip.

Among/Between

The words *among* and *between* are often used incorrectly. Fortunately, the rules that distinguish one term from the other are easy (and important) to remember.

Rule 1: *Between* is used with two, and only two, named objects.

> Prenylmagnesium chloride . . . reacts with low chemoselectivity *between* an aldehyde and a ketone. (From Bartolo and Woerpel, 2018)
>
> Spectra were obtained in the Raman shift range *between* 400 and 4000 cm^{-1}. (From Kizil et al., 2002)

Rule 2: *Among* is used with three or more named objects.

> In this study, we present evidence for the first time of a three-way interaction *among* starch, protein, and lipid. (Adapted from Zhang and Hamaker, 2003)

Rule 3: Rules 1 and 2 apply even if the objects are not explicitly stated.

> For the length assay, comparisons were made *between* treatments at the same concentration using a Student's *t* test. (From Lichtenberg et al., 2019)
>
> Rosell et al.[3] reported the highest lauric acid concentrations *among* coconut oil samples from Sri Lanka and the Philippines. (Adapted from Laureles et al., 2002)

Because/Since and Although/While

Two frequently misused words are *since* and *while*. These words are commonly used to mean *because* and *although*, respectively. In formal writing, for scientific and expert audiences, *since* and *while* should be used to connote time:

Incorrect	*Since* pesticides are designed to be toxic for living organisms, their toxicity to humans is often significant.[8]
Correct	*Because* pesticides are designed to be toxic for living organisms, their toxicity to humans is often significant.[8] (From Pereira et al., 2019)
Incorrect	*While* there exist empirical rules for determining conditions favorable for protein crystallization, protein crystallization remains predominantly a trial-and-error process.

Correct *Although* there exist empirical rules for determining conditions favorable for protein crystallization, protein crystallization remains predominantly a trial-and-error process. (Adapted from Curtis et al., 2001)

Tips 1–5 provide suggestions for using *because*, *since*, *although*, and *while* accurately. We also comment on the word *whereas*.

Tip 1: *Because* is generally used to indicate a cause-and-effect relationship.

> Knowledge of the SO_2 level in beer is important *because* SO_2 complexes with aldehydes. (From Vesely et al., 2003)
>
> *Because* of their high sugar content, arils were not able to be freeze-dried. (Adapted from Díaz-Mula et al., 2019)

Tip 2: *Since* is generally used to indicate a passage of time. For example, *since* indicates "during the period after" or "from a point in the past until the present."

> There is a need to verify whether ozone-depleting compound emissions have truly decreased *since* the passage of the Montreal Protocol.

Tip 3: *Although* is generally used to indicate a contrast or something unexpected.

> A four-membered transition state is also possible, *although* this pathway is less likely.

Tip 4: *While* is generally used to indicate "during the time that" and "at the same time that."

> The NCs should be dispersed in the lower density nonpolar hexane phase, *while* the HF, butanol, and water remain in the acetonitrile phase. (Adapted from Langof et al., 2002)

Tip 5: *Whereas* is used to indicate a difference between two facts or ideas that have different characteristics.

> Wind energy and solar energy are comparatively clean, *whereas* fossil fuels lead to air pollution.

Compose/Comprise

Although *compose* and *comprise* look and sound somewhat similar, they are used differently, with different word combinations.

Consider these sentences, which have the same meaning:

The human body *is composed of* 206 bones.

The human body *comprises* 206 bones.

Tip 1: *Comprise* is used by itself.

Comprise is a free-standing verb, used by itself, never with the word "of" following it or some form of the verb "be" preceding it.

Incorrect	This results in decreased conjugation of the oligothienyl units that *are comprised of* the backbones of these ligands.
Correct	This results in decreased conjugation of the oligothienyl units that *comprise* the backbones of these ligands. (Adapted from Weinberger et al., 2001)
Incorrect	The images show that the Au films *are comprised of* large (approximately 1000 Å wide) crystalline (111)-terminated grains.
Correct	The images show that the Au films *comprise* large (approximately 1000 Å wide) crystalline (111)-terminated grains. (Adapted from Han et al., 2001)

Tip 2: *Compose* is used with other words, not as a stand-alone word.

Compose is used with this combination of words: *(be) composed of*. Notice that *composed* (ending in *ed*) is preceded by some form of the verb *be* and followed by *of*.

Incorrect	Pomegranate PAs *composed* a rich mixture of flavan-3-ols: catechin, epicatechin, gallocatechin, epigallocatechin, epigallocatechin gallate, and epicatechin gallate.
Correct	Pomegranate PAs *were composed of* a rich mixture of flavan-3-ols: catechin, epicatechin, gallocatechin, epigallocatechin, epigallocatechin gallate, and epicatechin gallate. (From Díaz-Mula et al., 2019)
Incorrect	Each compound *composes* an amino sugar group and a nitrogen-free sugar group bound directly to a 14-membered-ring aglycon moiety.
Correct	Each compound *is composed of* an amino sugar group and a nitrogen-free sugar group bound directly to a 14-membered-ring aglycon moiety. (Adapted from McClellan et al., 2002)

Tip 3: The *(be) composed of* word combination can be altered slightly when the "be" form is not the main verb of the sentence. In these cases, the "be" form can be dropped, but it is always understood.

Be is implied	In this paper, we report on a method based on hexachloroplatinic acid impregnation on carbon followed by an electrochemical reduction step *composed of* several current pulses.[7,8]
Be is included	In this paper, we report on a method based on hexachloroplatinic acid impregnation on carbon followed by an electrochemical reduction step *that is composed of* several current pulses.[7,8] (Adapted from Adora et al., 2001)
Be is implied	The proven validity of the differences method for this kind of system will enable more complex systems to be studied, like those *composed of* particular metal-IDA complexes with some resin and oxyanions such as selenite and tellurite.
Be is included	The proven validity of the differences method for this kind of system will enable more complex systems to be studied, like those *that are composed of* particular metal-IDA complexes with some resin and oxyanions such as selenite and tellurite. (Adapted from Atzei et al., 2001)

Farther/Further

Farther and *further* are commonly confused words. Most often, *further* signals quantity or degree and *farther* signals literal or figurative distance.

Tip 1: *Further* (adjective) indicates *additional* or *more*.

These experiments provide *further* evidence that prenylmagnesium chloride and allylmagnesium chloride react at similar rates. (From Bartolo and Woerpel, 2018)

However, three molecules including SF_5NNF, CF_3SO_2F, and $CF_3OC(O)F$ need *further* assessment. (From Yu et al., 2018)

Tip 2: *Further* (adverb) indicates *additionally* or *to a greater degree*.

The precursor and mature BACE1 proteins were *further* characterized for their ability to interact with a substrate-based transition state BACE1 peptide inhibitor. (From Wang et al., 2005)

The pH was immediately adjusted to 8.7, and the refolding mixture was *further* incubated with slow stirring at room temperature for 4 h. (From Wang et al., 2005)

Tip 3: *Further* (verb) is used to mean *to help* or *to advance* (something).

To further our fundamental understanding of the underlying science, . . .

Tip 4: *Furthermore* is often used as a transition devise to indicate "in addition" (see Table 6.6 in Chapter 6).

> Furthermore, we describe experiments that suggest this transition state is responsible for the rapid and unselective reactions of allylic Grignard reagents. (From Bartolo and Woerpel, 2018)

Tip 5: *Farther* (adjective) means *more distant*.

> In the third case, the upper surface was moved a *farther* 0.74 Å to the right. (From Neitola and Pakkanen, 2001)

Tip 6: *Farther* (adverb) means *at* or *to a greater distance* or *more distant point. Farther* is often accompanied by words such as these: apart, away, from, and toward.

> When the concentration is higher, the metal ions move *farther* toward the maximum field. (From Fujiwara et al., 2004)
>
> The magnitude becomes smaller because the chromophore is *farther* apart from the rim of the cavity of the chiral macrocyclic host.[6,7] (Adapted from Park et al., 2002)

Fewer/Less

The words *fewer* and *less* are commonly confused. A few simple rules will help you use these words correctly.

Rule 1: Use *fewer* with items that can be counted.

> These crystals have *fewer defects* than those grown from solution.
>
> *Fewer* than *four chlorine atoms* surround each platinum atom.

Rule 2: Use *less* with quantities that cannot be counted.

> The preferred addition requires *less activation* than the HCl addition to ethylene.
>
> The other compounds possessed significantly *less potency*.

Rule 3: Use *less* with number and unit of measure combinations (because the combinations are regarded as singular).

> Data collection time was *less than 3 h* per crystal.
>
> Barriers of *less than 7 kcal/mol* are predicted.
>
> The oxide was isolated in high yield but with *less than 5%* ee.

Its/It's

Its and *It's* have totally different meanings but similar spellings.

Tip 1: *Its* is the possessive form of *it*.

> Concerns about the toxicity of imazalil derive from *its* potential effects on reproductive and neurobehavioral parameters.[17] (From Pereira et al., 2019)
>
> The SF_3N molecule involves the S≡N triple bond and *its* $\Delta H_{f,\ 298.15}$ is −83.3 kcal/mol. (Adapted from Yu et al., 2018)

Tip 2: *It's* is the contracted form of *it is*.

It's is inappropriate when writing for an expert audience. *It's* is sometimes acceptable in writing for a general audience, as in, for instance, *Science News* and the Latest News section of *Science*.

Tip 3: Writers should avoid using *it is* (even in its un-contracted form) when it creates "excess verbiage" (Raunikar, 2020a, section 5.1.3).

Avoid	*It is* a method that has been described previously.[3]
Preferred	*This* method has been described previously.[3]

Precede/Proceed

Precede and *proceed* are sometimes confused by writers, even though the words are not interchangeable. Hence, it is important to use them correctly.

Tip 1: *Precede* (verb): To come, exist, or occur *before* in time, order, rank, or position:

> Reductive activation of dioxygen *precedes* C–H bond cleavage. (Adapted from Crespo et al., 2006)
>
> The decrease in the concentration of IMP *preceded* a simultaneous increase in the concentrations of both inosine and hypoxanthine in meat samples. (From Tikk et al., 2006)
>
> The oxidative addition is *preceded* by decoordination of the anion.[11] (From Ahlquist et al., 2006)

Tip 2: *Preceding* (adjective): Previous; existing or coming before something in time, order, rank, or position:

> As mentioned in the *preceding* section, the reaction of . . .
>
> In agreement with the *preceding* experiment, the average concentration . . .

Note that the adjective *preceding* commonly precedes words and phrases such as those in Table 16.6.

Table 16.6. Words and Phrases That Commonly Follow the Adjective *Preceding*

conformational change	experiment	residue	stage
cycle	paper	results	step(s)
discussion	paragraph	scheme	study
example	position	section	treatment

Tip 3: *Proceed* (verb): To go forward or onward, to continue, to begin an action or a process:

> The thermal decomposition of 2,1-benzisoxazole will *proceed* at the lower temperature. (Adapted from Lifshitz et al., 2006)
>
> When the carbonyl group is relatively unhindered, the addition *proceeds* through a six-membered transition state with allylic transposition. (From Bartolo and Woerpel, 2018)

Principal/Principle

Principal and *principle* are commonly misused. Although *principle* is used more often than *principal* in chemistry writing, both words are common.

Tip 1: *Principle* (noun): a basic truth or assumption.

Principle also refers to a standard; a guiding rule, code of conduct, method of operation; a rule or law concerning the functioning of natural phenomena or mechanical processes:

> The remarkable selectivity of ions by macrocycles illustrates the *principle* of molecular recognition. (From Parra et al., 2006)
>
> The basic *principle* of semiconductor photocatalysis involves photogenerated electrons and holes migrating to the surface. (From Wang et al., 2006)
>
> *Principle* of Ebola Virus Glycoprotein and Whole Virion Detection Using Enhanced Strip Biosensor (section heading in Hu et al., 2017)

Additional forms of "principle" common in chemistry writing:

- *In principle*: with regard to the basics; according to what is supposed to be true, though not proven; in theory:

  More calculations are, *in principle*, needed to corroborate this assumption.

In principle, $CH_3COCH_2COCH_3$ may exist as an asymmetric (C_s) or symmetric (C_{2v}) structure in the ground state. (From Chen et al., 2006)

- *First principles*: a calculation is said to be from *first principles* (or ab initio) if it relies on basic and established laws of nature without additional assumptions or special models. The term is often used in titles and as a two-word modifier.

first-principles calculations	first-principles molecular dynamics study
first-principles kinetics	first-principles simulations
first-principles methods	first-principles study

- *Principal* (adjective): most important; main; chief; first in rank, worth, or degree:

The elimination of HO_2 from the alkylperoxy radicals is the *principal* reaction pathway for the R + O_2 reactions at temperatures above the transition region. (From Taatjes, 2006)

Principal components analysis (PCA) is a statistical procedure.

Principal is often used with words and phrases such as those listed in Table 16.7.

Table 16.7. Words and Phrases Commonly Used with *Principal*

principal advantages	principal factors	principal reaction
principal characteristics	principal goals	principal resonance peaks
principal driving force	principal indices	principal role
principal electronic parameters	principal ions	principal setup
principal emission sources	principal product	principal step

Than/Then

Than and *then* have very different meanings; nonetheless, writers often mistakenly use them interchangeably. Note that *then* is used only sparingly in expert scientific writing because the order of events (e.g., first, second, then) is generally implied and does not need to be stated explicitly.

Tip 1: *Than* (conjunction): used to show comparisons.

Than is preceded by words like fewer, less, lower, more, greater, higher. *Than* is also used with "rather" (*rather than*, similar in meaning to "instead of" and "in place of") to show a contrast.

In all pomegranate samples, EGCG and ECG were present but at low concentrations (less *than* 8%). (From Díaz-Mula et al., 2019)

In addition, some plant chitinases exhibit other biological functions *rather than* antifungal activity, such as cryoprotective activity *(14)*. (From Liu et al., 2020)

Tip 2: *Then* (adverb): next or after that.

> The oven temperature was held at 70 °C for 1 min, ramped to 150 °C at a rate of 50 °C/min, increased to 200 °C at 6°C/min, *then* followed by a final ramp to 280 °C at 16 °C/min. (From Maestroni et al., 2018)

[Remember that *then*, as used here, is used sparingly in expert writing.]

That/Which

The difference between *that* and *which* may appear to be subtle, but the improper use of these two words can alter the interpretation of a sentence. To help you use these words correctly, consider the following rules:

Rule 1: A *that* clause is essential to the meaning of the sentence.

A *that* clause, referred to as a "restrictive clause" (Raunikar, 2020b, section 5.2.2), introduces information (about the word or phrase just before it) that is essential to the meaning of the sentence. If the *that* clause were deleted, the sentence would lose its intended meaning. Consider these examples:

> The mass spectra showed no peaks *that* would suggest contamination.
>
> Transcriptomic analysis revealed several novel genes *that* are involved in anthocyanin biosynthesis in peach flesh.[19] (Adapted from Yue et al., 2018)

Rule 2: A *which* clause adds non-essential information.

A *which* clause, referred to as a "non-restrictive clause" (Raunikar, 2020b, section 5.2.2), adds information, but this information is not essential for understanding the sentence. Note that the extra information, introduced by *which*, is set off by a comma:

> The analyte was extracted and resuspended in an adequate volume of solvent, *which* resulted in a more concentrated analyte solution.
>
> The laser energy was set to 2.2 mJ, *which* corresponds to a fluence of 44 kJ/m^2.

Three "tests" can be used to check whether you are using *that* and *which* correctly.

- Ask yourself if the information following *that* or *which* is essential or extra. If the information is essential for communicating your intended message, use *that*. If the information could be deleted without changing your main message, use *which*.

Essential Information	Most aldehydes, except formaldehyde, form two geometrical isomers of the derivatives *that* are represented by two peaks in the chromatogram. (From Vesely et al., 2003)

Extra Information	PAL catalyzes the biosynthesis of cinnamic acid, *which* is the first step in Scheme 1. (Adapted from Yue, 2018)

- Check to verify if the word you have chosen (*that* or *which*) works with a comma immediately before it (and possibly at the end of the clause). *That*, used to introduce a restrictive clause, does not come after a comma. *Which*, used to introduce a non-restrictive clause, comes after a comma.

Incorrect	Pennsylvania, that is my home state, is known as the Keystone State.
Correct	Pennsylvania, which is my home state, is known as the Keystone State.

- Imagine inserting the phrase *by the way* after *which*. If the phrase sounds right, then you are probably using *which* correctly to signal extra information; if it sounds wrong, you probably should use *that*.

Sounds right	I like McIntosh apples, which (*by the way*) are red and green.
(*which* is correct)	I like McIntosh apples, *which* are red and green.
Sounds wrong	The results, which (*by the way*) support the mechanism, are shown in Table 1.
(*that* is correct)	The results that support the mechanism are shown in Table 1.

Part 7: Overall Writing Effectiveness

Effective writing is both fluid and unambiguous; we focus on these attributes of writing in the final section of this chapter.

Fluid Writing

Sometimes you may find sentences in your writing that seem awkward or choppy, even though they are grammatically correct. Follow these rules to make your writing flow more smoothly.

Rule 1: Use fewer words *before* the verb than *after* the verb. (Verbs are italicized in these examples.)

Awkward	One of the most common and costly diseases in poultry *is* coccidiosis.
Better	Coccidiosis *is* one of the most common and costly diseases in poultry. (Adapted from Peippo et al., 2004)
Awkward	A rapid and simple method for the screening of residues of these two coccidiostatic compounds in poultry and eggs *is presented* in this paper.
Better	This paper *presents* a rapid and simple method for the screening of residues of these two coccidiostatic compounds in poultry and eggs. (Adapted from Peippo et al., 2004)

Rule 2: State given information before new information.

In sentences that contain both new information and given information (i.e., information that has already been mentioned), state the given information first. In this way, the reader encounters familiar ideas first, making the sentence easier to read. Consider these examples. In the awkward example, the given information (italicized) comes after the new information, making the second sentence harder to understand.

Awkward	Anticoccidial drugs are extensively used in the poultry industry to control infection. Polyether ionophores, such as narasin and salinomycin, are *the most commonly used coccidiostats in poultry*.
Better	Anticoccidial drugs are extensively used in the poultry industry to control infection. *The most commonly used coccidiostats in poultry* are polyether ionophores such as narasin and salinomycin. (Adapted from Peippo et al., 2004)

Note, too, that different words are used to refer to the given information in each example: "coccidiostats" is another way of saying "anticoccidial drugs." By using different words with the same meaning, the writer creates linkages between sentences without becoming overly repetitive.

Unambiguous Writing

Novice writers often use the pronouns *this*, *that*, and *it* in ways that lead to ambiguity because readers cannot be certain what *this*, *that*, or *it* refers to. To minimize ambiguity in your writing, avoid the use of such pronouns by themselves (e.g., *this* versus *this plant*) unless the noun to which the pronoun refers is crystal clear.

Set A

Ambiguous — As a mosquito repellant, eucamalol has been reported to outperform *N,N*-diethyl-*m*-toluamide in terms of potency.[12,13] The presence of eucamalol in *S. insularis* explains the widespread use of *this* to fend off mosquitoes[4] and further exemplifies the potential of plants to provide biodegradable alternatives to synthetic insecticides.

Unambiguous — As a mosquito repellant, eucamalol has been reported to outperform *N,N*-diethyl-*m*-toluamide in terms of potency.[12,13] The presence of eucamalol in *S. insularis* explains the widespread use of *this plant* to fend off mosquitoes[4] and further exemplifies the potential of plants to provide biodegradable alternatives to synthetic insecticides. (Adapted from Fattorusso et al., 2004)

Set B

Ambiguous — A 65-µm poly(dimethylsiloxane)/divinyl benzene (PDMS/DVB) fiber coating (Supelco, Bellefonte, PA) was used in this method. *This* was selected for its ability to retain the derivatizing agent and for its affinity for the PFBOA-aldehyde oxime.

Unambiguous — A 65-µm poly(dimethylsiloxane)/divinyl benzene (PDMS/DVB) fiber coating (Supelco, Bellefonte, PA) was used in this method. *This fiber coating* was selected for its ability to retain the derivatizing agent and for its affinity for the PFBOA-aldehyde oxime. (Adapted from Vesely et al., 2003)

Set C

Ambiguous — A qualitative model aimed at explaining the observed field-induced adsorption of negatively charged gold particles on a similarly charged silica surface is proposed. *It* is based on the idea that the adsorbed ion distribution at the particle surface is influenced by the applied electric field.

Unambiguous — A qualitative model aimed at explaining the observed field-induced adsorption of negatively charged gold particles on a similarly charged silica surface is proposed. *The model* is based on the idea that the adsorbed ion distribution at the particle surface is influenced by the applied electric field. (Adapted from Kloepper et al., 2004)

References

Adora, S.; Soldo-Olivier, Y.; Faure, R.; Durand, R.; Dartyge, E.; Baudelet, F. Electrochemical Preparation of Platinum Nanocrystallites on Activated Carbon Studied by X-ray Absorption Spectroscopy. *J. Phys. Chem. B* **2001**, *105* (43), 10489–10495. DOI: org/10.1021/jp004532d

Aga, D. CAREER: Immunochemical Techniques for Investigations on the Occurrence and Fate of Agrochemicals in the Environment, 2002 (NSF CHE 0233700).

Ahlquist, M.; Fabrizi, G.; Cacchi, S.; Norrby, P-O. The Mechanism of the Phosphine-Free Palladium-Catalyzed Hydroarylation of Alkynes. *J. Am. Chem. Soc.* **2006**, *128* (39), 12785–12793. DOI: org/10.1021/ja061543x

Alley, M. *The Craft of Scientific Writing*, 3rd ed.; Springer, 1996.

Alley, M. *The Craft of Scientific Writing*, 4th ed.; Springer, 2018.

Almela, C.; Laparra, J. M.; Velez, D.; Barbera, R.; Farre, R.; Montoro, R. Arsenosugars in Raw and Cooked Edible Seaweed: Characterization and Bioaccessibility. *J. Agric. Food Chem.* **2005**, *53* (18), 7344–7351. DOI: org/10.1021/jf050503u

American Chemical Society. *The Chemical Professional's Code of Conduct.* https://www.acs.org/content/acs/en/careers/career-services/ethics/the-chemical-professionals-code-of-conduct.html (accessed 2021-03-12).

American Chemical Society. *Ethical Guidelines to Publication of Chemical Research.* https://pubs.acs.org/page/policy/ethics/index.html (accessed 2021-03-12).

Ampollini, L.; Katz, E. F.; Bourne, S.; Tian, Y.; Novoselac, A.; Goldstein, A. H.; Lucic, G.; Waring, M. S.; DeCarlo, P. F. Observations and Contributions of Real-Time Indoor Ammonia Concentrations during HOMEChem. *Environ. Sci. Technol.* **2019**, *53* (15), 8591–8598. DOI: 10.1021/acs.est.9b02157

Andersen, H. C. Diagrammatic Formulation of the Kinetic Theory of Fluctuations in Equilibrium Classical Fluids. III. Cluster Analysis of the Renormalized Interactions and a Second Diagrammatic Representation of the Correlation Functions. *J. Phys. Chem. B* **2003**, *107* (37), 10234–10242. DOI: org/10.1021/jp034754m

Anholt, R. R. H. *Dazzle 'Em with Style: The Art of Oral Scientific Presentation*, 2nd ed.; Elsevier, 2006.

Aranda, A.; Jiménez-Marti, E.; Orozco, H.; Matallana, E.; del Olmo, M. Sulfur and Adenine Metabolisms Are Linked, and Both Modulate Sulfite Resistance in Wine Yeast. *J. Agric. Food Chem.* **2006**, *54* (16), 5839–5846. DOI: org/10.1021/jf060851b

Asatryan, R.; Pal, Y.; Hachmann, J.; Ruckenstein, E. Roaming-like Mechanism for Dehydration of Diol Radicals. *J. Phys. Chem. A* **2018**, *122* (51), 9738–9754. DOI: 10.1021/acs.jpca.8b08690

Ashbrook, S. E.; Berry, A. J.; Wimperis, S. ^{17}O Multiple-Quantum MAS NMR Study of Pyroxenes. *J. Phys. Chem. B* **2002**, *106* (4), 773–778. DOI: org/10.1021/jp0130362
Atzei, D.; Ferri, T.; Sadun, C.; Sangiorgio, P.; Caminiti, R. Structural Characterization of Complexes between Iminodiacetate Blocked on Styrene-Divinylbenzene Matrix (Chelex 100 Resin) and Fe(III), Cr(III), and Zn(II) in Solid Phase by Energy-Dispersive X-ray Diffraction. *J. Am. Chem. Soc.* **2001**, *123* (11), 2552–2558. DOI: org/10.1021/ja0003728
Balijepalli, A. S.; McNeely, J. H.; Hamoud, A.; Grinstaff, M. W. Guidelines for β-Lactam Synthesis: Glycal Protecting Groups Dictate Stereoelectronics and [2+2] Cycloaddition Kinetics. *J. Org. Chem.* **2020**, *85* (19), 12044–12057. DOI: 10.1021/acs.joc.0c00510
Banik, G. M.; Baysinger, G.; Kamat, P. V.; Pienta, J. J., Eds.; *ACS Guide to Scholarly Communication*; American Chemical Society, 2020. DOI: https://doi.org/10.1021/acsguide
Bandara, S.; Froese, D. G.; St. Louis, V. L.; Cooke, C. A.; Calmels, F. Postdepositional Mercury Mobility in a Permafrost Peatland from Central Yukon, Canada. *ACS Earth Space Chem.* **2019**, *3* (5), 770–778. DOI: 10.1021/acsearthspacechem.9b00010
Bartolo, N. D.; Woerpel, K. A. Mechanistic Insight into Additions of Allylic Grignard Reagents to Carbonyl Compounds. *J. Org. Chem.* **2018**, *83* (17), 10197–10206. DOI: 10.1021/acs.joc.8b01430
Beal, H.; Trimbur, J. *A Short Guide to Writing about Chemistry*, 2nd ed.; Pearson Education, 2001.
Bekker, M. Z.; Kreitman, G. Y.; Jeffery, D. W.; Danilewicz, J. C. Liberation of Hydrogen Sulfide from Dicysteinyl Polysulfanes in Model Wine. *J. Agric. Food Chem.* **2018**, *66* (51), 13483–13491. DOI: 10.1021/acs.jafc.8b04690
Benedek, N. CAREER: Ionic and Thermal Transport Properties of Complex Oxides from First Principles, 2015 (NSF DMR 1454096).
Besser, J. M.; Brumbaugh, W. G.; Kemble, N. E.; May, T. W.; Ingersoll, C. G. Effects of Sediment Characteristics on the Toxicity of Chromium(III) and Chromium(VI) to the Amphipod, *Hyalella azteca*. *Environ. Sci. Technol.* **2004**, *38* (23), 6210–6216. DOI: org/10.1021/es049715i
Biber, D.; Conrad, S.; Reppen, R. Corpus-Based Approaches to Issues in Applied Linguistics. *Appl. Linguist.* **1994**, *15* (2), 169–189. DOI: org/10.1093/applin/15.2.169
Biber, D.; Conrad, S.; Reppen, R. *Corpus Linguistics: Investigating Structure and Use*; Cambridge University Press, 1998.
Black, I.; Heiss, C.; Azadi, P. Comprehensive Monosaccharide Composition Analysis of Insoluble Polysaccharides by Permethylation To Produce Methyl Alditol Derivatives for Gas Chromatography/Mass Spectrometry. *Anal. Chem.* **2019**, *91* (21), 13787–13793. DOI: 10.1021/acs.analchem.9b03239
Boesten, W. H. J.; Seerden, J.-P. G.; de Lange, B.; Dielemans, H. J. A.; Elsenberg, H. L. M.; Kaptein, B.; Moody, H. M.; Kellogg, R. M.; Broxterman, Q. B. Asymmetric Strecker Synthesis of α-Amino Acids via a Crystallization-Induced Asymmetric Transformation Using (*R*)-Phenylglycine Amide as Chiral Auxiliary. *Org. Lett.* **2001**, 3 (8), 1121–1124. DOI: org/10.1021/ol007042c
Bol, A. A.; Meijerink, A. Luminescence Quantum Efficiency of Nanocrystalline ZnS:Mn^{2+}. 2. Enhancement by UV Irradiation. *J. Phys. Chem. B* **2001**, *105* (42), 10203–10209. DOI: org/10.1021/jp010757s
Booth, W. C.; Colomb, G. G.; Williams, J. M.; Bizup, J.; FitzGerald, W. T. *The Craft of Research*, 4th ed.; University of Chicago Press, 2016.
Bridi, R.; Atala, E.; Núñez Pizarro, P.; Montenegro, G. Honeybee Pollen Load: Phenolic Composition and Antimicrobial Activity and Antioxidant Capacity. *J. Nat. Prod.* **2019**, *82* (3), 559–565. DOI: 10.1021/acs.jnatprod.8b00945

Broere, D. L. J.; Mercado, B. Q.; Bill, E.; Lancaster, K. M.; Sproules, S.; Holl, P. L. Alkali Cation Effects on Redox-Active Formazanate Ligands in Iron Chemistry. *Inorg. Chem.* **2018**, *57* (16), 9580–9591. DOI: 10.1021/acs.inorgchem.8b00226

Burner, J.; West, B. J.; Mayer, P. M. What Will Photo-Processing of Large, Ionized Amino-Substituted Polycyclic Aromatic Hydrocarbons Produce in the Interstellar Medium? *J. Phys. Chem. A* **2019**, *123* (24), 5027–5034. DOI: 10.1021/acs.jpca.9b02395

Cain, C. N. NSF Graduate Research Fellowship Program Proposal: Chemistry—Chemical Measurement and Imaging (2019).

Cantrill, S. J.; Youn, G. J.; Stoddart, J. F.; Williams, D. J. Supramolecular Daisy Chains. *J. Org. Chem.* **2001**, *66* (21), 6857–6872. DOI: org/10.1021/jo010405h

Cao, X.-L. Plasticizer di-(2-ethylhexyl)phthalate (DEHP) in Vegetables and Fruits. In *Technical Program Archives: Fall 2019 American Chemical Society National Meeting & Expo*, San Diego, CA; Poster AGFD 61.

Cao, Z.; Zhao, L.; Zhang, Y.; Ren, M.; Zhang, Y.; Liu, X.; Jie, J.; Wang, Z.; Li, C.; Shen, M.; Bu, Q. Influence of Air Pollution on Inhalation and Dermal Exposure of Human to Organophosphate Flame Retardants: A Case Study During a Prolonged Haze Episode. *Environ. Sci. Technol.* **2019**, *53* (7), 3880–3887. DOI: 10.1021/acs.est.8b07053

Carroll, J.; Jonsson, E. N.; Ebel, R.; Hartman, M. S.; Holman, T. R.; Crews, P. Probing Sponge-Derived Terpenoids for Human 15-Lipoxygenase Inhibitors. *J. Org. Chem.* **2001**, *66* (21), 6847–6851. DOI: org/10.1021/jo015784t

Chang,Y. C.; Xiong, B.; Xu, Y.; Ng, C.-Y. Quantum Spin–Orbit Electronic State Selection of Atomic Transition Metal Vanadium Cation for Chemical Reactivity Studies. *J. Phys. Chem. A* **2019**, 123 (12), 2310–2319. DOI: 10.1021/acs.jpca.9b00511

Chen, H.; Chow, A. T.; Li, X.-W.; Ni, H.-G.; Dahlgren, R. A.; Zeng, H.; Wang, J.-J. Wildfire Burn Intensity Affects the Quantity and Speciation of Polycyclic Aromatic Hydrocarbons in Soils. *ACS Earth Space Chem.* **2018**, *2* (12), 1262–1270. DOI: 10.1021/acsearthspacechem.8b00101

Chen, X.-B.; Fang, W.-H.; Phillips, D. L. Theoretical Studies of the Photochemical Dynamics of Acetylacetone: Isomerization, Dissociation, and Dehydration Reactions. *J. Phys. Chem. A* **2006**, *110* (13), 4434–4441. DOI: org/10.1021/jp057306i

Cheng, A.-C.; Huang, T.-C.; Lai, C.-S.; Kuo, J.-M.; Huang, Y.-T.; Lo, C.-Y.; Ho, C.-T.; Pan, M.-H. Pyrrolidine Dithiocarbamate Inhibition of Luteolin-Induced Apoptosis through Up-regulated Phosphorylation of Akt and Caspase-9 in Human Leukemia HL-60 Cells. *J. Agric. Food Chem.* **2006**, *54* (12), 4215–4221. DOI: org/10.1021/jf060269n

Cheng, W.; Xie, Y.; Yang, Z.; Sun, S.; Zhang, M.-Z.; Ding, Y.; Zhang, W. General Strategy for in Situ Generation of a Coumarin-Cu^{2+} Complex for Fluorescent Water Sensing. *Anal. Chem.* **2019**, *91* (9), 5817–5823. DOI: 10.1021/acs.analchem.9b00132

Chirollo, C.; Vassallo, A.; Piaz, F. D.; Lamagna, B.; Tortora, G.; Neglia, G.; De Tommasi, N.; Severino, L. Investigation of the Persistence of Penicillin G and Dihydrostreptomycin Residues in Milk of Lactating Buffaloes (Bubalus bubalis) Using Ultra-High-Performance Liquid Chromatography and Tandem Mass Spectrometry. *J. Agric. Food Chem.* **2018**, *66* (25), 6388–6393. DOI: 10.1021/acs.jafc.8b00229

Clancy, C. M. R.; Krogmeier, J. R.; Pawlak, A.; Rozanowska, M.; Sarna, T.; Dunn, R. C.; Simon, J. D. Atomic Force Microscopy and Near-Field Scanning Optical Microscopy Measurements of Single Human Retinal Lipofuscin Granules. *J. Phys. Chem. B* **2000**, *104* (51), 12098–12101. DOI: org/10.1021/jp0030544

Coffin, S.; Huang, G.-Y.; Lee, I.; Schlenk, D. Fish and Seabird Gut Conditions Enhance Desorption of Estrogenic Chemicals from Commonly Ingested Plastic Items. *Environ. Sci. Technol.* **2019**, *53* (8), 4588–4599. DOI: 10.1021/acs.est.8b07140

Cooper, D. J.; Husband, F. A.; Mills, E. N. C.; Wilde, P. J. Role of Beer Lipid-binding Proteins in Preventing Lipid Destabilization of Foam. *J. Agric. Food Chem.* **2002**, *50* (26), 7645–7650. DOI: org/10.1021/jf0203996

Crespo, A.; Martí, M. A.; Roitberg, A. E.; Amzel, L. M.; Estrin, D. A. The Catalytic Mechanism of Peptidylglycine α-Hydroxylating Monooxygenase Investigated by Computer Simulation. *J. Am. Chem. Soc.* **2006**, *128* (39), 12817–12828. DOI: org/10.1021/ja062876x

Crismore, A.; Farnsworth, R. Metadiscourse in Popular and Professional Science Discourse. In *The Writing Scholar: Studies in Academic Discourse*; Nash, W., Ed.; Sage, 1990; pp 118–136.

Curtis, R. A.; Blanch, H. W.; Prausnitz, J. M. Calculation of Phase Diagrams for Aqueous Protein Solutions. *J. Phys. Chem. B* **2001**, *105* (12), 2445–2452. DOI: org/10.1021/jp003087j

Davis, H. B.; Tyson, J. F.; Pechenik, J. A. *A Short Guide to Writing about Chemistry*; Longman/Pearson Education, 2010.

Dellinger, B.; Pryor,W. A.; Cueto, R.; Squadrito, G. L.; Hegde, V.; Deutsch, W. A. Role of Free Radicals in the Toxicity of Airborne Fine Particulate Matter. *Chem. Res. Toxicol.* **2001**, *14* (10), 1371–1377. DOI: org/10.1021/tx010050x

Demko, Z. P.; Sharpless, K. B. Preparation of 5-Substituted 1*H*-Tetrazoles from Nitriles in Water. *J. Org. Chem.* **2001**, *66* (24), 7945–7950. DOI: org/10.1021/jo010635w

Dethoff, E. A.; Weeks, K. M. Effects of Refolding on Large-Scale RNA Structure. *Biochemistry* **2019**, *58* (28), 3069–3077. DOI: 10.1021/acs.biochem.8b01219

Dhanaraj, J.; Jagannathan, R.; Kutty, T. R. N.; Lu, C.-H. Photoluminescence Characteristics of Y_2O_3:Eu^{3+} Nanophosphors Prepared Using Sol-Gel Thermolysis. *J. Phys. Chem. B* **2001**, *105* (45), 11098–11105. DOI: org/10.1021/jp0119330

D'hooghe, M.; Van Brabandt, W.; De Kimpe, N. New Synthesis of Propargylic Amines from 2-(Bromomethyl)aziridines. Intermediacy of 3-Bromoazetidinium Salts. *J. Org. Chem.* **2004**, *69* (8), 2703–2710. DOI: org/10.1021/jo035759i

Díaz-Mula, H. M.; Tomás-Barberán, F. A.; García-Villalba, R. Pomegranate Fruit and Juice (cv. Mollar), Rich in Ellagitannins and Anthocyanins, Also Provide a Significant Content of a Wide Range of Proanthocyanidins. *J. Agric. Food Chem.* **2019**, *67* (33), 9160–9167. DOI: 10.1021/acs.jafc.8b07155

Dimitrov, S. D.; Azzouzi, M.; Wu, J.; Yao, J.; Dong, Y.; Shakya Tuladhar, P.; Schroeder, B. C.; Bittner, E. R.; McCulloch, I.; Nelson, J.; Durrant, J. R. Spectroscopic Investigation of the Effect of Microstructure and Energetic Offset on the Nature of Interfacial Charge Transfer States in Polymer: Fullerene Blends. *J. Am. Chem. Soc.* **2019**, *141* (11), 46344643. DOI: 10.1021/jacs.8b11484

Ding, X.; Zhang, Y.-Q.; He, Q.-F.; Yu, Q.-Q.; Wang, J.-Q.; Shen, R.-Q.; Song, W.; Wang, Y.-S.; Wang, X.-M. Significant Increase of Aromatics-Derived Secondary Organic Aerosol during Fall to Winter in China. *Environ. Sci. Technol.* **2017**, *51* (13), 7432–7441. DOI: 10.1021/acs.est.6b06408

Dyer, D. CAREER: Career Plan for Research in Polar Organic Materials with an Integrated Emphasis on Industrial Problem Solving, 2001 (NSF CHE 0094195).

Fabjan, N.; Rode, J.; Košir, I. J.; Wang, Z.; Zhang, Z.; Kreft, I. Tartary Buckwheat (Fagopyrum tataricum Gaertn.) as a Source of Dietary Rutin and Quercitrin. *J. Agric. Food Chem.* **2003**, *51* (22), 6452–6455. DOI: org/10.1021/jf034543e

Fairbrother, H. CAREER: Exploring the Mechanisms of Organic Surface Modification Processes, 2000 (NSF CHE 9985372).

Fattorusso, E.; Santelia, F. U.; Appendino, G.; Ballero, M.; Taglialatela-Scafati, O. Polyoxygenated Eudesmanes and *trans*-Chrysanthemanes from the Aerial Parts of *Santolina insularis*. *J. Nat. Prod.* **2004**, *67* (1), 37–41. DOI: org/10.1021/np0302221

Finley, J. W.; Ip, C.; Lisk, D. J.; Davis, C. D.; Hintze, K. J.; Whanger, P. D. Cancer-Protective Properties of High-Selenium Broccoli. *J. Agric. Food Chem.* **2001**, *49* (5), 2679–2683. DOI: org/10.1021/jf0014821

Finney, N. CAREER: New Approaches to the Development of Fluorescent Chemosensors for Heavy Metal Ions, 1999 (NSF CHE 9876333).

Fredriksson, S.-Å.; Hulst, A. G.; Artursson, E.; de Jong, A. L.; Nilsson, C.; van Baar, B. L. M. Forensic Identification of Neat Ricin and of Ricin from Crude Castor Bean Extracts by Mass Spectrometry. *Anal. Chem.* **2005**, *77* (6), 1545–1555. DOI: org/10.1021/ac048756u

Friedland, A. J.; Folt, C. L.; Mercer, J. L. *Writing Successful Science Proposals*, 3rd ed.; Yale University Press, 2018.

Frydrych, R.; Ślepokura, K.; Bil, A.; Gregoliński, J. Mixed Macrocycles Derived from 2,6-Diformylpyridine and Opposite Enantiomers of *trans*-1,2-Diaminocyclopentane and *trans*-1,2-Diaminocyclohexane. *J. Org. Chem.* **2019**, *84* (9), 5695–5711. DOI: 10.1021/acs.joc.9b00614

Fujiwara, M.; Chie, K.; Sawai, J.; Shimizu, D.; Tanimoto, Y. On the Movement of Paramagnetic Ions in an Inhomogeneous Magnetic Field. *J. Phys. Chem. B* **2004**, *108* (11), 3531–3534. DOI: org/10.1021/jp0303523

Galarreta, B.; Hernández, Y.; Lagos, L.; Veliz, L. Accessible Mycotoxin Identification by Label-Free SERS Nanoaptasensors in Solution. In *Technical Program Archives: Spring 2019 American Chemical Society National Meeting & Expo,* Orlando, FL; Poster ANYL 353.

García-Villalba, R.; Selma, M. V.; Espín, J. C.; Tomás-Barberán, F. A. Identification of Novel Urolithin Metabolites in Human Feces and Urine after the Intake of a Pomegranate Extract. *J. Agric. Food Chem.* **2019**, *67* (40), 11099–11107. DOI: 10.1021/acs.jafc.9b04435

Garner, A. Strategies for Targeting Aberrant MicroRNA Activity in Cancer. In *Technical Program Archives: Fall 2019 American Chemical Society National Meeting & Expo*, San Diego, CA; Paper MEDI 171.

Gastel, B.; Day, R. A. *How to Write and Publish a Scientific Paper*, 8th ed.; Greenwood, 2016.

Gershon, H.; Lykkeberg, A.; Goren, F.; Mason, S. Identifying Fraudulent Natural Products: A Perspective on the Application of Carbon-14 Analysis. *J. Agric. Food Chem.* **2019**, *67* (49), 13393–13399. DOI: 10.1021/acs.jafc.9b01821

Ghodke, P. P.; Bommisetti, P.; Nair, D. T.; Pradeepkumar, P. I. Synthesis of N^2-Deoxyguanosine Modified DNAs and the Studies on Their Translesion Synthesis by the *E. coli* DNA Polymerase IV. *J. Org. Chem.* **2019**, *84* (4), 1734–1747. DOI: 10.1021/acs.joc.8b02082

Goldstein, I. J. Lectin Structure-Activity: The Story Is Never Over. *J. Agric. Food Chem.* **2002**, *50* (22), 6583–6585. DOI: org/10.1021/jf0201879

Guan, S.; Rougier, A.; Viraphong, O.; Denux, D.; Penin, N.; Gaudon, M. Two-Step Synthesis of VO_2 (M) with Tuned Crystallinity. *Inorg. Chem.* **2018**, *57* (15), 8857–8865. DOI: 10.1021/acs.inorgchem.8b00753

Gudmundsdottir, A. CAREER: Photolysis of Alkylazides in Solution and in Crystals, 2001 (NSF CHE 0093622).

Gunther, M. J.; Pavlović, R. Z.; Fernandez, J. P.; Zhiquan, L.; Gallucci, J.; Hadad, C. M.; Badjić, J. D. Stereo- and Regioselective Synthesis of Molecular Baskets. *J. Org. Chem.* **2019**, *84* (7), 4392–4401. DOI: 10.1021/acs.joc.9b00330

Guo, Y.; Zhu, X.; Li, N.; Yang, J.; Yang, Z.; Wang, J.; Yang, B. Molecular Sensitivities of Substrate-Supported Gold Nanocrystals. *J. Phys. Chem. C* **2019**, *123* (12), 7336–7346. DOI: 10.1021/acs.jpcc.8b12096

Hafez, A. M.; Taggi, A. E.; Dudding, T.; Lectka, T. Asymmetric Catalysis on Sequentially Linked Columns. *J. Am. Chem. Soc.* **2001**, *123* (44), 10853–10859. DOI: org/10.1021/ja016556j

Hageman, K. J.; Istok, J. D.; Field, J. A.; Buscheck, T. E.; Semprini, L. In Situ Anaerobic Transformation of Trichlorofluoroethene in Trichloroethene-Contaminated Groundwater. *Environ. Sci. Technol.* **2001**, *35* (9), 1729–1735. DOI: org/10.1021/es001577j

Hageman, K. J.; Simonich, S. L.; Campbell, D. H.; Wilson, G. R.; Landers, D. H. Atmospheric Deposition of Current-Use and Historic-Use Pesticides in Snow at National Parks in the Western United States. *Environ. Sci. Technol.* **2006**, *40* (10), 3174–3180. DOI: org/10.1021/es060157c

Han, S. M.; Ashurst, W. R.; Carraro, C.; Maboudian, R. Formation of Alkanethiol Monolayer on Ge(111). *J. Am. Chem. Soc.* **2001**, *123* (10), 2422–2425. DOI: org/10.1021/ja993816c

Harpp, K. CAREER: Teaching through Research: Plume-Ridge Interactions in the Galápagos as Focus of an Integrated Interdisciplinary Chemistry Curricular Initiative, 1998 (NSF CHE 9733597).

Hemida, M.; Coates, L. J.; Lam, S.; Gupta, V.; Macka, M.; Wirth, H.-J.; Gooley, A. A.; Haddad, P. R.; Paull, B. Miniature Multiwavelength Deep UV-LED-Based Absorption Detection System for Capillary LC. *Anal. Chem.* **2020**, *92* (20), 13688–13693. DOI: 10.1021/acs.analchem.0c03460

Henderson, L. Units of Measure. In *ACS Guide to Scholarly Communication*; Banik, G. M., Baysinger, G., Kamat, P., Pienta, N., Eds.; American Chemical Society, 2020. DOI: https://doi.org/10.1021/acsguide.50404

Hergenrother, P. CAREER: Combinatorial Chemistry in the Classroom and Laboratory: Identification of Novel Small Molecule Ligands for Apoptotic Proteins, 2002 (NSF CHE 0134779).

Hill, S. S.; Soppelsa, B. F.; West, G. K. Teaching ESL Students to Read and Write Experimental-Research Papers. *TESOL Quarterly* **1982**, *16* (3), 333–347. DOI: 10.2307/3586633

Hirai, M.; Ajito, S.; Takahashi, K.; Iwasa, T.; Li, X.; Wen, D.; Kawai-Hirai, R.; Ohta, N.; Igarashi, N.; Shimizu, N. Structure of Ultrafine Bubbles and Their Effects on Protein and Lipid Membrane Structures Studied by Small- and Wide-Angle X-ray Scattering. *J. Phys. Chem. B* **2019**, *123* (16), 3421–3429. DOI: 10.1021/acs.jpcb.9b00837

Hoffmann, R. *The Metamict State*; University of Central Florida Press, 1987.

Hofmann, A. H. *Scientific Writing and Communication: Papers, Proposals, and Presentations*, 4th ed.; Oxford University Press, 2020.

Houp, K. W.; Pearsall, T. E.; Tebeaux, E.; Dragga, S. *Reporting Technical Information*, 11th ed.; Oxford University Press, 2006.

Houser, R. P. CAREER: Models for Metal Active Sites in Proteins, 2001 (NSF CHE 0094079).

Hu, J.; Jiang, Y.-Z.; Wu, L.-L.; Wu, Z.; Bi, U.; Wong, G.; Qiu, X.; Chen, J.; Pang, D.-W.; Zhang, Z.-L. Dual-Signal Readout Nanospheres for Rapid Point-of-Care Detection of Ebola Virus Glycoprotein. *Anal. Chem.* **2017**, *89* (24), 13105–13111. DOI: 10.1021/acs.analchem.7b02222

Huang, J.; Kang, S.; Ma, M.; Guo, J.; Cong, Z.; Dong, Z.; Yin, R.; Xu, J.; Tripathee, L.; Ram, K.; Wang, F. Accumulation of Atmospheric Mercury in Glacier Cryoconite over Western China. *Environ. Sci. Technol.* **2019**, *53* (12), 6632–6639DOI: 10.1021/acs.est.8b06575

Huffman, T. R. NSF Graduate Research Fellowship Program Proposal: Chemistry—Chemical Synthesis (2019).

Hyland, K. Academic interaction: Where's it all going? In *Specialised English: New Directions in ESP and EAP Research and Practice*; Hyland, K., Wong, L. L. C., Eds.; Routledge, 2019; pp 91–106.

Hyland, K. *Hedging in Scientific Research Articles*; John Benjamins, 1998.

Hyland, K.; Salager-Meyer, F. Scientific Writing. In *Annual Review of Information Science and Technology*; Cronin, B. Ed., Vol. 42; Information Today, 2008; pp 297–338.

Ito, M.; Matsuumi, M.; Murugesh, M. G.; Kobayashi, Y. Scope and Limitation of Organocuprates, and Copper or Nickel Catalyst-Modified Grignard Reagents for

Installation of an Alkyl Group onto *cis*-4-Cyclopentene-1,3-diol Monoacetate. *J. Org. Chem.* **2001**, *66* (17), 5881–5889. DOI: org/10.1021/jo010474w
Jacoby, M. It's Time to Recycle Lithium-Ion Batteries. *C&EN Global Enterprise* **2019**, *97* (28), 2932. DOI: 10.1021/cen-09728-cover
Janotti, A. CAREER: Defects and Polarons in Complex Materials, 2017 (NSF DMR 1652994).
Jaoui, M.; Szmigielski, R.; Nestorowicz, K.; Kolodziejczyk, A.; Sarang, K.; Rudzinski, K. J.; Konopka, A.; Bulska, E.; Lewandowski, M.; Kleindienst, T. E. Organic Hydroxy Acids as Highly Oxygenated Molecular (HOM) Tracers for Aged Isoprene Aerosol. *Environ. Sci. Technol.* **2019**, *53* (24), 14516–14527. DOI: 10.1021/acs.est.9b05075
Jarvis, R. M.; Goodacre, R. Discrimination of Bacteria Using Surface-Enhanced Raman Spectroscopy. *Anal. Chem.* **2004**, *76* (1), 40–47. DOI: org/10.1021/ac034689c
Jia, H.; Du, A.; Zhang, H.; Yang, J.; Jiang, R.; Wang, J.; Zhang, C. Y. Site-Selective Growth of Crystalline Ceria with Oxygen Vacancies on Gold Nanocrystals for Near-Infrared Nitrogen Photofixation. *J. Am. Chem. Soc.* **2019**, *141* (13), 5083–5086. DOI: 10.1021/jacs.8b13062
Johnson, J. S. CAREER: Discovery of New Enantioselective Dipolar Cycloadditions from Small Ring Heterocycles, 2003 (NSF CHE 0239363).
Jozefaciuk, G.; Muranyi, A.; Fenyvesi, E. Effect of Cyclodextrins on Surface and Pore Properties of Soil Clay Minerals. *Environ. Sci. Technol.* **2001**, *35* (24), 4947–4952. DOI: org/10.1021/es010083z
Jozefaciuk, G.; Muranyi, A.; Fenyvesi, E. Effect of Randomly Methylated β-Cyclodextrin on Physical Properties of Soils. *Environ. Sci. Technol.* **2003**, *37* (13), 3012–3017. DOI: org/10.1021/es026236f
Jung, H.; Zhou, B.; Schimpf, A. Photoreduction of Iron Oxide Nanocrystals. In *Technical Program Archives: Fall 2019 American Chemical Society National Meeting & Expo*, San Diego, CA; Paper INOR 364.
Kalenius, E; Malola, S.; Matus, M. F.; Kazan, R.; Bürgi, T.; Häkkinen. H. Experimental Confirmation of a Topological Isomer of the Ubiquitous $Au_{25}(SR)_{18}$ Cluster in the Gas Phase. *J. Am. Chem. Soc.* **2021**, *143* (3), 12731277. DOI: 10.1021/jacs.0c11509
Kinsel, G. CAREER: Analytical Applications of MALDI Mass Spectrometry to the Study of Surface-Protein Interactions, 1999 (NSF CHE 9876249).
Kizil, R.; Irudavaraj, J.; Seetharaman, K. Characterization of Irradiated Starches by Using FT-Raman and FTIR Spectroscopy. *J. Agric. Food Chem.* **2002**, *50* (14), 3912–3918. DOI: org/10.1021/jf011652p
Kloepper, K. D.; Onuta, T.-D.; Amarie, D.; Dragnea, B. Field-Induced Interfacial Properties of Gold Nanoparticles in AC Microelectrophoretic Experiments. *J. Phys. Chem. B* **2004**, *108* (8), 2547–2553. DOI: org/10.1021/jp035644i
Koga, S.; Rieder, A.; Ballance, S.; Uhlen, A. K.; Veiseth-Kent, E. Gluten-Degrading Proteases in Wheat Infected by *Fusarium graminearum*—Protease Identification and Effects on Gluten and Dough Properties. *J. Agric. Food Chem.* **2019**, *67* (40), 11025–11034. DOI: 10.1021/acs.jafc.9b03869
Kohen, A. CAREER: Protein Dynamics and Hydrogen Tunneling in Enzymatic Catalysis, 2002 (NSF CHE 0133117).
Krechmer, J.; Lopez-Hilfiker, F.; Koss, A.; Hutterli, M.; Stoermer, C.; Deming, B.; Kimmel, J.; Warneke, C.; Holzinger, R.; Jayne, J.; Worsnop, D.; Fuhrer, K.; Gonin, M.; de Gouw, J. Evaluation of a New Reagent-Ion Source and Focusing Ion–Molecule Reactor for Use in Proton-Transfer-Reaction Mass Spectrometry. *Anal. Chem.* **2018**, *90* (20), 12011–12018. DOI: 10.1021/acs.analchem.8b02641
Kuechle, K. J.; Webb, E. B.; Mengel, D.; Main, A. R. Factors Influencing Neonicotinoid Insecticide Concentrations in Floodplain Wetland Sediments across Missouri. *Environ. Sci. Technol.* **2019**, *53* (18), 10591–10600. DOI: 10.1021/acs.est.9b01799

Kugel, R. W.; Ault, B. S. Infrared Matrix-Isolation and Theoretical Study of the Reactions of Ruthenocene with Ozone. *J. Phys. Chem. A* **2019**, *123* (27), 5768–5780. DOI: 10.1021/acs.jpca.9b02374

Langof, L.; Ehrenfreund, E.; Lifshitz, E. Continuous-Wave and Time-Resolved Optically Detected Magnetic Resonance Studies of Nonetched/Etched InP Nanocrystals. *J. Phys. Chem. B* **2002**, *106* (7), 1606–1612. DOI: org/10.1021/jp013720g

Laureles, L. R.; Rodriguez, F. M.; Reaño, C. E.; Santos, G. A.; Laurena, A. C.; Mendoza, E. M. T. Variability in Fatty Acid and Triacylglycerol Composition of the Oil of Coconut (*Cocos nucifera* L.) Hybrids and Their Parentals. *J. Agric. Food Chem.* **2002**, *50* (6), 1581–1586. DOI: org/10.1021/jf010832w

Lee, J. CAREER: Mechanistic Studies of Nucleotide Reactivity, 2001 (NSF CHE 0092215).

Lee, M. V.; Kaur, S.; Saad, O. M. Conjugation Site Influences Antibody-Conjugated Drug PK Assays: Case Studies for Disulfide-Linked, Self-Immolating Next-Generation Antibody Drug Conjugates. *Anal. Chem.* **2020**, *92* (18), 12168–12175. DOI: 10.1021/acs.analchem.0c00773

Li, C.-J.; Meng, Y. Grignard-Type Carbonyl Phenylation in Water and Under an Air Atmosphere. *J. Am. Chem. Soc.* **2000**, *122* (39), 9538–9539. DOI: org/10.1021/ja001699b

Li, X.; Lai, J.; Deng, Y.; Song, J.; Zhao, G.; Dong, S. Supramolecular Adhesion at Extremely Low Temperatures: A Combined Experimental and Theoretical Investigation. *J. Am. Chem. Soc.* **2020**, *142* (51), 21522–21529. DOI: 10.1021/jacs.0c10786

Liang, S.; Zheng, W.; Zhu, L.; Duan, W.; Wei, C.; Feng, C. One-Step Treatment of Phosphite-Laden Wastewater: A Single Electrochemical Reactor Integrating Superoxide Radical-Induced Oxidation and Electrocoagulation. *Environ. Sci. Technol.* **2019**, *53* (9), 5328–5336. DOI: 10.1021/acs.est.9b00841

Lichtenberg, S. S.; Tsyusko, O. V.; Palli, S. R.; Unrine, J. M. Uptake and Bioactivity of Chitosan/Double-Stranded RNA Polyplex Nanoparticles in *Caenorhabditis elegans*. *Environ. Sci. Technol.* **2019**, *53* (7), 3832–3840. DOI: 10.1021/acs.est.8b06560

Lifshitz, A.; Tamburu, C.; Suslensky, A.; Dubnikova, F. Decomposition of Anthranil. Single Pulse Shock-Tube Experiments, Potential Energy Surfaces and Multiwell Transition-State Calculations. The Role of Intersystem Crossing. *Phys. Chem. A* **2006**, *110* (27), 8248–8258. DOI: org/10.1021/jp056339v

Liu, J.; Alvarez, J.; Ong, W.; Roman, E.; Kaifer, A. E. Phase Transfer of Hydrophilic, Cyclodextrin-Modified Gold Nanoparticles to Chloroform Solutions. *J. Am. Chem. Soc.* **2001**, *123* (45), 11148–11154. DOI: org/10.1021/ja003957a

Liu, M.; Gong, Y.; Sun, H.; Zhang, J.; Zhang, L.; Sun, J.; Han, Y.; Huang, J.; Wu, Q.; Zhang, C.; Li, Z. Characterization of a Novel Chitinase from Sweet Potato and Its Fungicidal Effect against *Ceratocystis fimbriata*. *J. Agric. Food Chem.* **2020**, *68* (29), 7591–7600. DOI: 10.1021/acs.jafc.0c01813

Liu, Y.; Cheng, P.; Li, T.; Wang, R.; Li, Y.; Chang, S.-Y.; Zhu, Y.; Cheng, H.-W.; Wei, K.-H.; Zhan, X.; Sun, B.; Yang, Y. Unraveling Sunlight by Transparent Organic Semiconductors toward Photovoltaic and Photosynthesis. *ACS Nano* **2019**, *13* (2), 1071–1077. DOI: 10.1021/acsnano.8b08577

Llompart, M.; Pazos, M.; Landín, P.; Cela, R. Determination of Polychlorinated Biphenyls in Milk Samples by Saponification—Solid-Phase Microextraction. *Anal. Chem.* **2001**, *73* (24), 5858–5865. DOI: org/10.1021/ac0106546

Locke, L. F.; Spirduso, W. W.; Silverman, S. J. *Proposals That Work: A Guide for Planning Dissertations and Grant Proposals,* 6th ed.; Sage, 2014.

Lorigan, G. A. CAREER: Investigating Membrane Proteins with Magnetic Resonance Spectroscopy, 2002 (NSF CHE 0133433).

Lyon, L. A. CAREER: Bioresponsive Hydrogel Nanoparticles, 2000 (NSF CHE 9984012).

Lyons, L.; Carmichael, E.; McRoberts, C.; Aubry, A.; Thomson, A.; Reynolds, C. K. Prediction of Lignin Content in Ruminant Diets and Fecal Samples Using Rapid Analytical Techniques. *J. Agric. Food Chem.* **2018**, *66* (49), 13031–13040. DOI: 10.1021/acs.jafc.8b03808

Maestroni, B.; Abu Alnaser, A.; Ghanem, I.; Islam, M.; Cesio, V.; Heinzen, H.; Kelly, S.; Cannavan, A. Validation of an Analytical Method for the Determination of Pesticide Residues in Vine Leaves by GC-MS/MS. *J. Agric. Food Chem.* **2018**, *66* (25), 6421–6430. DOI: 10.1021/acs.jafc.8b00453

Mahaffy, P. Discovery of Organics on Mars with the SAM Experiment on the Curiosity Rover. In *Technical Program Archives: Spring 2019 American Chemical Society National Meeting & Expo*, Orlando, FL; Poster ANYL 17.

Manae, M. A.; Hazra, A. Triplet Decay Dynamics in Sulfur-Substituted Thymine: How Position of Substitution Matters. *J. Phys. Chem. A* **2019**, *123* (51), 10862–10867. DOI: 10.1021/acs.jpca.9b08214

Marra, R.; Lombardi, N.; d'Errico, G.; Troisi, J.; Scala, G.; Vinale, F.; Woo, S. L.; Bonanomi, G.; Lorito, M. Application of Trichoderma Strains and Metabolites Enhances Soybean Productivity and Nutrient Content. *J. Agric. Food Chem.* **2019**, *67* (7), 1814–1822. DOI: 10.1021/acs.jafc.8b06503

Mas-Torrent, M.; Rodríguez-Mias, R. A.; Solà, M.; Molins, M. A.; Pons, M.; Vidal-Gancedo, J.; Veciana, J.; Rovira, C. Isolation and Characterization of Four Isomers of a C60 Bisadduct with a TTF Derivative. Study of Their Radical Ions. *J. Org. Chem.* **2002**, *67* (2), 566–575. DOI: org/10.1021/jo010748f

Matsumoto, H.; Ichiyanagi, T.; Iida, H.; Ito, K.; Tsuda, T.; Hirayama, M.; Konishi, T. Ingested Delphinidin-3-rutinoside Is Primarily Excreted to Urine as the Intact Form and to Bile as the Methylated Form in Rats. *J. Agric. Food Chem.* **2006**, *54* (2), 578–582. DOI: org/10.1021/jf052411a

McCallum, E. S.; Cerveny, D.; Fick, J.; Brodin, T. Slow-Release Implants for Manipulating Contaminant Exposures in Aquatic Wildlife: A New Tool for Field Ecotoxicology. *Environ. Sci. Technol.* **2019**, *53* (14), 8282–8290. DOI: 10.1021/acs.est.9b01975

McClellan, J. E.; Murphy, J. P., III.; Mulholland, J. J.; Yost, R. A. Effects of Fragile Ions on Mass Resolution and on Isolation for Tandem Mass Spectrometry in the Quadrupole Ion Trap Mass Spectrometer. *Anal. Chem.* **2002**, *74* (2), 402–412. DOI: org/10.1021/ac015610b

McGuire, N. Spotlights on Recent JACS Publications. *J. Am. Chem. Soc.* **2020**, *142* (52), 21575–21576. DOI: 10.1021/jacs.0c13026

McMurry, J. *Organic Chemistry*, 6th ed.; Brooks/Cole-Thomson Learning, 2004.

Megadi, V. B.; Tallur, P. N.; Mulla, S. I.; Ninnekar, H. Z. Bacterial Degradation of Fungicide Captan. *J. Agric. Food Chem.* **2010**, *58* (24), 12863–12868. DOI: 10.1021/jf1030339

Miles, J. E. C.; Ramsewak, R. S.; Nair, M. G. Antifeedant and Mosquitocidal Compounds from *Delphinium* × cultorum Cv. Magic Fountains Flowers. *J. Agric. Food Chem.* **2000**, *48* (2), 503–506. DOI: org/10.1021/jf990354d

Mitchell, C. A.; Yu, L.; Ward, M. D. Selective Nucleation and Discovery of Organic Polymorphs through Epitaxy with Single Crystal Substrates. *J. Am. Chem. Soc.* **2001**, *123* (44), 10830–10839. DOI: org/10.1021/ja004085f

Mitsumi, M.; Murase, T.; Kishida, H.; Yoshinari, T.; Ozawa, Y.; Toriumi, K.; Sonoyama, T.; Kitagawa, H.; Mitani, T. Metallic Behavior and Periodical Valence Ordering in a MMX Chain Compound, $Pt_2(EtCS_2)_4I$. *J. Am. Chem. Soc.* **2001**, *123* (45), 11179–11192. DOI: org/10.1021/ja010900v

Monforte, A. R.; Martins, S. I. F. S.; Silva Ferreira, A. C. Impact of Phenolic Compounds in Strecker Aldehyde Formation in Wine Model Systems: Target and Untargeted Analysis. *J. Agric. Food Chem.* **2020**, *68* (38), 10281–10286. DOI: 10.1021/acs.jafc.9b02674

Motohashi, H.; Mikami, K. Nickel-Catalyzed Aromatic Cross-Coupling Difluoromethylation of Grignard Reagents with Difluoroiodomethane. *Org. Lett.* **2018**, *20* (17), 5340–5343. DOI: 10.1021/acs.orglett.8b02264

Murai, T.; Mutoh, N. Sequential Addition Reaction of Sulfanylmethyllithiums and Grignard Reagents to Thioformamides Leading to the Formation of 2-Phenyl-2-sulfanylethyl Tertiary Amines. *J. Org. Chem.* **2016**, *81* (18), 8131–8134. DOI: 10.1021/acs.joc.6b01857

Napolitano, A.; Di Donato, P.; Prota, G. Zinc-Catalyzed Oxidation of 5-*S*-Cysteinyldopa to 2,2'-Bi(2*H*-1,4-benzothiazine): Tracking the Biosynthetic Pathway of Trichochromes, the Characteristic Pigments of Red Hair. *J. Org. Chem.* **2001**, *66* (21), 6958–6966. DOI: org/10.1021/jo010320g

Neitola, R.; Pakkanen, T. A. Ab Initio Studies on the Atomic-Scale Origin of Friction between Diamond (111) Surfaces. *J. Phys. Chem. B* **2001**, *105* (7), 1338–1343. DOI: org/10.1021/jp003025t

Neufeld, B. H.; Tapia, J. B.; Lutzke, A.; Reynolds, M. M. Small Molecule Interferences in Resazurin and MTT-Based Metabolic Assays in the Absence of Cells. *Anal. Chem.* **2018**, *90* (11), 6867–6876. DOI: 10.1021/acs.analchem.8b01043

Neufeldt, S. CAREER: Combined Experimental and Computational Approach to Controlling Site Selectivity in Cross Coupling Chemical Reactions, 2019 (NSF CHM 1848090).

Ozen, B. F.; Mauer, L. J. Detection of Hazelnut Oil Adulteration Using FT-IR Spectroscopy. *J. Agric. Food Chem.* **2002**, *50* (14), 3898–3901. DOI: org/10.1021/jf0201834

Paradis, J. G.; Zimmerman, M. S. *The MIT Guide to Science and Engineering Communication*, 2nd ed.; MIT Press, 2002.

Park, J. W.; Song, H. E.; Lee, S. Y. Face Selectivity of Inclusion Complexation of Viologens with *β*-Cyclodextrin and 6-O-(2-Sulfonato-6-naphthyl)-*β*-cyclodextrin. *J. Phys. Chem. B* **2002**, *106* (29), 7186–7192. DOI: org/10.1021/jp020428f

Parra, R. D.; Yoo, B.; Wemhoff, M. Conformational Stability of a Model Macrocycle Tetraamide: An ab Initio Study. *J. Phys. Chem. A* **2006**, *110* (13), 4487–4494. DOI: 10.1021/jp060120e

Patrick, D. CAREER: Liquid Crystal Imprinting, 2000 (NSF CHE 9985428).

Pavia, D. L.; Lampman, G. M.; Driz, G. S.; Engel, R. G. *Introduction to Organic Laboratory Techniques: Small Scale Approach*; Saunders College Publishing, 1998; p 72.

Pearson, O. (née George, O.) NSF Graduate Research Fellowship Program Proposal: Chemistry—Macromolecular, Supramolecular, and Nanochemistry (2019).

Peippo, P.; Hagren, V.; Lövgren, T.; Tuomola, M. Rapid Time-Resolved Fluoroimmunoassay for the Screening of Narasin and Salinomycin Residues in Poultry and Eggs. *J. Agric. Food Chem.* **2004**, *52* (7), 1824–1828. DOI: org/10.1021/jf030716o

Pereira, I.; Banstola, B.; Wang, K.; Donnarumma, F.; Vaz, B. G.; Murray, K. K. Matrix-Assisted Laser Desorption Ionization Imaging and Laser Ablation Sampling for Analysis of Fungicide Distribution in Apples. *Anal. Chem.* **2019**, *91* (9), 6051–6056. DOI: 10.1021/acs.analchem.9b00566

Periyapperuma, K.; Pozo-Gonzalo, C.; MacFarlane, D. R.; Forsyth, M.; Howlett, P. C. High Zn Concentration Pyrrolidinium-Dicyanamide-Based Ionic Liquid Electrolytes for Zn^{2+}/Zn^0 Electrochemistry in a Flow Environment. *ACS Appl. Energy Mater.* **2018**, *1* (9), 4580–4590. DOI: 10.1021/acsaem.8b00742

Petronella, K. M. General Style Conventions. In *ACS Guide to Scholarly Communication*; Banik, G. M., Baysinger, G., Kamat, P., Pienta, N., Eds.; American Chemical Society, 2020. DOI: https://doi.org/10.1021/acsguide.50301

Pieper, K. J.; Martin, R.; Tang, M.; Walters, L. A.; Parks, J.; Roy, S.; Devine, C.; Edwards, M. A. Evaluating Water Lead Levels During the Flint Water Crisis. *Environ. Sci. Technol.* **2018**, *52* (15), 8124–8132. DOI: 10.1021/acs.est.8b00791

Plaper, A.; Jenko-Brinovec, S.; Premzl, A.; Kos, J.; Raspor, P. Genotoxicity of Trivalent Chromium in Bacterial Cells. Possible Effects on DNA Topology. *Chem. Res. Toxicol.* **2002**, *15* (7), 943–949. DOI: org/10.1021/tx010096q

Portillo-Cortez, K.; Martínez, A.; Dutt, A.; Santana, G. N719 Derivatives for Application in a Dye-Sensitized Solar Cell (DSSC): A Theoretical Study. *J. Phys. Chem. A* **2019**, *123* (51), 10930–10939. DOI: 10.1021/acs.jpca.9b09024

Prussin, A. J.; Garcia, E. B.; Marr, L. C. Total Concentrations of Virus and Bacteria in Indoor and Outdoor Air. *Environ. Sci. Technol. Lett.* **2015**, 2 (4), 84–88. DOI: 10.1021/acs.estlett.5b00050

Raunikar, C. Effective Writing & Word Usage. In *ACS Guide to Scholarly Communication*; Banik, G. M., Baysinger, G., Kamat, P., Pienta, N., Eds.; American Chemical Society, 2020a. DOI: https://doi.org/10.1021/acsguide.50101

Raunikar, C. Grammar, Punctuation, & Spelling. In *ACS Guide to Scholarly Communication*; Banik, G. M., Baysinger, G., Kamat, P., Pienta, N., Eds.; American Chemical Society, 2020b. DOI: https://doi.org/10.1021/acsguide.50201

Richardson, C. E. R.; Nolan, E. M.; Shoulders, M. D.; Lippard, S. J. A Sensitive, Nonradioactive Assay for Zn(II) Uptake into Metazoan Cells. *Biochemistry* **2018**, *57* (50), 6807–6815. DOI: 10.1021/acs.biochem.8b01043

Rodrigues, J. A. R.; Abramovitch, R. A.; de Sousa, J. D. F.; Leiva, G. C. Diastereoselective Synthesis of Cularine Alkaloids via Enium Ions and an Easy Entry to Isoquinolines by Aza–Wittig Electrocyclic Ring Closure. *J. Org. Chem.* **2004**, *69* (9), 2920–2928. DOI: org/10.1021/jo030287t

Rose-Petruck, C. CAREER: Ultrafast X-ray Imaging of Molecular Dynamics in Solution: A Research Program that Enhances Students' Learning, 2000 (NSF CHE 9984890).

Samuels, F. M. D. NSF Graduate Research Fellowship Program Proposal: Chemistry—Chemical Structure, Dynamics, and Mechanism (2019).

Sanov, A. CAREER: Structure and Dynamics of Negative Ions via Photoelectron Imaging Spectroscopy, 2002 (NSF CHE 0134631).

Seeram, N. P.; Adams, L. S.; Hardy, M. L.; Heber, D. Total Cranberry Extract versus Its Phytochemical Constituents: Antiproliferative and Synergistic Effects against Human Tumor Cell Lines. *J. Agric. Food Chem.* **2004**, *52* (9), 2512–2517. DOI: org/10.1021/jf0352778

Senda, T.; Ogasawara, M.; Hayashi, T. Rhodium-Catalyzed Asymmetric 1,4-Addition of Organoboron Reagents to 5,6-Dihydro-2(1*H*)-pyridinones. Asymmetric Synthesis of 4-Aryl-2-piperidinones. *J. Org. Chem.* **2001**, *66* (21), 6852–6856. DOI: org/10.1021/jo0103930

Song, Y.; Zhang, X.; Shen, Z.; Yang, W.; Wei, J.; Li, S.; Wang, X.; Li, X.; He, Q.; Zhang, S.; Zhang, Q.; Gao, B. Improving Brightness and Stability of Si-Rhodamine for Super-Resolution Imaging of Mitochondria in Living Cells. *Anal. Chem.* **2020**, *92* (18), 12137–12144. DOI: 10.1021/acs.analchem.9b04926

Spain, E. CAREER: Atomic Force Microscopy Studies of Transition Metal Chalcogenide Deposition Using Translationally Hot Atoms, 1997 (NSF CHE 9703345).

Spengler, J.; Anderle, F.; Bosch, E.; Grasselli, R. K.; Pillep, B.; Behrens, P.; Lapina, O. B.; Shubin, A. A.; Eberle, H.-J.; Knözinger, H. Antimony Oxide-Modified Vanadia-Based Catalysts: Physical Characterization and Catalytic Properties. *J. Phys. Chem. B* **2001**, *105* (44), 10772–10783. DOI: org/10.1021/jp012228u

Stachowiak, J. CAREER: Phase-separating Membrane Materials for Efficient and Specific Molecular Delivery to Cells, 2014 (NSF DMR 1352487).

Stockdill, J. CAREER: Harnessing the reactivity of neutral aminyl radicals for complex heterocycle synthesis, 2016 (NSF CHE 1554752).

Swales, J. M. *Genre Analysis: English in Academic and Research Settings*; Cambridge University Press, 1990.

Swales, J. M. *Research Genres: Exploration and Applications*; Cambridge University Press, 2004.

Swenson, R. E.; Sowin, T. J.; Zhang, H. Q. Synthesis of Substituted Quinolines Using the Dianion Addition of N-Boc-Anilines and α-Tolylsulfonyl-α, β-Unsaturated Ketones. *J. Org. Chem.* **2002**, *67* (26), 9182–9185. DOI: org/10.1021/jo0203387

Taatjes, C. A. Uncovering the Fundamental Chemistry of Alkyl + O_2 Reactions via Measurements of Product Formation. *J. Phys. Chem. A* **2006**, *110* (13), 4299–4312. DOI: org/10.1021/jp056997f

Tanaka, Y.; Kubosaki, S.; Osaka, K.; Yamawaki, M.; Morita, T.; Yoshimi, Y. Two Types of Cross-Coupling Reactions between Electron-Rich and Electron-Deficient Alkenes Assisted by Nucleophilic Addition Using an Organic Photoredox Catalyst. *J. Org. Chem.* **2018**, *83* (22), 13625–13635. DOI: 10.1021/acs.joc.8b02025

Tikk, M.; Tikk, K.; Tørngren, M. A.; Meinert, L.; Aaslyng, M. D.; Karlsson, A. H.; Andersen, H. J. Development of Inosine Monophosphate and Its Degradation Products during Aging of Pork of Different Qualities in Relation to Basic Taste and Retronasal Flavor Perception of the Meat. *J. Agric. Food Chem.* **2006**, *54* (20), 7769–7777. DOI: org/10.1021/jf060145a

Tobias, F.; McIntosh, J. C.; LaBonia, G. J.; Boyce, M. W.; Lockett, M. R.; Hummon, A. B. Developing a Drug Screening Platform: MALDI-Mass Spectrometry Imaging of Paper-Based Cultures. *Anal. Chem.* **2019**, *91* (24), 15370–15376. DOI: 10.1021/acs.analchem.9b03536

Tongraar, A.; Sagarik, K.; Rode, B. M. Effects of Many-Body Interactions on the Preferential Solvation of Mg^{2+} in Aqueous Ammonia Solution: A Born-Oppenheimer ab Initio QM/MM Dynamics Study. *J. Phys. Chem. B* **2001**, *105* (43), 10559–10564. DOI: org/10.1021/jp012103p

Truss, L. *Eats, Shoots & Leaves: The Zero Tolerance Approach to Punctuation*; Gotham Books, 2003.

Tuckerman, M. CAREER: Theoretical Investigations of Chemical Processes in Bulk Crystals and on Surfaces, 1999 (NSF CHE 9875824).

Turk, C. Do You Write Impressively? *Bulletin of the British Ecological Society* **1978**, *IX*, 5–10.

Tuulmets, A.; Nguyen, B. T.; Panov, D.; Sassian, M.; Järv, J. Kinetics of the Grignard Reaction with Silanes in Diethyl Ether and Ether-Toluene Mixtures. *J. Org. Chem.* **2003**, *68* (26), 9933–9937. DOI: org/10.1021/jo034941g

Underlin, E. N.; Böhm, M.; Madsen, R. Synthesis of Arabinoxylan Oligosaccharides by Preactivation-Based Iterative Glycosylations. *J. Org. Chem.* **2019**, *84* (24), 16036–16054. DOI: 10.1021/acs.joc.9b02529

Usugi, S.; Yorimitsu, H.; Shinokubo, H.; Oshima, K. Disulfidation of Alkynes and Alkenes with Gallium Trichloride. *Org. Lett.* **2004**, *6* (4), 601–603. DOI: org/10.1021/ol036391e

Valliant, J. F.; Schaffer, P.; Stephenson, K. A.; Britten, J. F. Synthesis of Boroxifen, A Nido-Carborane Analogue of Tamoxifen. *J. Org. Chem.* **2002**, *67* (2), 383–387. DOI: org/10.1021/jo0158229

Vebrosky, E. N.; Basirico, L. M.; Armbrust, K. L. Degradation of Dicloran in Irradiated Water-Sediment Systems. *J. Agric. Food Chem.* **2019**, *67* (27), 7609–7615. DOI: 10.1021/acs.jafc.9b01347

Veldhuyzen, W. F.; Shallop, A. J.; Jones, R. A.; Rokita, S. E. Thermodynamic versus Kinetic Products of DNA Alkylation as Modeled by Reaction of Deoxyadenosine. *J. Am. Chem. Soc.* **2001**, *123* (45), 11126–11132. DOI: org/10.1021/ja011686d

Vesely, P.; Lusk, L.; Basarova, G.; Seabrooks, J.; Ryder, D. Analysis of Aldehydes in Beer Using Solid-Phase Microextraction with On-Fiber Derivatization and Gas Chromatography/Mass Spectrometry. *J. Agric. Food Chem.* **2003**, *51* (24), 6941–6944. DOI: org/10.1021/jf034410t

Vieth, M.; Siegel, M. G.; Higgs, R. E.; Watson, I. A.; Robertson, D. H.; Savin, K. A.; Durst, G. L.; Hipskind, P. A. Characteristic Physical Properties and Structural Fragments of Marketed Oral Drugs. *J. Med. Chem.* **2004**, *47* (41), 224–232. DOI: org/10.1021/jm030267j

Viswanathan, G.; Oliynyk, A. O.; Antono, E.; Ling, J.; Meredig, B.; Brgoch, J. Single-Crystal Automated Refinement (SCAR): A Data-Driven Method for Determining Inorganic Structures. *Inorg. Chem.* **2019**, *58* (14), 9004–9015. DOI: 10.1021/acs.inorgchem.9b00344

Vyvyan, J. CAREER: Synthesis of Allelopathic Agents as Leads to New Agrochemicals, 2001 (NSF CHE 0094378).

Wait, E. E. NSF Graduate Research Fellowship Program Proposal: Chemistry—Chemical Theory, Models and Computational Methods (2019).

Walker, R. CAREER: Surface-Mediated Solvation at Hydrophobic and Hydrophilic Interfaces, 2001 (NSF CHE 0094246).

Walter, R. I. Absence of Detectable Freely Diffusing Radicals during the Formation of an Aromatic Grignard Reagent. *J. Org. Chem.* **2000**, *65* (16), 5014–5015. DOI: org/10.1021/jo0001028

Wang, N.; Lv, H.; Zhou, Y.; Zhu, L.; Hu, Y.; Majima, T.; Tang, H. Complete Defluorination and Mineralization of Perfluorooctanoic Acid by a Mechanochemical Method Using Alumina and Persulfate. *Environ. Sci. Technol.***2019**, *53* (14), 8302–8313. DOI: 10.1021/acs.est.9b00486

Wang, X.; Li, X.; Aya, S.; Araoka, F.; Ishida, Y.; Kikkawa, A.; Kriener, M.; Taguchi, Y.; Ebina, Y.; Sasaki, T.; Koshiya, S.; Kimoto, K.; Aida, T. Reversible Switching of the Magnetic Orientation of Titanate Nanosheets by Photochemical Reduction and Autoxidation. *J. Am. Chem. Soc.* **2018**, *140* (48), 16396–16401. DOI: 10.1021/jacs.8b09625

Wang, X. H.; Li, J.-G.; Kamiyama, H.; Moriyoshi, Y.; Ishigaki, T. Wavelength-Sensitive Photocatalytic Degradation of Methyl Orange in Aqueous Suspension over Iron(III)-doped TiO_2 Nanopowders under UV and Visible Light Irradiation. *J. Phys. Chem. B* **2006**, *110* (13), 6804–6809. DOI: org/10.1021/jp060082z

Wang, Y.-S.; Beyer, B. M.; Senior, M. M.; Wyss, D. F. Characterization of Autocatalytic Conversion of Precursor BACE1 by Heteronuclear NMR Spectroscopy. *Biochemistry* **2005**, *44* (50), 16594–16601. DOI: org/10.1021/bi051040o

Warren, T. CAREER: Metal-Ligand Multiple Bonding in Later, First Row Complexes, 2002 (NSF CHE 0135057).

Weaver, H. NSF Graduate Research Fellowship Program Proposal: Chemistry—Chemical Measurement and Imaging (2019).

Weinberger, D. A.; Higgins, T. B.; Mirkin, C. A.; Stern, C. L.; Liable-Sands, L. M.; Rheingold, A. L. Terthienyl and Poly-terthienyl Ligands as Redox-Switchable Hemilabile Ligands for Oxidation-State-Dependent Molecular Uptake and Release. *J. Am. Chem. Soc.* **2001**, *123* (11), 2503–2516. DOI: org/10.1021/ja0030008

Weston, D. P.; You, J.; Lydy, M. J. Distribution and Toxicity of Sediment-Associated Pesticides in Agriculture-Dominated Water Bodies of California's Central Valley. *Environ. Sci. Technol.* **2004**, *38* (10), 2752–2759. DOI: org/10.1021/es0352193

Whitaker, L.; Wheelhouse, K.; Kerr, W.; Reid, M. Reductive Etherification. In *Technical Program Archives: Fall 2019 American Chemical Society National Meeting & Expo*, San Diego, CA; Paper ORGN 76.

Xiao, J.-A.; Li, Y.-C.; Luo, Z.-J .; Cheng, X.-L.; Deng, Z.-X.; Chen, W.-Q.; Su, W.; Yang, H. Construction of Bispirooxindole Heterocycles via Palladium-Catalyzed Ring-Opening Formal [3 + 2]-Cycloaddition of Spirovinylcyclopropyl Oxindole and 3-Oxindole Derivatives. *J. Org. Chem.* **2019**, *84* (4), 2297–2306. DOI: 10.1021/acs.joc.8b02889

Yan, B.; Zhao, J.; Leopold, K.; Zhang, B.; Jiang, G. Structure-Dependent Response of a Chemiluminescence Nitrogen Detector for Organic Compounds with Adjacent Nitrogen Atoms Connected by a Single Bond. *Anal. Chem.* **2007**, *79* (2), 718–726. DOI: org/10.1021/ac061682x

Yan, Y.; Robinson, S.; Hendriks, K.; Sigman, M.; Sanford, M. Development of Highly Soluble Cyclopropenium Salts Leading to Highly Concentrated Non-Aqueous Flow Battery. In *Technical Program Archives: Fall 2019 American Chemical Society National Meeting & Expo*, San Diego, CA; Poster ORGN 157.

Yang, P.; Ji, Y.; Lu, J.; Huang, Q. Formation of Nitrophenolic Byproducts during Heat-Activated Peroxydisulfate Oxidation in the Presence of Natural Organic Matter and Nitrite. *Environ. Sci. Technol.* **2019**, *53* (8), 4255–4264. DOI: 10.1021/acs.est.8b06967

Yardley, R. Golden Opportunity: Evidence for MPC Isomers. *J. Am. Chem. Soc.* **2021**, *143* (4), 16811682. DOI: 10.1021/jacs.1c00937

Ye, Z.; Wang, X.; Xiao, L. Single-Particle Tracking with Scattering-Based Optical Microscopy. *Anal. Chem.* **2019**, *91* (24), 15327–15334. DOI: 10.1021/acs.analchem.9b02760

Yu, C. J.; Wan, Y.; Yowanto, H.; Li, J.; Tao, C.; James, M. D.; Tan, C. L.; Blackburn, G. F.; Meade, T. J. Electronic Detection of Single-Base Mismatches in DNA with Ferrocene-Modified Probes. *J. Am. Chem. Soc.* **2001**, *123* (45), 11155–11161. DOI: org/10.1021/ja010045f

Yu, X.; Hou, H.; Wang, B. A Priori Theoretical Model for Discovery of Environmentally Sustainable Perfluorinated Compounds. *J. Phys. Chem. A* **2018**, *122* (13), 3462–3469. DOI: 10.1021/acs.jpca.8b00606

Yue, Q.; Xu, L.; Xiang, G.; Yu, X.; Yao, Y. Characterization of Gene Expression Profile, Phenolic Composition, and Antioxidant Capacity in Red-Fleshed Grape Berries and Their Wines. *J. Agric. Food Chem.* **2018**, *66* (27), 7190–7199. DOI: 10.1021/acs.jafc.8b01323

Zhang, B.-B.; Yuan, S.; Ma, J.-P.; Zhou, Y.; Hou, J.; Chen, X.; Zheng, W.; Shen, H.; Wang, X.-C.; Sun, B.; Bakr, O. M.; Liao, L.-S.; Sun, H.-T. General Mild Reaction Creates Highly Luminescent Organic-Ligand-Lacking Halide Perovskite Nanocrystals for Efficient Light-Emitting Diodes. *J. Am. Chem. Soc.* **2019**, *141* (38), 15423–15432. DOI: 10.1021/jacs.9b08140

Zhang, G.; Hamaker, B. R. A Three Component Interaction among Starch, Protein, and Free Fatty Acids Revealed by Pasting Profiles. *J. Agric. Food Chem.* **2003**, *51* (9), 2797–2800. DOI: org/10.1021/jf0300341

Zhang, Y.-J.; Chen, S.; Radjenovic, P.; Bodappa, N.; Zhang, H.; Yang, Z.-L.; Tian, Z.-Q.; Li, J.-F. Probing the Location of 3D Hot Spots in Gold Nanoparticle Films Using Surface-Enhanced Raman Spectroscopy. *Anal. Chem.* **2019**, *91* (8), 5316–5322. DOI: 10.1021/acs.analchem.9b00200

Zhao, D. Y.; Manetsch, R. Studies towards the Total Synthesis of Anguidine. In *Technical Program Archives: Fall 2019 American Chemical Society National Meeting & Expo*, San Diego, CA; Poster ORGN 675.

Zhu, X.; Lin, Y.; Sun, Y.; Zhu, D.; Yan, Y. Lead Halide Perovskites for Photocatalytic Organic Synthesis. In *Technical Program Archives: Fall 2019 American Chemical Society National Meeting & Expo*, San Diego, CA; Poster ORGN 649.

Index

In page references, *ex* indicates *example*, *f* indicates *figure*, and *t* indicates *table*.

www.ingramcontent.com/pod-product-compliance
Ingram Content Group UK Ltd.
Pitfield, Milton Keynes, MK11 3LW, UK
UKHW051048150926
475595UK00005B/178

9 780190 098940